UNSETTLING EUROCENTRISM IN THE WESTERNIZED UNIVERSITY

The westernized university is a site where the production of knowledge is embedded in Eurocentric epistemologies that are posited as objective, disembodied and universal and in which non-Eurocentric knowledges, such as black and indigenous ones, are largely marginalized or dismissed. Consequently, it is an institution that produces racism, sexism and epistemic violence. While this is increasingly being challenged by student activists and some faculty, the westernized university continues to engage in diversity and internationalization initiatives that reproduce structural disadvantages and to work within neoliberal agendas that are incompatible with decolonization.

This book draws on decolonial theory to explore the ways in which Eurocentrism in the westernized university is both reproduced and unsettled. It outlines some of the challenges that accompany the decolonization of teaching, learning, research and policy, as well as providing examples of successful decolonial moments and processes. It draws on examples from universities in Europe, New Zealand and the Americas.

This book represents a highly timely contribution from both early career and established thinkers in the field. Its themes will be of interest to student activists and to academics and scholars who are seeking to decolonize their research and teaching. It constitutes a decolonizing intervention into the crisis in which the westernized university finds itself.

Julie Cupples is Professor of Human Geography and Cultural Studies at the University of Edinburgh, UK. Her current research is focused on indigenous and Afro-descendant media activism in Central America and Aotearoa New Zealand. She is the author of *Latin American Development* (2013) and co-author of *Communications/Media/Geographies* (2017) and *Shifting Nicaraguan Mediascapes: Authoritarianism and the Struggle for Social Justice* (2018).

Ramón Grosfoguel is Associate Professor of Ethnic Studies at the University of California, Berkeley, USA, and a senior researcher at the Maison des Sciences de l'Homme in Paris, France. He has published many articles on the political economy of the world system and on Caribbean migrations to Western Europe and the United States. He is author of *Colonial Subjects: Puerto Ricans in a Global Perspective* (2003) and co-editor of *Latin@s in the World System: Decolonization Struggles in the 21st Century US Empire* (2005) and *Decolonizing the Westernized University* (2016).

Routledge Research in New Postcolonialisms

Series Editor: Mark Jackson,
Senior Lecturer in Postcolonial Geographies, School of Geographical Sciences, University of Bristol, UK.

This series provides a forum for innovative, critical research into the changing contexts, emerging potentials, and contemporary challenges ongoing within postcolonial studies. Postcolonial studies across the social sciences and humanities are in a period of transition and innovation. From environmental and ecological politics, to the development of new theoretical and methodological frameworks in posthumanisms, ontology, and relational ethics, to decolonizing efforts against expanding imperialisms, enclosures, and global violences against people and place, postcolonial studies are never more relevant and, at the same time, challenged. This series draws into focus emerging transdisciplinary conversations that engage key debates about how new postcolonial landscapes and new empirical and conceptual terrains are changing the legacies, scope, and responsibilities of decolonising critique.

Postcolonialism, Indigeneity and Struggles for Food Sovereignty
Alternative food networks in the subaltern world
Edited by Marisa Wilson

Coloniality, Ontology, and the Question of the Posthuman
Edited by Mark Jackson

Unsettling Eurocentrism in the Westernized University
Edited by Julie Cupples and Ramón Grosfoguel

History, Imperialism, Critique
New Essays in World Literature
Edited by Asher Ghaffar

For more information about this series, please visit: https://www.routledge.com/Routledge-Research-in-New-Postcolonialisms/book-series/RRNP

UNSETTLING EUROCENTRISM IN THE WESTERNIZED UNIVERSITY

*Edited by Julie Cupples
and Ramón Grosfoguel*

Routledge
Taylor & Francis Group

LONDON AND NEW YORK

First published 2019
by Routledge
2 Park Square, Milton Park, Abingdon, Oxon OX14 4RN

and by Routledge
711 Third Avenue, New York, NY 10017

Routledge is an imprint of the Taylor & Francis Group, an informa business

© 2019 selection and editorial matter, Julie Cupples and Ramón Grosfoguel; individual chapters, the contributors.

The right of Julie Cupples and Ramón Grosfoguel to be identified as the authors of the editorial material, and of the authors for their individual chapters, has been asserted in accordance with sections 77 and 78 of the Copyright, Designs and Patents Act 1988.

British Library Cataloguing-in-Publication Data
A catalogue record for this book is available from the British Library

Library of Congress Cataloging-in-Publication Data
A catalog record has been requested for this book

ISBN: 978-1-138-06179-8 (hbk)
ISBN: 978-1-138-06180-4 (pbk)
ISBN: 978-1-315-16212-6 (ebk)

Typeset in Bembo
by Apex CoVantage, LLC

CONTENTS

ILLUSTRATIONS

Figures

Table

CONTRIBUTORS

Robert Aman is a Postdoctoral Research Fellow in Education at Linköping University, Sweden. He primarily conducts research on the relationship between education, the geopolitics of knowledge and various forms of exclusion and marginalisation, drawing on decolonial theories. His book *Decolonising Intercultural Education: Colonial Differences, the Geopolitics of Knowledge, and Inter-Epistemic Dialogue* was published by Routledge in 2017.

Trycia Bazinet is a white settler from Abitibi Lake in what is known as northern Quebec, Canada. She is a community organizer and PhD student in Indigenous and Canadian studies at Carleton University, Ottawa, on the unceded territory of the Algonquin Nations.

Johanna Bergström holds a PhD in Politics and International Studies from the European Union Erasmus Mundus Program, Globalisation, the EU and Multilateralism (GEM). Her research interests include gender equality, indigeneity, sustainability as well as social justice and she has mainly explored these issues in the Latin-American context. Previously she worked for UN Women in Guatemala and for various NGOs in Chile, Guatemala, Mexico and Venezuela.

Charisse Burden-Stelly is an Assistant Professor and Mellon Faculty Fellow of Africana Studies and Political Science at Carleton College, USA. She received a PhD in African Diaspora Studies from the University of California, Berkeley in 2016. Her areas of specialization include Black critical and political theory, epistemologies of Blackness, and the intersections of antiblackness and antiradicalism. Her current book project, titled *Epistemologies of Blackness,* will be published in November 2018. She has published in journals including *SOULS: A Critical Journal of Politics, Culture, and Society* and *The CLR James Journal.*

Sandra Camelo has a PhD in Cultural Studies from Goldsmiths, University of London, UK. She is also a participant of the Research Group in Cultural Studies at the Pontificia Universidad Javeriana in Bogotá, Colombia. She is interested in agonistic poetics, ethics and politics, decolonial and postcolonial studies, and critical ecologies.

Maricely Corzo Morales studied Fine Arts at Universidad Nacional de Colombia. She has a degree in Fine Arts and a PhD in Public Art from Universidad Politécnica de Valencia, and postgraduate studies in International Cooperation at Universidad de Valencia. She is currently a part-time professor at the Universidad Católica de Colombia.

Julie Cupples is Professor of Human Geography and Cultural Studies at the University of Edinburgh, UK. Her current research is focused on indigenous and Afro-descendant media activism in Central America and Aotearoa New Zealand. She is the author of *Latin American Development* (2013) and co-author of *Communications/Media/Geographies* (2017) and *Shifting Nicaraguan Mediascapes: Authoritarianism and the Struggle for Social Justice* (2018).

Lou Dear is a Research Assistant in English and Comparative Literature at the University of Glasgow, UK. They have published research on colonial/postcolonial literature and international education. Lou is currently editing a collection of essays (with Professor Martin Eve) entitled *The Abolition of the University* for the Open Library of the Humanities.

Christelle Gomis is a PhD student in History at the European University Institute, specializing in transnational race politics and Black internationalism. Her dissertation charts the educational struggles of Black Britons in England after 1945. Her research interests focus on decolonial feminisms, critical race theory and the history of colonialism (UK, France, Germany).

Ramón Grosfoguel is Associate Professor of Ethnic Studies at the University of California, Berkeley, USA, and a senior researcher at the Maison des Sciences de l'Homme in Paris, France. He has published many articles on the political economy of the world system and on Caribbean migrations to Western Europe and the United States. He is author of *Colonial Subjects: Puerto Ricans in a Global Perspective* (2003) and co-editor of *Latin@s in the World System: Decolonization Struggles in the 21st Century US Empire* (2005) and *Decolonizing the Westernized University* (2016).

Aitor Jimenez González is a Political Scientist and Attorney. His current research focuses on the constitutional and political construction of national pluralism under the frame of critical studies on race and law. He is currently completing a PhD at the University of Auckland, New Zealand.

Olivette Otele is a Reader in History at Bath Spa University, UK, and Fellow of the Royal Historical Society. Her research focuses on the links between history, memory, memorialisation and politics in relation to the transatlantic slavery. More broadly, she works on the history of people of African descent, identity, race and social cohesion in Britain and France. Dr Otele's forthcoming and latest publications include: *Afro-Europeans: A Short History* (2018) and "History of Slavery, Sites of Memory, and Identity Politics in Contemporary Britain" in *The States of Memory: International Comparative Perspectives* (eds) A. Gueye, J. Michel (2017). She is the recipient of several prestigious UK, French and European Union research grants, including EU-MSC-RISE grant *SLAFNET: A Dialogue between Europe and Africa*.

Lili Schwoerer is a PhD candidate in Sociology at the London School of Economics, UK. Her research focuses on situated knowledges and the marketization and internationalization of UK Higher Education.

Denisse Sepúlveda Sánchez is a Sociologist with a Master's Degree in Gender and Culture from the University of Chile. She is currently a PhD student in Sociology in the School of Social Sciences at the University of Manchester, UK. Her topics of interest are race, ethnicity, inequality, Mapuche People, social class, identities, gender and social stratification.

Francesca Sobande is a Lecturer in Marketing and Advertising at Edge Hill University, UK, with a background in sociology and politics. Her works addresses issues regarding identity, intersecting inequalities, digital media and the marketplace. Francesca's PhD thesis focuses on the online and media experiences of Black women in Britain. She has published work in the *European Journal of Cultural Studies,* as well as in edited collections including, *HBO's Original Voices: Race, Gender, Sexuality and Power* (eds) V. McCollum and G. Monteverde (2018).

Marcin B. Stanek is a PhD student in Geography at Durham University, UK. His doctoral research focuses on the politics of decolonial education in urban Bolivia. Previously he has researched geopolitics of knowledge mobility within the wider system of secondary international education.

Lucas Van Milders is a Lecturer in Politics and International Relations at Canterbury Christ Church University, UK. His research interrogates the epistemicidal dynamics in Western epistemology and seeks to disrupt those through interventions grounded in decolonial and autonomist standpoints. In addition, he also works on conjunctures of precarity in the post-industrial labour environment.

Simone Vegliò is a PhD candidate in the Department of Geography at King's College London, UK. He works on a project on urban transformations in Latin America which investigates the multiple ways in which cities have been used as strategic tools in order to "modernize" postcolonial countries. His research is mostly focused on Argentina, Mexico, and Brazil, historically, and he has a strong interest in contemporary urbanization worldwide. He has also worked on the figures of José Martí and Antonio Gramsci.

ACKNOWLEDGEMENTS

This book emerged from an event held at the University of Edinburgh in February 2016 entitled Decolonizing the Academy that was organized by Julie Cupples at the University of Edinburgh and led by Ramón Grosfoguel from University of California-Berkeley. We held a two-day seminar followed by a one-day conference that was attended by more than 70 researchers affiliated to universities in 13 different countries and who are engaged in a wide range of decolonial research projects. The event was funded by the University of Edinburgh's Global Development Academy and underscored the urgent need for spaces of decolonial dialogue on university campuses. We are grateful to all the people who came to the event and participated in important conversations about the persistence of Eurocentrism and the ways in which it can be unsettled. We are especially grateful to the 16 participants who agreed to contribute to this book project and thus take forward the conversations we had in Edinburgh. It has been a great pleasure to work with you all. Thanks too to Jan Penrose, Tom Slater, Farhana Sultana and Amanda Thomas for fantastic support and feedback, discussions about coloniality in our workplaces, and their efforts to make the westernized university a better place.

1

INTRODUCTION

Coloniality resurgent, coloniality interrupted

Julie Cupples

Introduction

In many parts of the world, university communications and marketing teams work hard to produce a continuous stream of promotional slogans and triumphalist narratives of success. Browse the website of any westernized university and you are likely to find stories of exciting scientific breakthroughs, upwards movement in one of a myriad of academic league tables and external grant successes. Such stories are often accompanied by glossy images of smiling successful students. Quite frequently the students who feature in marketing materials are ethnic minority ones in order to signal the extent to which this university has embraced diversity. In spite of the repeated chest thumping and displays of institutional confidence, the contemporary westernized university is currently enmeshed in a profound conjunctural crisis. As Stuart Hall and Doreen Massey (2010: 57) write, conjunctural crises "are moments of potential change, but the nature of their resolution is not given". There is no way of knowing how soon the current crisis will be resolved. But we do know that conjunctural crises deepen through processes of articulation, when discourses or events with no necessary connection begin to be connected (Grossberg, 1996). There is some evidence that points of articulation between the westernized university in crisis and broader social and cultural anxieties are also increasing. We are, however, living in moments of grave danger. The racist and white supremacist politics that accompany Brexit politics, the Trump presidency, anti-terror legislation, growing authoritarianism and the criminalization of protest have implications for universities. This book represents an intellectual intervention into this crisis with the explicit intention to contribute to the acceleration of the academy's urgent and necessary decolonization. It is apparent that the westernized university will be forced to negotiate an intensifying set of contradictions in the coming years, the resolution of which might not permit (colonial-capitalist) business as usual.

The westernized university is a site where learning and the production, acquisition and dissemination of knowledge are embedded in Eurocentric epistemologies that are posited as objective, disembodied and universal and in which non-Eurocentric knowledges such as black and indigenous knowledges are largely ignored, marginalized or dismissed. The westernized university does not only exist in so-called Western nations. As Ramón Grosfoguel (2012: 83) writes, the westernized university with its "disciplinary divisions" and its "racist/sexist canon of thought" is also to be found in "Dakar, Buenos Aires, New Delhi, Manila, New York, Paris or Cairo". The westernized university, like colonial conquest and slavery, is then a globalized phenomenon, but one that manifests itself in geographically specific forms, depending on, in part, whether the university is located in a site of labour-focused colonialism, a site of settler[1] colonialism (for the distinction, see Veracini, 2011) or in a colonial or imperial nation such as the UK that is home to large established communities of Commonwealth migrants and their descendants as well as to many international students from formerly colonized countries. These distinct colonial forms reproduce Eurocentrism in different ways and engender different kinds of decolonial responses.

The westernized institution, as it is currently constituted, is subject to a range of conjunctural challenges, some of which have neoliberalizing and (re)colonizing dimensions, and others that are anti-capitalist, feminist and decolonial in their orientation. The university is challenged externally by policymakers and taxpayers, a powerful sector of which believes that the main role of the university is to train young people for employment and to produce research that contributes to society in narrow instrumental ways or that supports economic growth. For this sector, it seems justifiable that students should individually finance their own studies through large interest-bearing student loans (or family wealth if they have it) and that academics should be required to demonstrate their teaching competence, their research productivity as well as the impacts of their research through a proliferating range of audit mechanisms, such as REF, TEF, NSS, TAS, QAA, ERA, PBRF and "pathways to impact"; what Fred Inglis (2000: 429) has referred to as the "mad rout of acronyms" that "blinds visions and stifles thought". Such pressures are not just external, as there is a managerial class within the university as well as a number of (usually elite, senior or securely employed) academics who also buy into such a vision and work to put it into practice. While the principles of academic freedom and free speech are often reasserted, academics who do not respond appropriately to such measures often find themselves targeted in a range of ways. Recent research and media reports reveal, for example, how bullying in academic settings is on the rise and is much higher than in other professions (see for example Keashly and Neuman, 2010; Lester, 2013; Shaw and Ratcliffe, 2014), while sexual harassment and misconduct are endemic problems that westernized universities are largely failing to address (Ahmed, 2016; Gluckman, 2017).

This neoliberalizing and instrumentalist framing is, however, intensely contested from within by faculty and students that believe that higher education is a public and collective good that should be financed by the public purse, that university is not (just) about job training but is about producing citizens capable of independent and critical thought that might disrupt rather than reproduce the neoliberal status quo, and that the forms of audit, metricization and surveillance to which academics are subject are intellectually impoverishing. Furthermore, they draw attention to the serious mental health crisis that characterizes our campuses that sometimes results in breakdowns and even suicides. For students, the mental health crisis on campus is clearly linked to indebtedness, tuition fee hikes and concerns about future economic and employment insecurity (Gani, 2016). Faculty on the other hand, who must teach increasingly anxious and indebted students, often struggle with their own failure to construct appropriately neoliberal academic subjectivities and to meet unrealistic performance targets in teaching, publication, grant income and leadership. Academics are, however, talking and writing about the neoliberalization of the university and its associated issues, such as casualization and outsourcing on the pages of academic books and journals, on blogs, on email discussion lists and to each other (see for example Aronowitz, 2001; Newfield, 2008; Collini, 2012; Cupples and Pawson, 2012; Giroux, 2014; Eagleton, 2015). This phenomenon has also produced a new academic genre of writing that Francesca Coin (2017) has referred to as "quit lit", written by academics who resign because of intolerable working conditions (see, for example, Ahmed, 2016; Morrish, 2017).

There is then a well-developed critique of the neoliberal academy, most of it produced by academics that feel professionally jaded or mentally destabilized by neoliberalization. Much of it is, however, characterized by a rather problematic sense of nostalgia, of a need to get back to an earlier postwar university, when academics had time to read, think and discuss, when the worth of academics was not measured in numbers of publications or grant targets, and when there were no student loans as the cost of higher education was mostly covered by the state. There is, however, another related but more substantial and serious challenge to the westernized university that is not accompanied by nostalgia but by a recognition that for certain kinds of people – especially women, people of colour and indigenous people – that both the pre- and post-neoliberal university is a site of elitism, pain, exclusion, coloniality and Eurocentric thinking. It underscores that nostalgia for an earlier, saner university is only really possible for those who enjoy white privilege, and to a lesser extent male privilege, who do not see coloniality on their university campuses and would deny it even exists. In fact, the westernized university always was and continues to be a sexist and racist institution in which the knowledges and worldviews of indigenous, African and Muslim people as well as scholarship by women and female perspectives, especially working class ones, are either excluded completely or kept on the margins of curricula, modes of governance and institutional practices. There is, as Kate Parizeau *et al.* (2016: 196) write, a connection between the absence of mental wellness on campus and the "privileges and exclusions of certain ways of working and being in the academy". As Lou Dear (this volume) notes, westernized universities were not set up to benefit the colonized, women or the working classes and many were built "by rich white men to benefit rich white men" and "to protect a class of social and cultural elites when elite was synonymous with white" (Iorio, 2017). Craig Wilder's (2013) study of the first universities in what is now the United States shows how they were founded on and benefited from the profits of slavery and some of them actively worked to destroy and domesticate indigenous cultures through evangelization. In the 19th and 20th centuries, European universities, as Enrique Dussel (2003: 12) notes, functioned as "theaters of 're-education', of brainwashing" that converted colonial elites into "puppets" that "repeated in the periphery what their eminent professors of the great metropolitan universities had propounded". As a result, Eurocentrism is as embedded in the universities of the Global South as it is in those of the Global North. Trained to reject their native epistemes, many eminent academics in the Global South are *not* decolonial thinkers.

Unable to shake off the legacies of these origins, the westernized university is, then, a site in which the concepts of universality and objectivity continue to be routinely mobilized to disguise the localized, provincial and historically specific nature of (Eurocentric) research. Indigenous scholars such as Linda Tuhiwai Smith (2012) and Rauna Kuokkanen (2007) have provided important insights into what Kuokkanen refers to as the Enlightenment university. They note that the Enlightenment university is colonial and monocultural in its orientation and while universities do increasingly seek to include women, indigenous people and people of colour in hiring practices and committee memberships, they do so in an assimilationist mode that does not include indigenous and other non-Eurocentric epistemes. Māori legal scholar Jacinta Ruru describes her legal training in a New Zealand university as follows:

> I really struggled with law in a lot of ways. It is a really hostile discipline for indigenous people, including Māori. It silences our stories and our relationships with the land. It does more than just silence. The law has been a major tool used by the colonialists to take our land, and take away our knowledge.
>
> *(cited in Day, 2017)*

Kuokkanen's work develops Derrida's notion of the gift, applying it to the university, revealing how the marginalization of indigenous epistemes is part of the neoliberal logic of market exchange that currently prevails in universities. In other words, we cannot decolonize the university very effectively while we continue to perpetuate neoliberal logics within it. It is imperative that the existing neoliberal critique is much more overtly articulated to the decolonial critique, particularly as the latter gains traction through various kinds of faculty and student activism and what Grosfoguel (2012: 81) calls "epistemic insurgency". This is one of the key contradictions that university managers have failed to address and if they don't do so in the near future, the tensions between their diversity and equity aspirations and ambitions to enhance the student experience, on the one hand, and the relentless pursuit of academic excellence defined according to a narrow set of largely white, male, neoliberal and Eurocentric criteria, on the other, will very quickly become impossible to manage.

This chapter serves as an introduction to a book project that began as a three-day event at the University of Edinburgh in February 2016 that sought to explore ways that we might decolonize the academy. Organized by myself as the Co-Director of the Global Development Academy and led by Ramón Grosfoguel from UC-Berkeley, the event attracted an overwhelming amount of interest, especially from young and early career BAME scholars, many of whom travelled quite long distances to join the event, including Australia, Canada, Chile, Colombia, France, Italy, Spain, Sweden and Switzerland, as well as from all over the UK. We were dramatically oversubscribed and as a result many people who expressed interest in attending had to be turned away. The interest in our event reveals the urgent need that many scholars have to engage with decolonial thought in order to reflect on their own experiences of racism, epistemic marginalization and coloniality in the westernized university or to advance their own decolonial research projects, or both. As this book underscores, many of us have been inspired by the intellectual debates associated with the so-called Modernity/Coloniality/Decoloniality (MCD) research paradigm, and to which Ramón is a key contributor. The MCD paradigm locates the start of modernity not with the European Enlightenment in the 17th century but with the conquest of America in the 15th century. It understands the conquest of America to have unleashed the global capitalist order still in place today; it sees colonialism, coloniality and capitalism as entangled, and it seeks to recover and foreground non-Eurocentric knowledges that have been subject to acts and processes of epistemicide wrought by the colonial power matrix (for an excellent introduction to these ideas see the 2007 special issue of the journal *Cultural Studies* on the MCD).

Recolonizing and decolonizing moves in New Zealand, the UK and Nicaragua

My academic career to date has been spread mainly across the UK, Aotearoa New Zealand and Nicaragua, with shorter visiting positions in Spain, Australia and the United States. I've worked in and with universities that approach the question of Eurocentrism in quite distinct ways and these approaches are shaped by their geographic location and by their colonial and postcolonial histories.

In universities in Aotearoa New Zealand, Māori academics have drawn our attention to the ways in which the exclusion of Māori knowledges from curricula does a disservice to Māori academics and students, who don't find their worldviews represented in curricula or institutional practices (see for example Smith, 2012). Māori scholars who, for example, emphasize the oral dissemination of research findings to their own communities through *hui* (meetings) held at *marae* (meeting house of a subtribe or hapu) rather than through high impact international journals find themselves at a disadvantage when applying for academic positions or for promotion. Others fail to apply for promotion to more senior positions, as the rewards system is based on self-promotion and many Māori do not feel it is culturally appropriate to boast of one's own achievements. For Māori undergraduate students, the situation can be even more challenging. In the late 1990s, a doctoral student in education, Hazel Phillips (2003), recruited a group of Māori students at the University of Canterbury, setting out to document in her thesis their struggles to succeed in a hostile, colonial and monocultural institution that repeatedly invisibilized or silenced their perspectives. Sadly, none of the research participants recruited at the start of the study completed their degrees.

I was witness to some of the hostility that Māori students encounter in 2003, when I attended a Treaty of Waitangi[2] workshop at the University of Canterbury with a number of my colleagues. In this workshop, an articulate Ngāi Tahu speaker explained to a group of Pākehā (of European descent) academics how Māori ways of knowing differed from Eurocentric ones and so if we were to be serious about honouring the Treaty in our classrooms, we would have to take on board this question of epistemological difference. She noted how western scientific taxonomic schemas sat uneasily with Māori, who couldn't play the British game animal-mineral-vegetable because Māori didn't divide things into the same set of categories as western scientists did – fish and jade might, for example, share a set of properties. Her talk hit a raw nerve among many of those present. One white male scientist claimed he could not teach geology according to any other knowledge system because it would simply be *wrong* and therefore irresponsible, while a white male chemist stood up waving his assimilationist fist in the air and exhorted the speaker to "tell your people to come and study science, it's not hard".

That was a decade and a half ago and there is no doubt that today the New Zealand university is subject to tangible decolonial pressures and transformations. There are conversations about how New Zealand universities should implement the Treaty of Waitangi in the workplace; applications for research grants and ethics forms must now include reflections on engagement with Mātauranga Māori (Māori knowledge); te reo (the Māori language) is sometimes spoken on campuses and at academic conferences; and scientists are increasingly collaborating with iwi (Māori tribes) to make sure their science is compatible with Māori ways of knowing and being. The thoroughly British graduation ceremony is often disrupted by a *haka* performed by the family members of a graduating Māori student. So Māori students and academics are challenging Eurocentrism and coloniality in highly visible ways and indigenous knowledges are increasingly found on curricula, although there is still a long way to go in that respect. In particular, there needs to be a discursive shift from viewing Māori students as vulnerable and who need special help (while strengthening and expanding necessary affirmative action programmes), as it is frequently Pākehā settlers who require education to prevent them from simplistically reproducing Eurocentrism in their curricula and pedagogical strategies and from acting microaggressively towards students that disrupt their settled view of the world.

Despite some positive changes, not only does coloniality persist but it reasserts itself in institutional spaces in quite astonishing ways. In 2017, while a visiting scholar in the Media Studies programme at Victoria University of Wellington, I went to visit a friend and colleague in the School of Geography, Environment and

FIGURE 1.1   Epistemic erasure in the westernized university

Photograph reproduced with permission of the artist.

Earth Sciences and was confronted with a painting of four white men[3] observing a landscape that appears as empty or inhabited that was hanging in the stairwell in one of the main entrances to the school (see Figure 1.1) and from which both Māori and women were completely absent. While the monocultural and sexist image and its prominent placing were hard to take, it was even more shocking to learn that the painting had been commissioned as recently as 2009, and so could not be justified on historical grounds (a justification usually mobilized by those in favour of keeping the portraits of dead white men on the walls of universities). I learned recently that after pressure from staff and students in the School, the painting was removed.

While British universities are also subject to substantive decolonial challenges, the challenges come not from indigenous peoples, who were settled there prior to conquest, but primarily from non-white British students whose parents or grandparents migrated to the UK in the postwar period from Britain's former colonies. European universities, especially those that consider themselves to be elite institutions, appear to be quite resilient to decoloniality, and their sense of elitism works to build this resilience. Elitism is expressed in different ways but includes membership of various groupings such as the Russell Group or the Coimbra Group, its high position in various ranking tables such as QS or Times Higher, the length of its history (the older the better!) *and* the way this history is narrated in official publications. The combination of a set of elite-sounding characteristics facilitates the reproduction of coloniality. For example, the official slogan of the University of Edinburgh, where I currently work, is "Influencing the World since 1583" and much is made of the University of Edinburgh's role in the Scottish Enlightenment, in spite of the fact that for the first three centuries of Edinburgh's four-century history, women were not admitted and the Enlightenment involved the codification of knowledge in thoroughly Eurocentric and frequently racist ways, especially in the ways that it confers humanity, rationality and liberty to some people but not to others (see Emejulu, 2017; Gomis, this volume). In 2017, apparently without any process of critical reflection, a new international scholarship scheme has been named the Enlightenment scholarships, even though Enlightenment knowledges have been thoroughly devastating for indigenous and other colonized peoples (Louis, 2007; Clement, 2017). As I tell my (overwhelmingly, but not exclusively, white and privileged) students, coloniality is everywhere in our institution and in my teaching I seek to encourage its identification and interrogation. The classroom next to my office is named after Charles Darwin, a man who believed that the "civilized races of man will almost certainly exterminate and replace the savage races through the world" (Darwin, 1888: 159). Like a number of rooms on campus, the "Old Library" in my building is filled with black and white portraits of dead white men, along with two colour photos of two (both living) former female professors of geography. All nine full professors in the Institute of Geography are white, and only one is female. Each day that I am at work, I must walk under a massive globe that hangs above the stairwell that celebrates the British Empire. I have colleagues that firmly believe and often state that objectivity and universality are desirable and possible research aspirations, and others that take funding from oil companies. And we even have a Geographer Royal![4]

Furthermore, everyday intellectual, pedagogical and social interactions and encounters are also frequently characterized by Eurocentric thinking. In 2016, I co-organized an academic workshop on hazard and disaster communication in Central America funded by the UK government's Global Challenges Research Fund (GCRF). The GCRF provides extraordinarily large publicly funded research grants to address global challenges and "intractable problems" in the Global South and the grants must be ODA-compliant. While the funds once received can be put to decolonial purposes, the calls mobilize the dominant and highly contested notion that development expertise resides in the Global North and can be delivered by British academics to Global South countries that have been unable to solve their own problems. Pat Noxolo (2017) describes the GCRF as a means through which understandings of the UK as an imperial power can be re-activated. Given the desire to decolonize our own project, we were talking in our workshop about the urgent need to bring indigenous knowledges into dialogue with scientific ones, so that hazard and risk communication would be more sensitive to indigenous peoples living near highly active volcanoes and would accommodate the more diverse range of meanings that indigenous Guatemalans attach to volcanoes. I was sharing some of the ways in which my former graduate students and colleagues in New Zealand and I had engaged in horizontal and reciprocal forms of knowledge exchange in our work and with non-human ontologies (Guyatt, 2005; Johnson and Murton, 2007; Johnson et al., 2007; Jardine-Coom, 2009; Cupples, 2012; Palomino-Schalscha, 2012). One of my colleagues, in an outburst that reminded me of that of the chemist in 2003, stated that he would not be adding his name "to any academic paper that said the world began on the back of a turtle". With one such intervention, discussions about how we might decolonize our knowledges and our research practices and think from and with indigenous epistemes were brought to an abrupt close as the enunciator tried to reinstate "the myth of the unsituated knowledge of the Cartesian ego-politics of knowledge" by dismissing indigenous knowledges as inferior, irrelevant and suspicious (see Grosfoguel, 2013: 76).

In Nicaragua, I have a long-term and well-established research relationship with URACCAN (University of the Autonomous Regions of the Nicaraguan Caribbean Coast). URACCAN is a community and intercultural university, created specifically by Black Creole and indigenous intellectuals, to provide culturally relevant (i.e non-Eurocentric) teaching and research for the Caribbean Coast region that seeks to tackle the racism, discrimination and poverty that prevents black and indigenous students from reaching their potential (see Cupples and Glynn, 2014). URACCAN organizes its departments and curricula in quite different and non-hierarchical ways, and it is not uncommon to find cultural, scientific and spiritual dimensions in a horizontal dialogue in a course or research programme. Indigenous and Afrodescendant worldviews, experiences and perspectives form the basis of degrees, and the university produces graduates who are committed to working for and with their local communities. URACCAN is one of a number of intercultural universities that have emerged across Latin America. These universities take a variety of forms; some of them are created by the state, some receive state support, some emerge from alliances with existing institutions and some are created directly by indigenous groups and social movements (for an overview of these institutions, see Mato, forthcoming). That the westernized university is a site of epistemic violence is something well understood by black and indigenous scholars around the world and a point repeatedly made by the contributors to this volume. The creation of intercultural universities is a key decolonial response to this state of affairs and something from which westernized universities should learn.

Institutional colonialities

For those who experience epistemic violence and marginalization directly and for those attuned to its identification, coloniality is then everywhere in the westernized university. It is in our buildings, names and monuments. Many US universities, including the universities of North Carolina-Chapel Hill (UNC) and Virginia, were built by slaves. UNC also has a highly controversial statue known as Silent Sam that commemorates a confederate soldier. Most Canadian universities sit on stolen and unceded Aboriginal land (see Bazinet, this volume). Rhodes University in South Africa is named after Cecil Rhodes, a white supremacist British colonizer, prime minister of the Cape Colony and diamond magnate, who believed that the Anglo-Saxon race was "the best, the most human, the most honourable race the world possesses" and that "Africa is still lying ready for us and it is our duty to take it" (cited in Perry, Berg and Krukones, 2009: 21). There is a statue of Cecil Rhodes at Oriel College at Oxford and until 2015 there was also one at the University of Cape Town

(UCT). The statue at UCT was removed after a high profile student-led campaign, Rhodes Must Fall, which called for the removal of the statue and the decolonization of higher education in South Africa. There was a similar movement that followed at Oxford, but university administrators at Oxford have refused to move the statue, in part as a result of pressure from wealthy donors, who threatened to withdraw their donations if the statue was removed (Espinoza and Rayner, 2016). Oxford University also has a statue of and a library named after slave owner Charles Codrington. In Barbados, where Codrington owned slaves, an International Baccalaureate school and a theological college still bear his name. The Wills Memorial Building at Bristol University is named after slave owner Henry Overton Wills. Liverpool University has a hall of residence named after William Gladstone, a former prime minister who was in favour of slavery. Queen Mary-University of London recently removed two plaques dedicated to Leopold II, the Belgian genocidal colonial ruler of what is now known as Democratic Republic of Congo, after the student campaign Leopold Must Fall, named after Rhodes Must Fall. Yale University recently agreed to change the name of Calhoun College, named after slavery supporter John Calhoun, while Princeton University refused to remove the name of segregationist Woodrow Wilson from the School of Public and International Affairs. Massey University in New Zealand is named after William Massey, also a former prime minister and white supremacist who once said "Nature intended New Zealand to be a white man's country, and it must be kept as such" (Massey, 1921: 7). Linnaeus University in Sweden is named after 18th century botanist Carl Linnaeus, who developed the hierarchy of being that placed Europaeus Albus (white people) at the top and Afer Niger (black people) at the bottom (Linnaeus, 1758). UCL has a research institute, a collection and a lecture theatre named after eugenicist Francis Galton and also celebrates other scholars who were associated with Galton and the eugenics movement, including Galton's cousin Charles Darwin and his student Carl Pearson. As UCL student Mahmoud Arif asked at a recent event to discuss the legacy of eugenics at UCL, "Why do we celebrate Francis Galton when he hated people like us?" (UCLTV, 2014; see also Tobias Coleman, 2014).

Coloniality and Eurocentrism do not just exist in the built environment and in historical naming practices; they can be found in the narratives that are used to promote the university – in marketing and branding strategies, in curricula, in publication and citation practices, in hiring and promotion practices, in the outsourcing of cleaning and catering services and most importantly in the everyday experiences of students and faculty of colour. It is not uncommon for students to complete an entire degree and not read a single black or indigenous scholar. Racial inequality on campus is also manifest in the woeful absence of black professors. In 2013, it became apparent from a HESA survey that out of a total of more than 18,000 professorships in the UK, only 85 of them were held by black academics and only 17 by black women (Berliner, 2013; Johnson, 2014). There have been very small increases since 2014, but there are still many British universities that don't have a single black professor of any gender (Black British Academics, 2016; Williams, 2013). While many British universities do successfully recruit BAME students, many so-called elite universities fail to do so. Through freedom of information requests to Oxbridge colleges, Labour MP David Lammy was able to reveal a depressing picture, with the vast majority of places going to white wealthy students from the South East of England. Lammy's data showed that a third of Oxford colleges and six Cambridge colleges had failed to admit a single black student in the previous academic year (see Adams and Bengtsson, 2017; Lammy, 2017). There are also *no* black academics in senior management roles in the UK (Adams, 2017). The absence of black professors and of black and indigenous scholars on curricula results in a highly constrained education. Black and indigenous students do not find their worldviews represented in curricula and they are denied role models and mentors, while white students complete degrees believing there is only one knowledge system worth studying. It is not only the numbers that are shocking, but also the experiences of BAME students and academics. Divya Tolia-Kelly's (2017) long-term research project with black female academics in UK Geography departments reveals how they often feel lonely, isolated, marginalized and out of place, are subject to everyday microaggressions and are accused of being oversensitive or troublemakers when they call out racism and sexism on campus. Shirley Ann-Tate (2014), one of the small number of black female professors in the UK, describes how the "silent working of [white] networks" that denies black academic access to the "organizational knowledge practice that is essential for success" constitutes obstacles to the career progression of black academics in British universities. Cambridge student, Ore Ogunbiyi (2017) has poignantly described the emotional and intellectual exhaustion involved in attending Cambridge University "as a black girl". She provides advice to other black students who must negotiate the racism, microaggressions, misunderstanding and hostility along with the dominance of Eurocentric curricula that constitute the black experience at Cambridge, telling them to "bathe every essay in black girl magic and write in resistance to the Eurocentrism of academia that did not see you coming".[5]

The disciplining of decoloniality

Ore Ogunbiyi's article reveals the kind of decolonial tactics that are also present in the westernized university. The intensity of the conjunctural crisis is revealed in a range of decolonial activisms led by students that are growing in strength and connecting with one another in important ways. Both closely and loosely articulated movements led mostly by students to decolonize the university are proliferating – Rhodes Must Fall, Leopold Must Fall, Why is My Curriculum White?, Why Isn't My Professor Black?, Dismantling the Masters House, Liberate my Degree, Silence Sam. Decolonial movements on university campuses are connecting positively with movements outside of the academy, such as the reparations movement, the call for the removal of confederate and other offensive and racist monuments outside of university campuses, the debates about institutional racism in police forces and judiciaries, and the ongoing challenges to museum curators, television and film producers, arts festivals or literary prizes to adequately acknowledge actors, artists and writers of colour and anti-colonial histories. These modes of articulation outside the academy are positive and help decolonial movements to gain traction and wider social acceptance.

Some universities have responded positively to these decolonial challenges. Brown University in the US has implemented a number of initiatives to deal with problematic and uncomfortable pasts, including the creation of the Center for the Study of Slavery and Justice (Saner, 2017), while Birmingham City University has introduced an undergraduate degree in Black Studies, the first to exist in the UK. As one of the course founders Kehinde Andrews (2016) writes, this course emerges in response to "the unrepresentative and outdated knowledge and experiences being reproduced in British universities" to which many students cannot relate and also seeks to tackle broader conjunctural factors through a pedagogical approach that permits "a much wider examination of the problem of a neocolonial economic order that destroys poorer countries to the point where people are willing to risk their children's lives on makeshift boats". Of course, as Andrews notes, all students, and not just those racialized as black, would benefit from the democratization of the knowledge that the Black Studies degree promotes.

Those irritated by the suggestion that our universities are sites of epistemic violence often point to the principles of academic freedom or free speech and state that there are no barriers to the questioning and critiquing of colonialism, coloniality and legacies of slavery. On one level, this is clearly correct. As academics, we are free to develop our own courses, write our own syllabi, compose our reading lists, publish decolonial thought, and we can often get internal and external funding to organize events on the colonial underpinnings of our institutions and curricula. Indeed, this book emerges from such an event. Because "the university" is not just its administrators, but also its faculty and students, and because far from being an ivory tower it operates within the existing ideological terrain, the administrators have been forced to accommodate a number of decolonial demands. Some indigenous and minority students experience racism and hostility, but also gain important forms of social mobility; they mobilize new and empowering claims to indigeneity, and develop skills to challenge racisms (see Sepúlveda; Aman, both this volume). Universities do, however, sometimes move to block decolonial content and initiatives. In 2015, UCL came under fire for rejecting a proposed MA in Black Studies that resulted in the termination of a temporary employment contract held by one of only five black philosophers in the UK (Lusher, 2015). In 2017, an indigenous journalist and scholar resigned from the University of North Dakota after his proposal to deliver a lecture series on indigenous resistance to the Dakota Access pipeline was rejected (Democracy Now, 2017). When universities do accommodate decolonial demands, they frequently do so in a way that ensures that they remain marginal and do not bring about structural change in the institution as a whole. As both Francesca Sobande and Lou Dear (this volume) argue, they usually incorporate these without undertaking the wholesale structural reforms that the university requires and without standing up to the externally imposed policies that place a number of students under suspicions of radicalization and place many more in a state of long-term indebtedness.

Consequently, decolonial progress in one area is often accompanied by reasser-
tion of coloniality in another. That some of us are free to decolonize our curricula
exists in parallel with the freedom to continue to exclude black and indigenous
thought and scholarship. Further, decolonization can easily be co-opted or disciplined by the westernized university in quite insidious ways. Along with the commodification of degrees and the cultures of audit and surveillance, other processes
also work to constrain and contain decolonization. These include the STEM agenda,
the diversity agenda, the internationalization agenda (see Dear, this volume) and the
implementation on campus of anti-terror and anti-radicalization legislation.

Both Smith (2012) and Kuokkanen (2007) acknowledge the importance of
interdisciplinary intellectual trends such as Gender Studies and Cultural Studies
that have opened up spaces for indigenous intellectuality (see Burden-Stelly, this
volume, for an alternative perspective on the interdisciplines and how in the westernized university radical and critical perspectives are readily domesticated), but it
is important to note that these are areas often threatened with defunding or closure
by rampant managerialism, by the STEM agenda, and programme and department
mergers enacted in the name of efficiency that undermine the critical work being
done. In the decade I spent working at the University of Canterbury (UC) in
New Zealand, I witnessed the downsizing, closure and threatened closure of several
interdisciplinary programmes where incipient engagement with Māori knowledges
was underway and where Māori and Pacific Island students were gaining access
to intellectual perspectives and debates that helped them to make sense of racial
inequality in New Zealand, including American Studies, Theatre and Film Studies, Gender Studies, Religious Studies and Cultural Studies. Cultural Studies was
the only one of these programmes that managed to fight off closure. The closure
of Religious Studies in 2006–07 resulted in the laying off of UC's only Muslim
scholar of Islamic Studies and the downsizing and subsequent closure of American
Studies to the laying off of UC's only African American scholar.

One institutional response to decolonial demands is to talk a lot about diversity
and internationalization. Universities often sign up to diversity programmes such
as the Race Equality Charter and Athena Swan and showcase their diversity and
global focus in externally facing publications. Official publications publish photo
shoots with non-white colleagues to celebrate the university's international diversity or they engage in forms of exoticization or folkloricization, where once a year
those from overseas are invited to share food and even dress up in their national
costumes. What Lucas Van Milders (this volume) refers to as "the empty rhetoric of
diversity" can therefore further entrench racism and coloniality. As Lili Schwoerer
(also this volume) notes, the language of diversity can gloss over racism rather than
work to remove it (see also Ahmed, 2012). Kavita Bhanot (2015) has challenged this
institutional focus on diversity that keeps white privilege intact and writes:

> The concept of diversity only exists if there is an assumed neutral point from
> which 'others' are 'diverse.' Putting aside for now the straight, male, middle-
> classness of that 'neutral' space, its dominant aspect is whiteness. Constructed

by a white establishment, the idea of 'diversity' is neo-liberal speak. It is the new corporatized version of multiculturalism. It is about management, efficiency, box-ticking.

One highly problematic "diversity" initiative is a focus on "implicit bias" and universities often require that faculty complete implicit bias training in a way that does not disrupt neoliberalism or coloniality. In a tweet retweeted more than 3000 times, Sanjay Srivastava (2017) captures the problem with these initiatives that "presume that the problem lies in the minds of search committee members, which excuses the university from spending resources on structural fixes".[6] Diversity work often places substantial burdens on faculty of colour, who are frequently called to sit on committees so that they are more diverse, with this work being time-consuming, exhausting (because it means trying to educate racist and sexist colleagues), and not adequately recognized when it comes to promotion. Such practices allow managers and administrators to tick the diversity box without changing anything, while those who call out racism and sexism are sometimes accused of stifling free speech (see Ferber, 2017; Dutt-Ballerstadt, 2017).

In the UK, diversity and internationalization policies are implemented in conjunction with anti-terror legislation. British government policies rooted in concerns about growing Islamic fundamentalism and terrorism have delegated the functions of the UK Border Agency (UKBA) to universities. Universities now face a statutory requirement to participate in and enact the UK's counter-terrorist strategy, monitoring non-EU student attendance in ways, as Matt Jenkins (2014) writes, that transform classrooms into border spaces, lecturers into border agents, and students into highly surveilled border-crossers. Furthermore, the deeply flawed Prevent strategy, which forces schools and universities to enact policies and actions to prevent people being drawn into terrorism, has resulted in some quite extraordinary outcomes, including the cancellation of a conference on Islamophobia at Birkbeck College organized by the Islamic Human Rights Commission (IHRC, 2014).

The conjunctural analysis is crucial here, as the decolonial challenges to the westernized university are taking place in a global conjuncture characterized by the rise of what Boaventura de Sousa Santos (2007) refers to as social fascism and abyssal logics. Beyond the academy, as noted, we see the resurgence of racist and imperial thinking and the proliferation of what Frantz Fanon (1967) referred to as zones of non-being (see Gordon, 2007; De Sousa Santos, 2007; Grosfoguel, 2016, this volume) both in the Global South and in Europe. This conjunctural shift can be witnessed in the growing indifference towards human suffering from hegemonic powers. Racism and coloniality entangled with capitalism underpin the avoidable deaths of migrants in the Mediterranean, the needless deaths of residents in the Grenfell Tower fire tragedy in London, the denial of US citizenship to US citizens in New Orleans and Puerto Rico in the wake of devastating hurricanes and the widespread dependence on food banks in the UK and the US. In this context, a common sense takes hold among large sectors of the population that believe it is indeed possible to "take back control" or "make America great again", slogans that are articulated to the racist exclusion of certain kinds of people and the reproduction of white privilege. Such common sense requires the mobilization of a set of discourses that assert that colonialism was a good thing and the Empire can be resurrected. As a result of these conjunctural dynamics, not only do students and faculty that demand curricula decolonization or the removal of offensive racist statues struggle to enrol support from their colleagues and classmates, but they are often subject to a wider social backlash that tends to play out on social media as well as in established media sites.

In the UK in both tabloid and broadsheet newspapers, decolonizing initiatives at universities are often roundly rejected or ridiculed. There is no shortage of headlines expressing outrage at the idea of decolonization on university campuses. For example, in *The Daily Mail* rightwing historian Dominic Sandbrook described moves to remove the name of William Gladstone from a university hall of residence as "a distressing sign of the intolerance plaguing our universities", while *The Sunday Express* ran with the erroneous headline "Newsnight guest DEFENDS calls to ban Plato and Kant because the Enlightenment is 'racist'"[7] (Stromme, 2017). Not infrequently, the political backlash leads to online abuse enacted towards those trying to democratize and decolonize institutions. Cambridge student and women's officer for the students union Lola Olufemi, who was one of a number of students calling on the English department to decolonize and diversify the English curriculum and include more BAME authors, was subject to vicious racist and sexist abuse after *The Daily Telegraph* covered the story including her photo on the front page of the newspaper, singling her out for criticism and incorrectly stating in the headline that she was calling for the removal of white male authors (Osamede Okundaye, 2017). In 2017, *Third World Quarterly*, a well-respected academic journal that hitherto had had an anti-colonial ethos, published an article that asserted that colonialism was beneficial and that many formerly colonized nations would benefit from recolonization. Not only was the article factually wrong and deeply offensive, it did not even meet even the most basic academic publishing standards, including peer review, and was later retracted (for good overviews of the *Third World Quarterly* controversy, see Paradkar, 2017; Prashad, 2017). Articles in *The Times* on the controversy and the retraction (which were quite sympathetic to the author and to his ideas) produced copious comments from readers defending colonialism (see Biggar, 2017; Whipple, 2017). Hamid Dabashi (2017) captures the ways that responses to this article reveal the resilience of white male privilege in universities and how what he calls "bourgeois etiquette and liberal politesse" obscures their colonial thinking. A Syracuse University professor, Farhana Sultana, led a successful and widely supported petition calling for the article's retraction, not for retraction's sake or to suppress free speech (the author is of course free to express his pro-colonial views in other sites), but to ensure that academic journals uphold academic publishing standards, integrity and ethics. As a result of this action, she was also subject to horrific online abuse (see Flaherty, 2017). Unfortunately, when faculty and students are attacked online by white supremacists, they often receive insufficient support from their own institutions (Ferber, 2017). Fighting for decolonization, especially for non-white students and faculty, takes courage.

Fortunately, there is no shortage of courageous student activists and scholars seeking to do so, and a number of them have contributed to this book. Many of the chapter contributors are young and early career; some are in the final stages of PhD projects. They are positioned in diverse ways in relation to colonial legacies, and their work is drawn from experience both in Europe and the Americas. They share a commitment to tackling the epistemic violence that is central to the university experience and are all seeking to identify the ways in which Eurocentric hierarchies are reproduced as well as the ways in which they are being challenged and dismantled. As a result, the book documents sources of suffering, spaces of hope and modes of political intervention that can inform our own struggles. Chapters 2, 3 and 4 tackle the ways in which neoliberalism and coloniality are entangled in the westernized university and call upon scholars not to neglect persistent racisms in their critiques of neoliberalism. In Chapter 2, Lou Dear uses the scholarship of Sylvia Wynter to critique the westernized university's internationalization agenda and in particular the formation of overseas branch campuses. According to Dear, such initiatives are informed by plantation logics and colonial histories that reproduce economic exploitation and epistemic homogeneity. In Chapter 3, Lucas Van Milders discusses how neoliberalization and racialization are entangled in the contemporary university through a hegemonic whiteness that obscures and excludes. Van Milders calls for an intersectional standpoint that can exploit the cracks within the white neoliberal university. Chapter 4 by Lili Schwoerer tackles the problematic and linear discourse of the university having fallen from its former critical and autonomous position to a neoliberal and marketized one. Through in-depth research with students and faculty in Sociology at Cambridge University in 2016, Schwoerer assesses the possibilities that exist for anti-colonial agency and for dismantling the racist-colonial structures of power-knowledge. Chapters 5 and 6 deal with the challenges posed by, and facing, Black Studies in the westernized university. Chapter 5 by Charisse Burden-Stelly focuses on the US context and the development of Black Studies in the 1960s during the Cold War. Despite its anti-imperial and anti-racist potential, it was subject to academic disciplining within the westernized university in an anti-Marxist political context that undermined its radical potential. In Chapter 6, Francesca Sobande responds to the need for radical modes of knowledge production and explores how Black feminist and intersectional approaches are key to dismantling and undermining discourses of objectivity and universality. Sobande also explores the potential offered by the new media environment and digital tools for expanding Black feminist knowledge. Chapters 7, 8 and 9 consider strategies for decolonizing particular fields of study with a focus on development studies, urban studies and law. In Chapter 7, Trycia Bazinet draws our attention to the conceptual blindspots in international development education in Canada that do not even begin to come to terms with settler colonial violence within Canada. In Chapter 8, Simone Vegliò explores the possibilities that exist for decolonizing urban studies and how we might work with both postcolonial and decolonial perspectives to theorize cities in a way that avoids the North/South division. In Chapter 9, Aitor Jimenez turns his attention to the entrenched coloniality of legal education in the westernized university and discusses some of the ways in which Latin American universities are embarking on decolonial legal practices and working in innovative ways with indigenous communities in the context of the new constitutionalism. Mapuche scholar, Denisse Sepúlveda explores in Chapter 10 what it means to be Mapuche in the Chilean university. While the Chilean university is embedded in Eurocentric thinking and Mapuche students must negotiate hostility and racism, it is also a site in which they begin to assert their identities as Mapuche, overcoming in part some of the negative experiences of discrimination that they and their parents were exposed to and that resulted in cultural loss. Chapters 11 and 12 focus on what scholars in westernized universities can learn from indigenous decolonial concepts from Latin America, such as *buen vivir* and interculturality. In Chapter 11, Johanna Bergström draws on fieldwork with Mayan women in Guatemala, looking at how they theorized "development" and "nature" and urges western feminists to pay more attention to these ways of knowing. In Chapter 12, Robert Aman calls for an interepistemic approach to interculturality, a core decolonial concept in both education and in indigenous social movements in the Andean region, in order to facilitate the claiming of indigenous ways of being through educational experiences. In Chapter 13, Sandra Camelo focuses on indigenous language practices that have been subject to colonial violence and explores the decolonial and (po)ethical initiatives to revitalize these and promote healing. In Chapter 14, Maricely Corzo poses the possibility of decolonizing artistic practices through knowledge hybridization and community engagement. Chapters 15 and 16 both focus on coloniality and racism in the French education system. In Chapter 15, Olivette Otele shows how the deficiencies in the French education system and the failure to move beyond colonial education articulate with forms of xenophobia in French society. In Chapter 16, through a discussion of France's history of chattel slavery, Christelle Gomis discusses the unmarked whiteness of the French education system and stresses the need to tell alternative histories such as that of Haitian revolution to splinter dominant Eurocentric narratives. In Chapter 17, Marcin Stanek also contemplates the implications of school education with a set of reflections on the epistemic politics of secondary education in Europe. The final chapter is authored by Ramón Grosfoguel, who outlines the work of two theorists, Frantz Fanon and especially Boaventura de Sousa Santos, whose scholarship and insights are fundamental for the decolonization of knowledge production in the westernized university.

Given the existing conjuncture, it is urgent to connect our decolonial struggles across campuses and with non-university spaces and call for structural rather than cosmetic change in our institutions. We need to find ways to convert our universities into spaces of hope and possibility, rather than sites of violence, anxiety and exclusion and in which we can overcome the constraints posed by market mechanisms and embedded racisms and sexisms to engage in more radical decolonial work. This book provides a set of interventions and conversations on how we might do so.

Notes

1 Of course, the concept of "settler" is as Māori legal scholar Moana Jackson (Sykes and Jackson, 2016) once noted a euphemism.
2 The Treaty of Waitangi, signed in 1840 between the British Crown and a number of Māori chiefs, is the founding document of New Zealand. Repeated Crown violations of the Treaty led to the establishment of the Waitangi Tribunal in 1975 so that Māori land loss, suffering and grievances could be addressed.
3 The men are Charles Cotton, physical geographer and the founder of the Geography Department; Harold Wellman, geologist who was the first to identify the Alpine Fault; Bob Clarke, geologist and founder of the Geology Department; and Frank Evision, who founded the Institute of Geophysics, and the artist was Bob Kerr. I am grateful to Amanda Thomas of Victoria University of Wellington for sharing her insights on the painting, its history and its institutional effects with me.
4 A position conferred to a senior white male professor in 2015, after a 118-year gap.
5 See also the collection of essays in Gutiérrez y Muhs et al., 2012.
6 https://twitter.com/hardsci/status/941471410739806208.
7 This headline came after Kehinde Andrews said on BBC Newsnight on 9 January "Of course the Enlightenment was racist, its philosophers and anthropologists were obsessed with white supremacy". A link to the statement can be found at https://twitter.com/BBCNewsnight/status/818597784030842880.

References

Adams, R. (2017) British universities employ no black academics in top roles, figures show. *The Guardian* 19 January. Available at: www.theguardian.com/education/2017/jan/19/british-universities-employ-no-black-academics-in-top-roles-figures-show (Accessed 15 December 2017).
Adams, R. and Bengtsson, H. (2017) Oxford accused of 'social apartheid' as colleges admit no black students. *The Guardian* 19 October. Available at: www.theguardian.com/education/2017/oct/19/oxford-accused-of-social-apartheid-as-colleges-admit-no-black-students (Accessed 3 December 2017).
Ahmed, S. (2012) *On Being Included*. Durham: Duke University Press.
Ahmed, S. (2016) Resignation. [blog] *feministkilljoys* 30 May. Available at: https://feministkilljoys.com/2016/05/30/resignation/ (Accessed 17 December 2017).
Andrews, K. (2016) At last, the UK has a black studies university course. It's long overdue. *The Guardian* 20 May. Available at: www.theguardian.com/commentisfree/2016/may/20/black-studies-university-course-long-overdue (Accessed 17 December 2017).
Aronowitz, S. (2001) *The Knowledge Factory: Dismantling the Corporate University and Creating True Higher Learning*. Boston: Beacon Press.
Berliner, W. (2013) Where are all the black professors? *The Guardian* 23 July. Available at: www.theguardian.com/education/2013/jul/23/teaching-students-higher-education-networks (Accessed 3 December 2017).
Bhanot, K. (2015) Decolonise, not diversify. *Media Diversified* 30 December. Available at: https://mediadiversified.org/2015/12/30/is-diversity-is-only-for-white-people/ (Accessed 2 December 2017).
Biggar, N. (2017) Don't feel guilty about our colonial history. *The Times* 30 November. Available at: www.thetimes.co.uk/article/don-t-feel-guilty-about-our-colonial-history-ghvstdhmj (Accessed 20 December 2017).
British Black British Academics (2016) HESA statistics professors by race and gender. Available at: http://blackbritishacademics.co.uk/focus/hesa-statistics-professors-by-race-and-gender/ (Accessed 3 December 2017).

Clement, V. (2017) Beyond the sham of the emancipatory Enlightenment: Rethinking the relationship of Indigenous epistemologies, knowledges, and geography through decolonizing paths. *Progress in Human Geography* (online early).

Coin, F. (2017) On quitting. *Ephemera: Theory and Politics in Organization* 17(3): 705–719.

Collini, S. (2012) *What are Universities for?* London: Penguin.

Cupples, J. (2012) Wild globalization: The biopolitics of climate change and global capitalism on Nicaragua's Mosquito Coast. *Antipode: A Journal of Radical Geography* 44(1): 10–30.

Cupples, J. and Glynn, K. (2014) Indigenizing and decolonizing higher education on Nicaragua's Atlantic Coast. *Singapore Journal of Tropical Geography* 35(1): 56–71.

Cupples, J. and Pawson, E. (2012) Giving an account of oneself: The PBRF and the neoliberal university. *New Zealand Geographer* 68(1): 14–23.

Dabashi, H. (2017) Moral paralysis in American academia. *Al Jazeera* 28 September. Available at: www.aljazeera.com/indepth/opinion/moral-paralysis-american-academia-1709280916 38481.html (Accessed 21 December 2017).

Darwin, C. (1888) *The Descent of Man*. New York: Appleton.

Day, S. (2017) If the hills could sue: Jacinta Ruru on legal personality and a Māori worldview. *The Spinoff* 27 November. Available at: https://thespinoff.co.nz/atea/atea-otago/27-11-2017/if-the-hills-could-sue-jacinta-ruru-on-legal-personality-and-a-maori-worldview/ (Accessed 18 December 2017).

Democracy Now (2017) Native professor quits after U. of North Dakota blocks him from teaching about Standing Rock. 27 October. Available at: www.democracynow.org/2017/10/27/headlines/native_professor_quits_after_u_of_north_dakota_blocks_him_from_teaching_about_standing_rock (Accessed 29 October 2017).

De Sousa Santos, B. (2007) Beyond abyssal thinking: From global lines to ecologies of knowledges. *Review (Fernando Braudel)* 30(1): 45–89.

Dussel, E. (2003) *The Philosophy of Liberation*. Eugene: Wipf and Stock.

Dutt-Ballerstadt, R. (2017) When free speech dismantles diversity initiatives. *Counterpunch* 31 October. Available at: www.counterpunch.org/2017/10/31/when-free-speech-dismantles-diversity-initiatives/ (Accessed 17 December 2017).

Eagleton, T. (2015) The slow death of the university. *The Chronicle of Higher Education* 6 April. Available at: www.chronicle.com/article/The-Slow-Death-of-the/228991 (Accessed 6 December 2017).

Emejulu, A. (2017) Another university is possible. [blog] *Verso* 12 January. Available at: www.versobooks.com/blogs/3044-another-university-is-possible (Accessed 6 December 2017).

Espinoza, J. and Rayner, G. (2016) Cecil Rhodes statue to remain at Oxford University after alumni threaten to withdraw millions. *The Telegraph* 29 January. Available at: www.telegraph.co.uk/education/universityeducation/12128151/Cecil-Rhodes-statue-to-remain-at-Oxford-University-after-alumni-threatens-to-withdraw-millions.html (Accessed 2 May 2018).

Fanon, F. (1967) *Black Skin, White Masks*. New York: Grove Press.

Ferber, A. (2017) White supremacists go to college: New tactics, same old white supremacy. [blog] *Mobilizing Ideas* 4 December. Available at: https://mobilizingideas.wordpress.com/2017/12/04/white-supremacists-go-to-college-new-tactics-same-old-white-supremacy/ (Accessed 17 December 2017).

Flaherty, C. (2017) A dangerous withdrawal? *Times Higher Ed* 9 October. Available at: www.insidehighered.com/news/2017/10/09/pro-colonialism-article-has-been-withdrawn-over-threats-journal-editor (Accessed 21 December 2017).

Gani, A. (2016) Tuition fees 'have led to surge in students seeking counselling'. *The Guardian* 13 March. Available at: www.theguardian.com/education/2016/mar/13/

tuition-fees-have-led-to-surge-in-students-seeking-counselling (Accessed 26 November 2017).

Giroux, H.A. (2014) *Neoliberalism's War on Higher Education*. Chicago: Haymarket Books.

Gluckman, N. (2017) A complete culture of sexualization: 1,600 stories of harassment in higher ed. *Chronicle of Higher Education* 12 December. Available at: www.chronicle.com/article/A-Complete-Culture-of/242040/ (Accessed 17 December 2017).

Gordon, L. (2007) Through the hellish zone of nonbeing: Thinking through Fanon, disaster, and the damned of the earth. *Human Architecture: Journal of The Sociology of Self-Knowledge* 5: 5–12.

Grosfoguel, R. (2012) The dilemmas of ethnic studies in the United States: Between liberal multiculturalism, identity politics, disciplinary colonization, and decolonial epistemologies. *Human Architecture: Journal of The Sociology of Self-Knowledge* 10(1): 81–90.

Grosfoguel, R. (2013) The structure of knowledge in westernized universities: Epistemic racism/sexism and the four genocides/epistemicides of the long 16th century. *Human Architecture: Journal of the Sociology of Self-Knowledge* 11(1): 73–90.

Grosfoguel, R. (2016) What is racism? *Journal of World Systems Research* 22(1): 9–15.

Grossberg, L. (1996) On postmodernism and articulation: An interview with Stuart Hall. In D. Morley and K. Chen (eds) *Stuart Hall: Critical Dialogues in Cultural Studies*. London: Routledge, pp.131–150.

Gutiérrez y Muhs, G., Y., Flores Niemann, C.G. González and A.P. Harris (eds) (2012) *Presumed Incompetent: The Intersections of Race and Class for Women in Academia*. Boulder: Utah State University Press.

Guyatt, V. (2005) Reconciling Multiple Identities: Māori Women Scientists in Aotearoa/New Zealand. Unpublished MSc thesis, Department of Geography, University of Canterbury.

Hall, S. and Massey, D. (2010) Interpreting the crisis. *Soundings* 44: 57–71.

IHRC (2014) Birkbeck College buckles to far right and cancels conference [Press release] Islamic Human Rights Commission 12 December. Available at: www.ihrc.org.uk/activities/press-releases/11302-press-release-birkbeck-college-buckles-to-far-right-cancels-islamophobia-conference-booking (Accessed 3 December 2017).

Inglis, F. (2000) A malediction upon management. *Journal of Education Policy* 15(4): 417–429.

Iorio, V. (2017) Student: Unraveling white privilege at Trinity. *Hartford Courant* 19 July. Available at: www.courant.com/opinion/op-ed/hc-op-fresh-talk-iorio-white-privilege-trinity-college-0719-20170718-story.html (Accessed 15 December 2017).

Jardine-Coom, L. (2009) When men and mountains meet: Rāiamoko, western science and political ecology in Aotearoa/New Zealand. Unpublished MSc thesis, Department of Geography, University of Canterbury.

Jenkins, M. (2014) On the effects and implications of UK Border Agency involvement in higher education. *The Geographical Journal* 180: 265–270.

Johnson, J.T., Cant, G., Howitt, R. and Peters, E. (2007) Creating anticolonial geographies: Embracing indigenous knowledges and rights. *Geographical Research* 45: 117–120.

Johnson, J.T. and Murton, B. (2007) Re/placing native science: Indigenous voices in contemporary constructions of nature. *Geographical Research* 45(2): 121–129.

Johnson, S. (2014) Just 85 professors (Be the change mix). [video] *Dismantling the Master's House* 8 December. Available at: www.dtmh.ucl.ac.uk/videos/just-85-professors-change-mix/ (Accessed 3 December 2017).

Keashly, L. and Neuman, J.H. (2010) Faculty experiences with bullying in higher education: Causes, consequences, and management. *Administrative Theory and Praxis* 32(1): 48–70.

Kuokkanen, R. (2007) *Reshaping the University: Responsibility, Indigenous Epistemes, and the Logic of the Gift*. Vancouver: University of British Columbia Press.

Lammy, D. (2017) Seven years have changed nothing at Oxbridge. In fact, diversity is even worse. *The Guardian* 20 October. Available at: www.theguardian.com/commentisfree/2017/oct/20/oxford-cambridge-not-changed-diversity-even-worse-admissions (Accessed 3 December 2017).

Lester, J. (2013) (ed) *Workplace Bullying in Higher Education*. London: Routledge.

Linnaeus, C. (1758) *Systema Naturae*. Stockholm: Laurentius Salvius.

Louis, R.P. (2007) Can you hear us now? Voices from the margin: Using indigenous methodologies in geographic research. *Geographical Research* 45(2): 130–139.

Lusher, A. (2015) Black academic claims he was denied university job over his plans to 'put white hegemony under the microscope'. *The Independent* 22 May. Available at: www. independent.co.uk/news/uk/home-news/black-academic-claims-he-was-denied-university-job-over-his-plans-to-put-white-hegemony-under-the-10270949.html (Accessed 17 December 2017).

Massey, W.F. (1921) Message from the Prime Minister. *The Ashburton Guardian* 3 January. Available at: https://paperspast.natlib.govt.nz/newspapers/ashburton-guardian/1921/1/3/7 (Accessed 24 November 2017).

Mato, D. (forthcoming) Intercultural universities and modes of learning. In J. Cupples, M. Palomino-Schalscha and M. Prieto (eds) *The Routledge Handbook of Latin American Development*. London: Routledge.

Morrish, L. (2017) Why audit culture made me quit. *Times Higher Education* 2 March. Available at: www.timeshighereducation.com/features/why-audit-culture-made-me-quit (Accessed 17 December 2017).

Newfield, C. (2008) *Unmaking the Public University: The Forty Year Assault on the Middle Class*. Cambridge: Harvard University Press.

Noxolo, P. (2017) Decolonial theory in a time of the re-colonisation of UK research. *Transactions of the Institute of British Geographers* 42(3): 342–344.

Ogunbiyi, O. (2017) A letter to my fresher self: Surviving Cambridge as a black girl. *Varsity* 4 December. Available at: www.varsity.co.uk/features/14214 (Accessed 17 December 2017).

Osamede Okundaye, J. (2017) The 'decolonise' Cambridge row is yet another attack on students of colour. *The Guardian* 25 October. Available at: www.theguardian.com/commentisfree/2017/oct/25/decolonise-cambridge-university-row-attack-students-colour-lola-olufemi-curriculums (Accessed 17 December 2017).

Palomino-Schalscha, M. (2012) Indigeneity, autonomy and new cultural spaces: The decolonisation of practices, being and place through tourism in Alto Bío-Bío, Chile. Unpublished PhD thesis, University of Canterbury.

Paradkar, S. (2017) How an article defending colonialism was ever published is a mystery roiling academia. *The Toronto Star* 21 September. Available at: www.thestar.com/amp/news/gta/2017/09/21/how-an-article-defending-colonialism-was-ever-published-is-a-mystery-roiling-academia-paradkar.html (Accessed 17 December 2017).

Parizeau, K., Shillington, L., Hawkins, R., Sultana, F., Mountz, A., Mullings, B. and Peake, L. (2016) Breaking the silence: A feminist call to action. *The Canadian Geographer/Le Géographe canadien* 60(2): 192–204.

Perry, M., Berg, M. and Krukones, J. (2009) *Sources of European History: Since 1900*. Boston: Cengage.

Phillips, H. (2003) Te Reo Karanga o nga tauria Māori: Māori students, their voices, their stories at the University of Canterbury 1996–1998. Unpublished PhD thesis, Department of Education, University of Canterbury.

Prashad, V. (2017) Third World Quarterly row: Why some western intellectuals are trying to debrutalise colonialism. *Scroll.In* 21 September. Available at: https://scroll.in/

article/851305/third-world-quarterly-row-why-some-western-intellectuals-are-trying-to-debrutalise-colonialism (Accessed 17 December 2017).

Sandbrook, D. (2017) The story of students seeking to remove William Gladstone's name from their halls of residence sounds like a spoof – but in truth is a distressing sign of the intolerance plaguing our universities. *The Daily Mail* 17 November. Available at: www. dailymail.co.uk/debate/article-5091347/William-Gladstone-student-petition-distressing-sign.html#ixzz51XNR0Kzx (Accessed 17 December 2017).

Saner, E. (2017) Renamed and shamed: taking on Britain's slave-trade past, from Colston Hall to Penny Lane. *The Guardian* 29 April. Available at: www.theguardian.com/world/2017/apr/29/renamed-and-shamed-taking-on-britains-slave-trade-past-from-colston-hall-to-penny-lane (Accessed 17 December 2017).

Shaw, C. and Ratcliffe, R. (2014) Struggle for top research grades fuels bullying among university staff. *The Guardian* 16 December. Available at: www.theguardian.com/higher-education-network/2014/dec/16/research-excellence-framework-bullying-university-staff#comments (Accessed 13 November 2017).

Smith, L.T. (2012) *Decolonizing Methodologies: Research and Indigenous Peoples*, 2nd ed. London: Zed Books.

Stromme, L. (2017) Newsnight guest DEFENDS calls to put the teachings of KANT in SOAS curriculum into context. *Sunday Express* 10 January. Available at: www.express.co.uk/news/uk/752279/Kehinde-Andrews-Enlightenment-racist-BBC-Newsnight-SOAS (Accessed 17 December 2017).

Sykes, A. and Jackson, M. (2016) Transforming the debate: Decarceration not prison, just-ness not justice, constitutional transformation not treaty settlement. Workshop at Space, Race, Bodies II: Sovereignty and Migration in a Carceral Age, University of Otago, 6–8 May 2016.

Tate, S.A. (2014) Why Isn't My Professor Black? [video] Dismantling the Masters House project. Available at: www.dtmh.ucl.ac.uk/videos/isnt-professor-black-shirley-ann-tate/ (Accessed 3 December 2017).

Tobias Coleman, N.A. (2014) Eugenics: the academy's complicity. *Times Higher Education* 9 October. Available at: www.timeshighereducation.com/comment/opinion/eugenics-the-academys-complicity/2016190.article (Accessed 24 November 2017).

Tolia-Kelly, D.P. (2017) A day in the life of a geographer: 'Lone', black, female. *Area* 49(3): 324–328.

UCLTV (2014) Eugenics at UCL: We inherited Galton [YouTube] 9 October. Available at: www.youtube.com/watch?v=3e412C7Pmm8 (Accessed 24 November 2017).

Veracini, L. (2011) Introducing settler colonial studies. *Settler Colonial Studies* 1(1):1–12.

Whipple, T. (2017) Colonialism was a force for good, claimed Professor Bruce Gilley. *The Times* 12 October. Available at: www.thetimes.co.uk/article/colonialism-was-a-force-for-good-claimed-professor-bruce-gilley-of-portland-state-university-vpsnck8nk (Accessed 17 December 2017).

Wilder, C.S. (2013) *Ebony and Ivy: Race, Slavery, and the Troubled History of America's Universities*. New York: Bloomsbury.

Williams, R. (2013) The university professor is always white. *The Guardian* 28 January. Available at: www.theguardian.com/education/2013/jan/28/women-bme-professors-academia (Accessed 3 December 2017).

2

THE UNIVERSITY AS BRANCH PLANT INDUSTRY

Lou Dear

Introduction

Writing from the University of West Indies, Mona, Jamaica in 1968, Sylvia Wynter reflects on the institutional dimensions of colonial knowledge: "the University, like the society, is a 'branch plant industry of a metropolitan system'" (1968a: 24). Wynter describes the situation in post-independence Jamaica, with the British empire still in control:

> The end result is an arrangement by which, with independence attained, the majority of the West Indians were illiterate. The writer wanting market and audience had to go to England; as the West Indian emigrant wanting a living wage had to go to England. As the West Indian University, wanting skilled personnel, had to turn to England. The presence of [an English academic] in the Caribbean and the absence of the writer in London are part of the same historical process.
>
> *(1968a: 26)*

Wynter wrote this essay in the context of the perceived failings of the national liberation struggles of which she was part. Her work in the intervening years has identified and analysed the ways in which the anti-colonial struggle continues. The university is one locus of the replication and reproduction of the global colonial/capitalist condition.[1] I aim to revisit Wynter's analysis of the institution in the context of 1968 Jamaica to reflect on the contemporary Westernized university. Specifically, I will elaborate on Wynter's allusion to the university as a "branch plant industry" in the context of the Westernized university's internationalization agenda. This involves consideration of the history of educational institutions in settler colonial plantations.[2] There are concrete differences between the history of settler colonialism involving universities and the neo-colonial present. My intention in positioning the two together is to explore the continuities of "plantation logic" and the political possibilities of bringing history and present into dialogue. When discussing internationalization, I foreground my own context (the UK) for explanatory reasons, but the findings and implications are not just applicable in Europe. The neo-colonial trajectory of the Westernized university has social, political and economic implications for millions of international students and the societies from which they come.

The Westernized university as plantation

What constitutes the "Westernized" university? As Jack Goody (2006) notes, the notion that *institutionalized higher education* started with the founding of universities in Western Europe is inaccurate. The process of "Westernization", however, refers to a hegemonic development of European institutions in tandem with imperial state power. It is characterized by the study of difference and "the other" as objects rather than knowledge-producing subjects. As the Westernized university becomes a global institution, as "a place of authoritative knowledge, certified knowledge", it becomes "the heart of epistemic violence" (Pillay, 2015).

Much scholarship on the Westernized "neoliberal" university is abstracted from its evolution alongside the history of capitalism.[3] This coterminous evolution occurred in (at least) two ways. First, the imperial secular state was the university's patron, displacing the clerical patronage of the Middle Ages. The large number of universities established in Britain in the nineteenth century at the apex of the British empire's material wealth speaks to this shift. Second, there is a more complex entanglement at the level of ideas, pedagogy and curriculum. When discussing the historical significance of the Reformation, Friedrich Engels (1976) refers to English universities (colleges and public schools) as "Protestant monasteries". Max Weber (1992) theorized the link between the Reformation, the ideas and practices of Protestantism, and its relationship to the evolution of capitalism. The Protestant (and in particular Puritan) moral philosophy fermented in universities drove and transformed the British imperial plantations, and thus also contributed to the evolution of European capitalism. An element of this was the "calling" – the necessity to civilize the wilderness and Christianize society – which resulted in universities being both seedbeds and frontiers in Western colonial and capitalist expansion.

During the Reformation the tentative freedoms of the mediaeval and Renaissance universities evaporated, and they became instruments of state control and functionaries for the national Church (Maxwell, 1946: 2). Furthermore, under the British and Spanish empires, universities became frontiers and agents of European colonial expansion and control.[4] As Craig Wilder (2013: 33) argues,

> Colleges were imperial instruments akin to armories and forts, a part of the colonial garrison with the specific responsibilities to train ministers and missionaries, convert indigenous peoples and soften cultural resistance, and extend European rule over foreign nations.

Wilder (2013: 18) notes that the Spanish conquest of the Americas was "followed by intensive periods of institution building. Universities catalyzed cultural shifts in the capitals and hinterlands of the defeated Taino, Carib, Aztec, Mayan, and Incan empires".

Trinity College Dublin (1592) and Harvard College (1636) were founded by Cambridge-educated Puritans as an express feature of the British imperial plantation regime (Boran and Robinson-Hammerstein, 1998). Both institutions were mandated "to plant religion, civilitie, and true obedience in the hearts of this people" (Maxwell, 1946: 4). The universities worked in tandem with other forms of coercion to further the aims of the imperial state: "it was clear that education might be used in Ireland as a powerful auxiliary to military force, by breaking down the two great barriers to the spread of English influence: Catholicism and the Gaelic cultural tradition" (McDowell and Webb, 1982: 2, see also Mahaffy, 1903 and Maxwell, 1946). The Puritans focused on Harvard as a central locus of their theological aims to build a "city of peace" in the "wilderness" (Boran and Robinson-Hammerstein, 1998: 479). As part of the British colonization of the Americas, Harvard was followed by William and Mary (1693), Yale (1701), Codrington (1745) in Barbados, and New Jersey (1746). Wilder (2013: 17) notes that these colleges were "instruments of Christian expansionism, weapons for the conquest of indigenous peoples, and major beneficiaries of the African slave trade and slavery".

Universities were deeply entangled with the plantation regime. Wilder (2013: 8) elaborates on the relationship between colonial colleges and colonial slavery and the legacy of slavery in intellectual culture. The universities of Ireland and New England were crucial to the success of the plantations. The monarch used the university (together with the Inns of Court) "as the instrument to develop royal power of control effectively, to establish royal rule as the rule of law", justifying "the coordination and centralisation of power as a 'civilising' influence" (Boran and Robinson-Hammerstein, 1998: 487). Far from the protection of the metropolitan government, the plantation was precarious and potentially volatile. The case of Anne Hutchison – the radical spiritualist who threatened the internal cohesion of the Puritan community (1636–1638) – is evidence of this precariousness. Harvard College had to instil its "own consistent logic" and "It needed tried, old learning" to ensure stability and success (Boran and Robinson-Hammerstein, 1998: 501). It may seem anomalous that the universities were being used as the bulwark of "order" given that they were set up by intellectual rebels of the time. The Cambridge Puritan scholars who set up Trinity and later Harvard had become "unemployable in England" due to their religious radicalism (Boran and Robinson-Hammerstein, 1998: 483). When theorizing the modern episteme, Wynter (1984: 21) offers us the jester – the epistemic antagonist – and reminds us "every rebellion is therefore metamorphosed into a conservative state", from heresy to orthodoxy.

Despite the governing impetus to civilize "natives", this aim was a relegated prerogative of the establishment of universities. Trinity College Dublin was open to Irish students but "chiefly patronised by the settlers" (Maxwell, 1946: 6). Despite purporting to be a primary instrument in establishing civility, for a long time – in the case of both Trinity and Harvard – the colleges were "fairly useless to the 'native'" (Irish and Native American respectively) population (Boran and Robinson-Hammerstein, 1998: 492). In New England, a separate college for "native" people was proposed, but ultimately rejected by the settlers:

> In the first instance, the plantation/colonization enterprise was simply an opportunity for the good, educated, highly 'civilised' European Christians to demonstrate that they understood the will of God [...] That the focus was on the European Christian elite rather than the fate of the natives is clear from the Puritans' attitude to unsuccessful plantations.
>
> *(Morison, 1935: 161)*

The settler's account of why the plantations failed is illuminating. The Winthrop Papers, beginning in 1630, offer historical insight into the colonization of New England. They suggest that earlier plantations proved unsuccessful because

> their maine ende ... was carnall and not religious; they aymed cheifly at profitt and not the propagating of Religion; ... they vsed unfitt instrumentes, a multitude of Rude and misgoverned persons the very scomme of the lande; they did not establish the right forme of Government.
>
> *(Forbes, 1929–1992, cited in Boran and*
> *Robinson-Hammerstein, 1998: 500)*

There was a tension between patrician families engaged in commerce that depended on the stable existence of law and order, and the "unemployable English university scholars" with "prophetic Puritan minds" (Boran and Robinson-Hammerstein, 1998: 491). Both had an interest in cementing Christian order, either for the success of imperial conquest or for expansionist mercantile ambition. But what is of great interest is the relationship between those two divergent motivations and the universities' mediation of these forces. The institution was the crucial vessel required to progress imperial expansion at both the level of ideas and the practicalities of governmental control and resource exploitation. The university's role in containing, expanding and mandating power is interesting in the current context.

There is also a relationship between the university and epochal change, problematizing the "transfer of knowledge"; situating this phenomenon as an imperial exchange. Specifically, "*Translatio studii* [transfer of knowledge] was the crucial concomitant to *translatio imperii* [transfer of control]" (Boran and Robinson-Hammerstein, 1998: 480). But what remains in the current incarnation of the Westernized university of this type of epochal thinking? What are the implications for the Westernized university's strategic investment in the "knowledge-economy" as an expansionary logic: *translation studii* to *translation imperii*?

Wilder elaborates on the plantation universities and the production of knowledge and the intellectual cultures that grew out of these institutions. There was a relationship between the settler universities and the persecution of free black communities and schools. New England's colonizers (including the presidents of Connecticut's colleges and universities) "silenced debate about slavery and vehemently attacked abolitionism" (Wilder, 2013: 2). The colleges and universities progressed the imperial project through the perpetuation of "theories of racial difference and scientific claims on the superiority of white people. The academy refined these ideas and popularized the language of race, providing intellectual cover for the social and political subjugation of non-white peoples" (Wilder, 2013: 3).

Furthermore, the colleges crafted and propagated "a determinist and widely held vision that white people would become the sole possessors of the North American mainland" (Wilder, 2013: 10). So embroiled are such institutions, at the level of knowledge production, intellectual culture, financial and political power, that they must be considered an integral part of the settler colonial project. As Wilder (2013: 10) argues,

> The American college trained the personnel and cultivated the ideas that accelerated and legitimated the dispossession of Native Americans and the enslavement of Africans. Modern slavery required the acquiescence of scholars and the cooperation of academic institutions.

The examples of Harvard and Trinity afford some insight into how universities operated as an integral feature of the colonial plantation regime. Wilder develops this to analyse the ways in which educational institutions (their built environment, endowments and investments, employment practices, curriculum, and local political involvement) advanced the settler colonial project. In 1968, Wynter makes a similar link between the built environment of her university, the University of the West Indies, colonial history and contemporary knowledge production.

Wynter and the university in 1968

> Since 1948, the Mona campus of the University of the West Indies (UWI) has occupied a square mile of land in the parish of St Andrews, on the outskirts of the city of Kingston, in the island of Jamaica [...] from the mid-seventeenth century, the Mona and Papine sugar estates were centred within these boundaries. Remnants of their sugar work [...] can still be seen on campus. African slaves served the English and creole plantation owners [...] The English had replaced the Spanish [...] The Spanish themselves had overborne the Taino or Arawak people who preceded them.
>
> *(Brown, 2004: 1)*

During World War II, Mona was used as a military base and internment camp. Gradually transformed into university buildings, what remained of the architecture of slavery was not renovated:

> The development of the University's physical structures has interfered less with the old sugar estate remnants [...] An early decision to develop green space around the Mona Works ruins, was undoubtedly a factor in their having

remained relatively undisturbed. In 2001, however, the University elected to develop two of the major ruins [...] The boiling house/curing house complex is planned to become an upscale restaurant and bar.

(Brown, 2004: 45–46)

Those invested in UWI will decide how they wish to remember their history and organize their campus, but notably, the infrastructure of the plantation is now a University-run upscale restaurant and bar. Brown notes in her book about the history and heritage of the university: "At Mona, the past is always present" (2004: 45–46), but also, it seems, partly occluded. In this sense, it seems important to recall the history of what came to pass at Mona.

In 1655, Oliver Cromwell, Lord Protector of the Commonwealth of England, Scotland and Ireland, despatched a military force to colonize Jamaica. Burgeoning free trade and the sugar plantations had grown Britain's slave trade exponentially. Between 1680 and 1686, Britain transported a yearly average of 5,000 people to be enslaved, mostly from the African continent; by 1786 there were 610,000 of those people in Jamaica (Williams, 1944: 3). By 1761 there were over 9,024 enslaved persons labouring on the sugar plantations of St Andrews alone (Brown, 2004: 4). The total number of people enslaved to the British colonies between 1680 and 1786 was over two million (Pitman, 1917, cited in Williams, 1944: 33). Wynter (1971: 95) reminds us:

> The Caribbean area is the classic plantation area since many of its units were 'planted' with people, not in order to form societies, but to carry on plantations whose aim was to produce single crops for the market. That is to say, the plantation-societies of the Caribbean came into being as adjuncts to the market system; their peoples came into being as an adjunct to the product, to the single crop commodity the sugar cane which they produced.

The value of Jamaican sugar plantations to British capitalism, both as an import and export market, was enormous.

A list of enslaved persons on the St Andrews estates in 1783 indicates 172 at Mona (Brown, 2004: 15). Estate maps indicate the whereabouts of "slave villages", including the likely whereabouts of their huts. Historians also use botanical markers to place these villages: for example, the proximity to ackee and mango trees is an indication of their presence.

When Wynter writes a two-part article "about a little culture" in 1968, her analysis is fixed on the past-as-future as she reflects on the situation at UWI. What Wynter describes in terms of the university's relationship to society, the production of writing and culture, and the politics of knowledge production resonates with the critique of contemporary Westernized universities. It is the experience of exile and displacement that she believes should inform us as to the historical foundations of our shared heritage and shared future:

> Our condition is one of uprootedness. Our uprootedness is the original model of the total twentieth century disruption of man. It is not often appreciated

that West Indian man, qua African slave, and to a lesser extent, white inden-
tured labourer, was the first labour force that capitalism had totally at its
disposal. We anticipated by a century the dispossession that would begin in
Europe with the Industrial Revolution. We anticipated, by centuries, that
exile, which in our century is now common to all.

(Wynter, 1968a: 24)

The institution of racialized capitalism created the framework in which we are
all *in common but differently* and "without exception still 'enchanted', imprisoned,
deformed and schizophrenic in its bewitched reality" (Wynter, 1971: 95). But why
does Wynter respond by needing to "talk about a little culture"?

As a University we have attacked the distortion of the *economic* dominance;
yet we continue to reflect in our goals, curriculum, 'standards' the cultural
corollary of this economic arrangement.
    A new 'culture' for us is not a luxury, not and no longer the playmate of an
elite soul; it must be instead the agent of man's drive to survive the twentieth
century.

(Wynter, 1968b: 42)

The university will dictate cultural output because the Caribbean could not
support another institution of comparable size and thus "the University must commit itself to the cultural destiny of our territories" (Wynter, 1968b: 41). But here is
Wynter's injunction: "to which Jamaica, which Caribbean, is the University of the
West Indies to belong?" (1968b: 41). How might the university transcend the values
of the plantation on which it was built?

For if the history of Caribbean society is that of a dual relation between plantation and plot, the two poles which originate in a single historical process,
the ambivalence between the two has been and is the distinguishing characteristic of the Caribbean response. This ambivalence is at once the root cause
of our alienation; and the possibility of our salvation.

(Wynter, 1971: 95)

Katherine McKittrick (2013: 10–11) suggests that Wynter's distinction between
plantation and plot "provides the theoretical scaffolding that rethinks how our present spatial struggles around race, segregation and violence might be reimagined".
Plantation/plot dichotomy is explained thus: with the Western colonial experience the "oscillatory process" in which Man adapts to Nature morphs into "the
reduction of Man to Labour and Nature to Land" (Wynter, 1971: 99). At this point
Western Man saw himself as "the Lord and possessor of Nature", thus a process of
dehumanization and alienation was set in place (Wynter, 1971: 99). In old societies
with traditional values based on the old relation, resistance to this was possible. In
new societies, created for the market, there seemed (at first) no possibility of such
resistance. However, from early on (to maximize profit) the planters gave the slaves

plots of land on which to grow food and so on: "We suggest that this plot system was […] the focus of resistance to the market system and market values" (Wynter, 1971: 99). The African peasants transplanted to the plot the entire structure of values that had been created by the traditional societies of Africa: "Around the growing of yam, of food for survival, he created on the plot a folk culture – the basis of social order – in three hundred years" (Wynter, 1971: 99). Wynter (1971: 100) argues that this folk structure became a source of guerrilla resistance to the plantation system, after abolition: "The plantation was the superstructure of civilisation; and the plot was the roots of culture".

Wynter demonstrates what it means to honour the values of the plantation over the plot in an analysis of the output of the Creative Arts Centre, UWI. She is interested in "praxis" not "sermons"; what we *do* in universities, not what we *say* (Wynter, 1968b: 42). To that end, she observes the connection between writing and political consciousness; between culture and imperialism; between materialism and the negation of humanity; material culture and elitism; the "branch plant perspective"; and the choice of those working within the university.

Wynter takes the position that her history and present are so bound up in trauma that it is the function of culture – writing and art – to be aware of its own sickness; its own unreality. Furthermore, "To *reinterpret* this reality is to commit oneself to a constant revolutionary assault against it" (Wynter, 1968a: 24). Reinterpreting is a dual task of rejecting the epistemologies of Western imperial thought and also resisting and transforming the uprooting, itself a facet of imperial control.

Through the Western pursuit of knowledge "culture has become a mere appendage to the market mechanism; another industry among others" (Wynter, 1968a: 25). Wynter notes that the continuous uprooting of the writer is part of the mandating of imperial culture by the university. Producing writing in order to survive or satiate the market becomes self-destructive, another facet of negation. Writing for the market, for the "consumer-connoisseur" is another way to mandate power and hierarchy, because it is those with power who dictate market forces and direction (Wynter, 1968a: 25). Wynter links the market, elite culture and praxis which appeases these interests:

> The writer who consciously sets out to write for an elite is as much involved in the appeasing arts as the writer, who, wanting to write 'for the people' falls into a trap of writing for the consumer.
>
> *(1968a: 25)*

Wynter notes that she returned to Jamaica because she had no choice. And because of the arrangement of society her creative instincts are driven into exile: "We come to terms with our 'branch plant' existence, our suburban raison d'etre" (Wynter, 1968a: 26). She explains the "branch plant" as,

> a perspective that views the part for the whole; that adjusts new experience to fit an imported model, with a shift here and a shift there; that blinds its

horizons in order not to perceive the logical and ultimate connections, that would invalidate the original model that had formed his being and distorted his way of seeing. The 'branch plant' perspective of all the 'appeasing arts'; and of their corollary, 'acquiescent criticism'.

(1968a: 26)

For those ensconced within the university, Wynter posits a choice between critical and appeasing work. It is a choice not just for individuals working within the institution, but also for the institution as a whole:

the University is poised between the choice; and leaning towards the older, the easier, the gesture of silence. [...] Above all, this kind of silence, of the unsaid, has deprived the University [...] of a genuine sense of purpose.

Without the realization of this purpose [...] we shall continue to turn out an elite technocrat class (some with liberal slogans, some with Black Panther slogans) all seeking to take an elite place in any order (liberal, Communist, Black Power) which is based on an elite.

(Wynter, 1968b: 41)

The dangers of acquiescence to this situation, she claims, are attributable to a lack of awareness among academics, but also to deliberate rejection of such awareness (Wynter, 1968a: 32). Writing in 1968, Wynter encouraged her colleagues at Mona to reflect on history to write a more progressive future. In this spirit, I now consider the contemporary Westernized university in the context of the history of colonial plantation, thinking also about Wynter's "branch plant perspective":

Our purpose begins to formulate itself with our awareness of *the University, as the logical result of a common history* stretching over four centuries [my emphasis].

(1968b: 41)

Internationalization and the Westernized university

A cursory survey of literature on internationalization and the Westernized university reveals an uncritical tendency towards the language of empire. Consider this statement from Heriot-Watt University, Edinburgh, on its branch campuses:

Internationalism has always been a concept that Heriot-Watt approaches differently: many universities export their expertise but Heriot-Watt has taken its vision, root and branch, to centres across the world.

(Whitaker, 2016)

Sean Matthews – Director of Studies in the School of Modern Languages and Cultures, University of Nottingham Malaysia Campus – argues that despite being "weary as next sahib of the easy banality of equating any and all overseas educational activity as 'neo-colonial'", certain viewpoints suggest:"the direction of transmission, the mode of production of knowledge, primarily involves the UK's finest bringing enlightenment to the dark places of the earth" (2012a). British Foreign Secretary Boris Johnson (2013), brother of the former Minister for Universities and Sciences Jo Johnson, stated that branch plant universities in other countries could be considered "epigone descendants" of their British instigators. My intention is to consider the neo-colonial implications of contemporary internationalization in higher education, guided by Wynter's commentary on universities and plantation logic.

The internationalization of higher education at the national, sector and institutional levels "is the process of integrating an international, intercultural or global dimension into the purpose, functions or delivery of postsecondary education" (Huang, 2013: 2). The elements of internationalization that involve institutional expansion in countries other than the source include: articulation, capacity building, distance learning, hybrids, joint degrees, student and staff mobility, validation and branch campuses (University of Glasgow, 2017b) The latter are defined as:

> an entity that is owned, at least in part, by a foreign education provider; operated in the name of the foreign education provider; and provides an entire academic program, substantially on site, leading to a degree awarded by the foreign education provider.
>
> *(Garrett et al., 2016: 6)*

Internationalization also refers to the strategic recruitment of international students to attend "home" universities. It also has other dimensions, like for example, the diversification of curriculum and so on.

Historical context

The "internationalization" of higher education has a history. Futao Huang (2013: 14) suggests at least four phases: first, from the thirteenth to the eighteenth century, from the emergence of the nation state through the extension of Christendom in Europe and the Americas to the age of industrialization; second, the creation of uniform national culture and national higher education systems in Europe in the nineteenth century associated with "the advancement of the scientific revolution; the establishment of modern nation-states; colonisation of Africa, America, Asia and other continents; and the breakout of two world wars". The third phase from 1947 to 1971 concerns the context of the Cold War, characterized by the bifurcated interests and influence of the Soviet Union and the United States. The fourth phase beginning in the 1990s includes the great majority of countries (except the United States) increasing enrolment, the movement from elite to mass education (the UK and China) and post-massification and near universal access (Japan and Korea) (Huang, 2013: 17). Huang (2013: 15) notes that during the second phase:

> the real internationalisation of academic profession sometimes occurred in a negative way, aimed at the suppression of traditions and conventions of the

academic profession in colonies in particular. To illustrate this, the importa-
tion or transplantation of languages, teaching programmes, academic norms
and standards through military government or policies of colonisation were
sometimes used to accelerate the adoption of foreign imports and the decline
or the disappearance of national languages, cultures and academic traditions.

Tamson Pietsch picks up this thread in *Empire of Scholars*, opening with the 1912
Congress of the Universities of the British Empire: "the colonial delegates who
attended it are recognisable as agents of cultural imperialism: this helped extend
British dominion over indigenous lands and peoples, asserting their metropolitan
'expertise' in the colonies and fashioning loyal imperial subjects in the classroom"
(2013: 1). Congress delegates

> presumed the universality and superiority of 'Western' knowledge and cul-
> ture and established themselves as its local representatives, proudly proclaim-
> ing this position in the neo-gothic buildings they erected and the Latin
> mottos they adopted. Although they adapted old-world models, in their early
> years settler universities were very much local affairs. Fashioned by colonial
> politics and frequently funded by the state, they were small institutions that
> served the sons of the colonial elite.
>
> *(Pietsch, 2013: 5)*

Similar to today, "settler universities and the individuals who worked in them were both
local and global actors. They were rooted in specific social and political communities
and also wayfarers on international routes of scholarship" (Pietsch, 2013: 3). Pietsch's
(2013) work to discover imperial networks, postcolonial and transnational exchange,
and the social construction of scientific knowledge addresses this neglected area of
study. Nothing is new in this account of the negative effects of "internationalization"
– ergo colonialism – in the nineteenth century, other than it seems we have neglected
to make meaningful comparisons with the actions of Westernized universities today.

Contemporary context

We must consider the extent to which the current incarnation of transnational
higher education is different to that of universities which were established as a
means of colonization. This undertaking takes as its standpoint that the replication
of ideological apparatus within a system of domination negates the possibility of
ethical communication and cross-fertilization of ideas and education systems. We
must consider the drivers of international branch campuses and what constellations
of power are at play within these new relationships. Doing so entails an analysis of
the political economy of knowledge (and the institutions promulgating knowledge)
and the economic strategies and material mechanisms that drive these processes
(Rivera Cusicanqui, 2012: 95–109).

The growth of transnational education through internationalization and branch
campuses is significant. Towards the end of 2016, there were 251 branch campuses

in 72 countries around the world (Merola, 2016: 1). The number of branches has grown steadily in the past decade. The US, UK, Russia, France and Australia are the top five countries developing branch campuses. Although the US has the greatest number of branch campuses in absolute terms (78), the UK has the greatest proportion compared to the number of institutions in Britain: 39 international branches and 451 higher education institutions at source. Of interest in this current context, enrolment data demonstrates that some of the biggest branch campuses are "in countries and regions that have historic involvement and ties with the UK" (Merola, 2016: 2). London, Manchester, Liverpool and Glasgow – the historic cities of empire – are prominent overseas, with branch campuses of their universities.

The 2008 financial crisis was seen as an indication that Europe and the US were stagnating at the level of ideas (De Sousa Santos, 2014), a notion shared by the Rectors and Presidents of the nine universities who wrote the 2015 communiqué for the University of Vienna's 650th birthday:

> In a time of crisis [...] universities have a key role to play in Europe's revival. [...] In order to achieve this, they have to be active on the global scale.
>
> *(Vienna Communiqué, 2015, cited in*
> *Chou, Kamola and Pietsch, 2016: 1)*

But the idea of a Eurocentric "revival" driven by globalized European universities is problematic. Chou, Kamola and Pietsch argue that we need to understand the "transnational politics of higher education" rather than the "globalisation of higher education"; emphasizing "the way in which students, faculty, money, and institutions do not simply move across borders, but rather cut across them in ways that reinforce or alter the power hierarchy and particularity of the evolving international order" (Chou, Kamola and Pietsch, 2016: 2)

Since the 1980s universities have not just been party to the educational ideas caught up in the processes described above, but have increasingly become the agents of economic change and growth. In the Middle East, petro-states are investing in higher education as a tool towards economic diversification for post-fossil fuel growth (Chou, Kamola and Pietsch, 2016).

To elaborate the case of the UK, between 2008 and 2009 education exports – which include international branch campuses – brought in £14.1 billion to the UK economy, forecast to rise to almost £27 billion by 2025 (British Council, 2012: 10). The 2008 to 2009 figures are equivalent to 1.0 per cent of GDP and 8.4 per cent of total service exports (this compares to banking at 28 per cent) (British Council, 2012: 10). The five top host countries of international branch campuses, China, The United Arab Emirates, Singapore, Malaysia and Qatar, are either the richest countries in the world in GDP per capita, or experiencing the highest levels of economic growth.

Capital is a clear domestic driver for internationalization. The European University Association reports that public funding for UK higher education has fallen 28 per cent (nominal change) from 2010 to 2016 to less than 0.5 per cent GDP (European University Association, 2017). Universities have had to meet the shortfall and internationalization is one way of compensating for this loss of central government funding. International students contribute more than £7 billion to the UK economy (Universities UK, 2014: 1). In addition to escalating fees for RUK students across the UK,[5] international students bolster university budgets. At the University of Glasgow, an undergraduate in English Literature in 2017 who is a Scottish domiciled student pays £1,820 per year (the Student Awards Agency for Scotland will pay this if you are eligible) in comparison to an international student who will pay £14,900. At the University of Glasgow, international student numbers have risen from 3,420 in 2011 to 5,213 in 2015 (University of Glasgow, 2017a). Similarly, RUK students at Glasgow have risen from 1,799 in 2011 to 3,199 in 2015. International and RUK students are now subsidising the education of Scottish (and European Union) students. The British and Scottish governments have shifted the burden of funding higher education, not just to domiciled individual students, but also to those in search of a UK education from abroad. The exploitative dimension to this practice is clear. That this funding structure is necessary is undermined by Germany, who recently abolished tuition fees for home and international students (and has just four international branch campuses).

There is a further domestic driver of internationalization and particularly a driver of the international branch campus expansion. The current Conservative Party has for some time faced an internal ideological rift: between those supportive of the free movement of labour demanded by neoliberal economics and those Conservative nationalists who demand protectionist policies. Indeed, to the dismay of many in higher education, in late 2016, the government announced a cut in international student numbers to UK "home" universities (Fazackerley, 2016). However, the phenomenon of the international branch campus effectively solves this dispute within the Conservative Party, at least when it comes to immigration associated with education (a sector which has proven to be an easy target for Home Office).[6] Alongside overseas degree programmes (offered in partner institutions), branch campuses afford the maximum economic benefits from British educational institutions but without those individual students necessarily requiring a British visa. Indeed, the British government and complicit universities now extract large sums of money from students (often from Britain's former colonies) to pay for the education of British-born people and drive British economic growth.

The practical drivers of international expansion are a separate matter to the ethical questions surrounding whether Westernized universities *should* open branches in the first place. In his role as Vice Chancellor of Liverpool University, Drummond Bone co-founded Xi'an Jiaotong-Liverpool University, a branch campus at Suzhou, China. Bone was knighted in the 2008 Birthday Honours for services to Higher Education and the regeneration of the North West. The Bone Report on Internationalisation and Higher Education, commissioned by the Brown government in 2008, ostensibly called for a "paradigm shift from internationalisation as a recruitment-led activity to internationalisation as a comprehensive partnership activity" (Grant, 2013: 3). But the paper contains this statement:

> Two ethical issues are obvious – first the hegemony of western culture spread through trans-national education and secondly the effect it has on in-country capacity building. The latter is of course even more true of traditional recruitment of students from developing countries to the UK, where there is a significant loss of talent from the donor country. Arguably TNE [transnational education] reduces the loss, but it may actually increase the erosion of the host culture.
>
> *(Bone, 2008: 10)*

Bone's report offers no further comment on this topic. It is an area of concern more fully developed by academics and intellectuals in "host" countries. In 2011, Tan Sri Dato' Dzulkifli Abdul Razak stood down from the University Sains Malaysia after a decade as Vice Chancellor to lead a new initiative, Albukhary International University, a not-for-profit, philanthropic institution dedicated to opening the path to Higher Education for underprivileged students from around the world, and to 'Education for Being' as opposed to 'Education for Having' (Matthews, 2012a). The following is a summary of his reflections and critique of internationalization. Razak notes that the prevalence of the English language has the potential to diminish indigenous languages and cultures and that the overuse of English may lead to cultural homogenization. The pursuit of institutional reputation and rankings is arguably motivated more by desire for prestige than interest in educational cooperation. Global competition may diminish the diversity of institutional models of what constitutes quality higher education. The pursuit of "a single model of excellence" will have a homogenizing effect, and "may result in the concentration of scarce national resources in a few or single institution to the detriment of a diverse national system" (Razak, 2012: 5). The risk of brain drain will continue or even accelerate due to internationalization, depriving existing institutions the ambitious and talented students they need to advance, change and grow. Large-scale recruitment of international students can be at times "unethical" (Razak, 2012: 6). Asymmetrical institutional relationships may emerge due to internationalization; access to resources will lead to the pursuit of goals which most benefit powerful institutions to the detriment of the partners (privileging Western universities). Taken together, Razak's critique offers a clear insight into the ways in which internationalization is having a homogenizing effect at the epistemic level. Internationalization also risks further exacerbating unequal power relationships between developed and developing countries, institutions and students. There is, of course, a positive case to be made around the benefits of internationalized education. However, Colin Grant, Vice President (International) of the University of Southampton suggests that the "internationalisation of the curriculum" (i.e. the potentially transformative sharing of epistemic systems and practices) "is the area of internationalisation that receives least interest" from those involved (Grant, 2013: 12).

In response to Razak's comments, Matthews (2012b) notes "the familiar allegation that quibbling about values and epistemology butters no parsnips, much less brings water to the thirsty, food to the starving, electricity to the powerless". All imperialists, observes one commentator on Matthews' blog, think they are doing something good. Wynter questions the very premise of "international development". If the South is to be ascendant, "it might very well have to get rid of the concept of 'development' altogether" (Wynter, 1996: 299). In this way, development becomes an epistemological rather than an economic matter, and the question of where education fits into this becomes one of whether it challenges the present hegemonic mode of rationality. In other words, what does not "butter the parsnips" is a world order justified by a master discipline of economics that ensures money flows from the South to the North by way of 'Third World' debt. And what illustrates the ideology of the parsnip (as Derek Walcott's daffodils)[7] is an education system that exacerbates and facilitates debt – to nations, institutions and individuals – from the South to the North, and justifies it as an education.

Conclusion

On the relationship between capital and the latest incarnation of global illiberalism, Zsolt Kapelner (2016) develops a particularly interesting argument in the context of thinking about the university's entanglement with the colonial plantation, as frontiers and seedbeds. It is both critical and a cause for concern that the Westernized university's current pull is to authoritarian contexts. Among the top five host countries for international branch campuses the United Arab Emirates and Qatar are absolute monarchies, with no political opposition and no trade unions, and China is a one-party state, which forgoes political opposition. This is not to say that there are not hugely valuable educational institutions and knowledge systems in these countries. But Westernized universities' current proclivity towards authoritarian contexts for international branch campuses is provocative. In addition to the idea that universities are frontiers and seedbeds, universities may also turn out to be barometers. My principal concern here is not academic freedom, freedom of speech, equality and so on. One of the arguments advanced by Piya Chatterjee and Sunaina Maira in *The Imperial University* relates to the relationship of the academy to the military and the prison industrial complex:

> We argue that the state of permanent war that is core to US imperialism and racial statecraft has three fronts: military, cultural, and academic. Our conceptualization of the imperial university links these fronts of war, for the academic battleground is part of the culture wars that emerge in a militarized nation.
>
> *(2014: 7)*

The subject areas perused by international branch campuses are often very narrow and market driven, furthermore by "business" it is possible to discern military or petrochemical- related science and technology. If the humanities and social sciences are pursued at all, to what extent do syllabi depart from Walcott's daffodils – their own imperial disciplinary history – and the demands of the contemporary market? The argument made by Kapelner is that capital and illiberalism have a fecund relationship. My question is: what role are universities playing as frontiers and seedbeds? And if Westernized universities are barometers, what does their migration to authoritarian contexts suggest about their values, and what does it tell us about the evolving (and historically repetitious) future of these institutions and the societies in which they are based?

Internationalization in this context does not just refer to the opening of branch campuses by Westernized universities. In fact, much more lucrative are the millions of students, often from former colonies, who travel to "the West" to study. The neo-colonial manifestation of the university plantation is an epistemological, economic and political phenomenon. The "original" settler colonial universities were not wholly interested in educating the "natives". The contemporary neo-colonial university offers millions of international students a homogenized diet of knowledge as part of an exploitative economic, social and political arrangement. Education is only one frontier in the story of renewed forms of colonization by capital. As Wynter (1968: 42) reminds us, the political and cultural work we must *do* on this topic is not elite academic sport, "it must be instead the agent of man's drive to survive the twentieth century". Outing the logic of the plantation and affiliating ourselves to the values of the plot (outlined in this article by Razak) is part of resisting the "branch plant perspective". Once again – as relevant now as in 1968 – "the University [and those working within it] is poised between the choice" – between critical or appeasing work (Wynter, 1968b: 41).

Notes

1 Wynter consistently draws attention to the institutional dimensions of systemic oppression. Her account of the modern episteme notes the significance of the institutional/disciplinary transition in medieval Europe, from scholasticism to the *studia humanitatis*, which nationalized and privileged knowledge based on secular rationality (1984:19–70).
2 I will resist direct metaphorical readings of the university and the plantation, mindful of the work of Katherine McKittrick and others, who are using the concept of plantation – pasts, futures and logics – in very concrete political ways. That is not my intention here. See, McKittrick, 2013: 1–15.
3 See, for example, Brown, 2015; Giroux, 2014.
4 France and Portugal preferred higher education to take place only within their own national borders. Belgium banned higher education in its colonies outright.
5 Rest of the UK students include those required to pay up to £9,250 per year in fees. The Scottish government has prevented comparative fee raises for Scottish domiciled students.
6 This can be read in two ways. First, establishing immigration quotas for students is bureaucratically more straightforward than cutting immigration from industry. Second, higher education representatives and institutions have been weak in defending their interests against Home Office encroachments.
7 Walcott famously questioned why Caribbean students knew more about Wordsworth's daffodils than they did indigenous flowers.

References

Bone, D. (2008) *Internationalisation of HE: A Ten Year View*. London: UK Government.

Boran, E. and Robinson-Hammerstein, H. (1998) The promotion of 'civility' and the quest for the creation of a 'city of peace': The beginnings of Trinity College, Dublin and of Harvard College, Cambridge, Mass. *Paedagogica Historica* 34(2): 479–505.

British Council (2012) *The Shape of Things to Come: Higher Education Global Trends and Emerging Opportunities to 2020*. London: British Council.

Brown, S.F. (2004) *Mona Past and Present: The History and Heritage of the Mona Campus, University of West Indies*. Kingston: University of the West Indies Press.

Brown, W. (2015) *Undoing the Demos: Neoliberalism's Stealth Revolution*. New York: Zone Books.

Chatterjee, P. and Maira, S. (2014) *The Imperial University: Academic Repression and Scholarly Dissent*. Minneapolis: University of Minnesota Press.

Chou, M.H., I. Kamola and T. Pietsch (2016) (eds) *The Transnational Politics of Higher Education: Contesting the Global / Transforming the Local*. London: Routledge.

De Sousa Santos, B. (2014) *Epistemologies of the South: Justice against Epistemicide*. Boulder: Paradigm Publishers.

Engels, F. (1976) Historical significance of the reformation, *Marx Engels on Literature and Art*. Available at: www.marxists.org/archive/marx/works/subject/art/index.htm (Accessed 30 November 2016).

European University Association (2017) *Public Funding Observatory*. Available at: www.eua.be/publicfundingobservatory (Accessed 29 March 2017).

Fazackerley, A. (2016) UK considers plans to nearly halve international student visas. *The Guardian* 12 December. Available at: www.theguardian.com/education/2016/dec/12/uk-halve-international-student-visa-tougher-rules (Accessed 20 December 2016).

Forbes A.B. (ed) (1943) *The Winthrop Papers, Volume 2*. Boston: Massachusetts Historical Society.

Garrett, R., Kinser, K., Lane, J.E. and Merola, R. (2016) International branch campuses: Trends and developments 2016. *The Observatory on Borderless Higher Education*. Available at: www.obhe.ac.uk/documents/view_details?id=1035 (Accessed on 19 December 2017).

Giroux, H.A. (2014) *Neoliberalism's War on Higher Education*. Chicago: Haymarket Books.

Goody, J. (2006) *The Theft of History*. Cambridge: Cambridge University Press.

Grant, C.B. (2013) Losing our chains: Contexts and ethics of university internationalization. *Stimulus Paper Series*. Available at: www.bath.ac.uk/news/system/wp-content/uploads/2013/02/LFHE_-Grant_SP_v4.pdf (Accessed 20 December 2016).

Huang, F. (2013) The internationalization of the academic profession. In F. Huang, M. Finkelstein and M. Rostan (eds) *The Internationalization of the Academy: Changes, Realities and Prospects*. New York: Springer, pp. 1–21.

Johnson, B. (2013) Interviewed by Sir Drummond Bone for Balliol College, University of Oxford. Available at: www.youtube.com/watch?v=rKfbvlKyx6g (Accessed 14 December 2016).

Kapelner, Z. (2016) Capital's paradise: The rise of global illiberalism. *LeftEast* 9 December. Available at: www.criticatac.ro/lefteast/capitals-paradise-the-rise-of-global-illiberalism/ (Accessed 3 August 2017).

Mahaffy, J.P. (1903) *An Epoch in Irish History: Trinity College Dublin Its Foundations and Early Fortunes 1591–1660*. London: T. Fisher Unwin.

Matthews, S. (2012a) Return of the civilizing mission. *University of Nottingham Blogs / Knowledge Without Borders*. Available at: http://blogs.nottingham.ac.uk/kwbn/2012/07/06/return-of-the-civilizing-mission/ (Accessed 3 August 2017).

Matthews, S. (2012b) Resisting internationalisation: Thinking about some contradictions in transnational education, *University of Nottingham Blogs/Knowledge Without Borders*. Available at: https://blogs.nottingham.ac.uk/aworldincrisis/2012/03/28/resisting-international-isation-thinking-about-some-contradictions-in-transnational-education/ (Accessed 3 August 2017).

Maxwell, C. (1946) *A History of Trinity College Dublin 1591–1892*. Dublin: The University Press Trinity College.

McDowell, R.B. and Webb, D.A. (1982) *Trinity College Dublin 1592–1952*. Cambridge: Cambridge University Press.

McKittrick, K. (2013) Plantation futures. *Small Axe* 17(3): 1–15.

Merola, R. (2016) 250+ and counting: What does IBC growth tell us about the direction of the phenomenon and its place in CBHE? *Observatory on Borderless Higher Education*. Available at: www.obhe.ac.uk/documents/download?id=1052 (Accessed 19 December 2017).

Morison, S.E. (1935) *The Founding of Harvard College*. Cambridge, MA: Harvard University Press.

Pietsch, T. (2013) *Empire of Scholars: Universities, Networks and the British Academic World 1850–1939*. Manchester: Manchester University Press.

Pillay, S. (2015) Decolonizing the university. *Africa is a Country*. Available at: http://africasa-country.com/2015/06/decolonizing-the-university/ (Accessed 16 November 2016).

Pitman, F.W. (1917) *The Development of the British West Indies, 1700–1763*. New Haven: Yale University Press.

Razak, D.A. (2012) Internationalisation: Levelling the playing field, *The Observatory on Borderless Higher Education*. Available at: www.obhe.ac.uk/conferences/2012_global_forum_kuala_lumpur/2012gf_presentations/S10_Razak (Accessed 19 December 2017).

Rivera Cusicanqui, S. (2012) Ch'ixinakax utxiwa: A reflection on the practices and discourses of decolonization. *South Atlantic Quarterly* 111(1): 95–109.

University of Glasgow (2017a) QlikView student headcount profiles for the University of Glasgow. Available at: www.gla.ac.uk/services/planning/qv/ (Accessed 3 August 2017).

University of Glasgow (2017b) What types of collaboration are possible?. Available at: www.gla.ac.uk/services/academiccollaborations/typesofcollaboration/ (Accessed 15 December 2017).

Universities UK (2014) International students in higher education: The UK and its competition. Available at: www.universitiesuk.ac.uk/policy-and-analysis/reports/Documents/2014/international-students-in-higher-education.pdf (Accessed 3 August 2017).

Weber, M. (1992) *The Protestant Ethic and the Spirit of Capitalism*. Translated by T. Parsons. London and New York: Routledge.

Whitaker, A. (2016) Scotland's universities seek to boost their overseas influence. *Holyrood*. Available at: www.holyrood.com/articles/inside-politics/scotlands-universities-seek-boost-their-overseas-influence (Accessed 15 December 2017).

Wilder, C.S. (2013) *Ebony and Ivory: Race, Slavery, and the Troubled History of America's Universities*. New York: Bloomsbury.

Williams, E. (1944) *Capitalism and Slavery*. Chapel Hill: The University of North Carolina Press.

Wynter, S. (1968a) We must learn to sit down together and talk about a Little Culture: Reflections on West Indian writing and criticism, part one. *Jamaica Journal: Quarterly of the Institute of Jamaica* 2(4): 23–32.

Wynter, S. (1968b) We must learn to sit down together and talk about a Little Culture: Reflections on West Indian writing and criticism, part two. *Jamaica Journal: Quarterly of the Institute of Jamaica* 3(1): 27–42.

Wynter, S. (1971) Novel and history, plot and plantation. *Savacou* 5: 95–102.

Wynter, S. (1984) The ceremony must be found: After humanism. *Boundary* 2: 19–70.

Wynter, S. (1996) Is development a purely empirical concept, or also teleological?: A perspective from 'We the Underdeveloped'. In A. Y. Yansané (ed) *Prospects for Recovery and Sustainable Development in* Africa. London: Greenwood Press.

3

THE WHITE UNIVERSITY

A platform of subjectification/subjugation

Lucas Van Milders

Introduction

This chapter conceptualizes the contemporary university as a white institution that hegemonizes an entangled episteme of neoliberalization and racialization. As such, it argues that two strands of critique that have recently proliferated about the university are in urgent need of an intersectional standpoint that draws on the institutionalized cartography of each discursive practice (i.e. of neoliberalization and racialization) yet also moves beyond them. The main contention will be that both discourses entail a process that seeks to obfuscate the locus of enunciation through a set of disciplinary and institutional practices. These range from mechanisms of perpetual control that produce the subjectification/subjugation of the so-called entrepreneur of the self (*homo economicus*) to the invisible category of whiteness as a systemic tool of measurement and hierarchy that segregates white superiority and privilege from non-white inferiority and exploitation. In a broader sense, the argument builds on the widespread view that the discursive and material operations of global capitalism and the institutionalized reproduction of (neo)colonial racism ought to be understood in their historical and sociological entanglement. The latter is then perpetuated in the contemporary university – conceptualized as both neoliberalized and westernized – as an institution that consolidates this symbiosis through both its production of the perpetually indebted student-subject and the atomized precariat of exploited academic workers on the one hand and its erasure of difference through empty rhetoric about diversity and inclusion that centralizes whiteness as a hegemonic category of knowledge production on the other.

Structurally, the argument unfolds in three steps. First, the white university is conceptualized as grafting an institutionally obscured conception of humanity-as-whiteness onto the university as a site of knowledge production as well as subject formation. This starts from the imprint that colonial imperialism has left on pedagogical practices and institutions that consolidate the Western standpoint as a transcendental view from nowhere, which explores, measures, divides, categorizes, and hierarchizes the entire world as an enclosed site of knowledge exploitation. Through epistemicidal curricula and course modulations, difference is depoliticized into hollow commitments to widening diversity that predicate inclusion and participation on a covert acceptance of white hegemony and supremacy. Second, the operations of the white university are analysed through the mechanisms of perpetual control and subjectification/subjugation that have engulfed both students and academic workers. Regarding the former, students-as-consumers are reconstituted as indebted subjects that actively feed the perpetualized cycle of lifelong learning as never-ending work on the self. Regarding the latter, class war is covertly restored as an atomized precariat is increasingly entangled in practices of permanent surveillance and self-disciplination that further segregate precarious knowledge workers that engage in mass teaching from an academic elite of corporate managers and researchers of excellence. In the third and concluding section, I suggest that what propels both dynamics is the obscured positionality of whiteness that inserts an aspirational logic of belonging into a practice of perpetual subject formation. This requires an intersectional standpoint that exploits the cracks within the white university as sites of resistance. The latter range from a recollection of alternative grounds of knowledge cultivation to practices of refusal and counter-organization that operate within, outside, and beyond the contemporary university. This entails both the development of a heterogeneous positionality that reflects an intersectional standpoint as well as prefiguring the pluriversity-to-come.

The unbearable whiteness of the university

In order to advance a genealogy of the white university as a hegemonic institution of white supremacy, critical scrutiny of the epistemic conjuncture that underpins it is required. According to Said's well-known argument, a power/knowledge nexus infuses the incessant production of the Orient as politically, sociologically, militarily, ideologically, scientifically, and imaginatively inferior to the Occident (2003 [1978]: 3). More so, through Orientalist discourse, the white Western voice becomes hegemonic and produces the Orient as such: an inferior category that is in existential need of Western tutelage/intervention. However, as a number of decolonial scholars have argued (Mignolo, 2011; Dussel, 2013 [1998]; Grosfoguel, 2013; De Sousa Santos, 2014), the emergence of difference is not to be situated within the geopolitical expansion of European empires during the 18th and 19th century but within the European encounter with difference in 1492. Indeed, the conquest, colonization, enslavement, and exploitation of the Americas marked the inception of a hierarchical difference between the mutually constructed European self and the non-European other that would come to determine the colonial mode of engagement between the west and the rest for the following 500 years (Todorov, 1984; Chomsky, 1993; Hall, 1996). It is this epistemic conjuncture that involves a coloniality of power that privileges the European epistemic along racialized, gendered, classed, linguistic, and religious blinkers (Quijano, 2000; Grosfoguel, 2009). It has directly informed the formation of a pedagogical orientation and knowledge production which, in turn, facilitates the reproduction and maintenance of this entangled hierarchy. For Willinsky, this involves an educational history of imperialism:

> Colonial rule gave rise to a new class of knowledge workers in universities, government offices, industry, and professions devoted to colonial conquest by classification and categorisation. They traveled, formed learned societies, created experimental gardens, and established laboratories. They joined in building the military, political, religious, and economic structures of global empires.
>
> *(1998: 26)*

Although the university as such is a medieval institution, through its institutionalized practices of "show-and-tell" colonialism and catering towards the interests of the recently emerged bourgeois class, it effectively became both a handmaiden and beneficiary of empire. Accordingly, the European university soon proliferated itself as the laboratory of imperial design and categorization through the emergence and consolidation of social scientific disciplines in the 18th century. This process is, for instance, discernible in the work of French sociologists such as Auguste Comte, who exploited the vast magnitude of the French colonial empire to present what was essentially a comparative study of the societal and developmental differences between the European metropole and the non-European colony as an indicator of the "positive" quality of European society. This was also presented as the aspirational goal for "traditional" (i.e. non-European) societies (Martineau, 2000; Connell, 2007: 9). Indeed, as suggested above, within the practices of disciplinarity and categorization, colonial societies are converted into "laboratories of Western sciences" where imperial control can be exercised through a variety of political, governmental, economic, and pedagogical tools (Smith, 2012: 38). Is it indeed through the double meaning of disciplinarity – as involving both institutional practices of governmentality as well as scientific modes of knowledge production and classification – that the university's entanglement with authority is revealed (Foucault, 1991 [1975]; Kaplan, 1995). This point is analysed in more detail in the next section. It is, however, necessary at this point to interrogate which dynamic effectively propels the classificatory cartography of the colonial world. In his encounter with the "Indians", Christopher Columbus famously invoked a lack of common humanity as determining the difference between Europeans and the indigenous peoples he had "discovered" (Todorov, 1984: 42–43). This did not merely amount to hierarchical shades of humanity but rather a ruthless questioning of humanity as such – a disposition that has incessantly acted as a blueprint for the relations between colonized and colonizer.

To demonstrate how this racialized episteme informs the contemporary university, a couple of clarifications are warranted. First, it is important to foreground that this chapter primarily draws on the context of the UK whilst nonetheless remaining aware of the extent to which the argument developed here resonates beyond the UK. Second, the argument in no way seeks to exceptionalize as a whole the Western context (since this would effectively reproduce the aforementioned episteme). The point is rather to reflect on how the idea of the *westernized* university also relates to the idea of the *westernizing* university. The governmental, corporate, and managerial practices that are unfolding (and exacerbating) in contemporary universities everywhere correspond to both ideas. The white university might then initially appear as an outdated notion, as equality policies in the UK have fostered a so-called overrepresentation of students from minority ethnic groups compared to white students (Modood and Shiner, 1994; HEFCE, 2005). However, as research has shown that students from these backgrounds "are more likely to drop out, are less likely to gain good honours degrees and tend to leave universities with fewer job prospects" (Pilkington, 2013: 226), the problem seems to be an institutional one that does not concern demographic representation. I note also that staff from minority ethnic groups often encounter obstacles in gaining promotion and face the implicit taken-for-grantedness of the whiteness of senior staff. If anything, the gradual increase of representation for both students and staff in the UK has given rise to their depiction as space invaders that are not the "natural" bodies of academia (Puwar, 2004: 53).

The institutional racism of the university is then exposed in its anxiety for the increased representation of ethnic minorities, which also reinforces a "White middle-class desire for distance from the Other" (Crozier *et al.*, 2016: 40). The demographic changes in the university sector that have taken place over the last two decades can thereby be interpreted as constituting a challenge to white, middle class privilege and the values that reflect it. This does not solely concern a racialized conflict but also a classed one, where preservation of the middle class is under threat from lower-class students that are expected to remain a distanced other (Holt and Griffin, 2005). The next section further explores this operation through discourses of neoliberalization. What suffices to be stipulated at this point is that racialized barriers in the university sector (both for students and staff) effectively operate to re-centre and normalize white privilege. And this cannot be resolved by solely advancing non-white representation. What recent debates on the whiteness of the curriculum – such as the "Rhodes Must Fall" movement in South Africa and the "Why is My Curriculum White?" debates in the UK (Why is My Curriculum White? collective, 2015; Heleta, 2016) – have exposed is that university courses, textbooks, and the institutional assumptions on knowledge production that inform them are constructed around an epistemicidal dynamic that strategically appropriates non-Western histories and knowledges into a fabricated and Eurocentric conception of history (Paraskeva, 2016).

These disciplinary mechanisms then condition representation on the acceptance of an institutionally engrained privilege of whiteness. As Clegg *et al.* (2003: 164) write, "although the student body is now more diverse, non-white students are only legitimately 'included' when they conform to the expectations of the traditional middle class white academic". Indeed, it is through empty rhetoric about diversity, inclusion, and participation that whiteness is centralized as a hegemonic category of knowledge production and difference is effectively sanitized. More so, it is through racial inequity that educational policies on diversity and inclusion become implicated in their constitutive role vis-à-vis white supremacy. This does not so much refer to the social darwinism vulgarly advocated by racist movements but rather a political, economic, and cultural system in which whites overwhelmingly exert power and control material resources, in which conscious or unconscious ideas of white superiority and entitlement are widespread, and in which relations of white dominance and non-white subordination are re-enacted daily across a broad array of institutions and social settings. The university's gatekeeping institutions privilege whiteness as an exercise of power and condition inclusion on the acceptance of this privilege. Indeed, white supremacy operates covertly by acquiring a nearly invisible status. In fact, it is this invisibility of whiteness as a racialized category that white people consistently fail to acknowledge as they attribute their privilege (and indeed even their identity and so-called "colour-blindness" (Bartoli *et al.*, 2016)) to anything but their racialized whiteness. Despite the absence of an articulated intentionality, the outcomes of educational policies – which dramatically favour whiteness as its key beneficiary – clearly indicate that white supremacy is a central part of the educational system in the UK (Gillborn, 2005: 498). What is therefore urgently required is not merely a diagnosis of how the white university functions through the integration of white supremacy as a mechanism of power but also a more fundamental dislodging of whiteness from its position of invisible transcendence (i.e. supposedly being above racialization) – as it is this invisibility that determines its efficacy in maintaining white privilege.

The white university is prefigured by "whiteness as an unmarked marker of others' differentness – whiteness not so much void or formless as norm" (Frankenberg, 1993: 198). The invisibility or transcendence of whiteness effectively operates as a distributive mechanism that categorizes and hierarchizes otherness. It appropriates universality and conditions inclusion on the extent to which the other effectively recognizes this appropriation. This dynamic of white supremacy permeates the white university through its policies, practices, and organization (Law, 2017: 335). It is therefore an obfuscated or hidden positionality of whiteness that essentially propels the entire project of white supremacy. To explore how this takes place in practice, the next section engages the ramifications and permutations of neoliberal discourses that effectively facilitate the reproduction of whiteness in the contemporary university.

The university as a platform of subjectification/subjugation

The struggle against the white university has manifested itself into a plethora of narratives, movements, and practices that seek to unsettle, dismantle, or disrupt the continued reproduction of white supremacy in higher education (Tate and Bagguley, 2017), and start from demonstrating that "universities are middle-class institutions displaying, imposing, and reproducing White middle-class values" (Crozier *et al.*, 2016: 41).Yet what this also implies is that the white university is a site where not only raced and gendered but also classed antagonisms unfold. An important strategy in this context has therefore been the institutional adoption of tokenism that seeks to obfuscate the class antagonism and the broader entrenchment of the university in the operations of global capitalism through the aforementioned emphasis on participation and inclusion as well as modifications to the white curriculum. Although these developments are intrinsically praiseworthy, a discriminatory emphasis on (a misconstrued conception of) identity politics as a privileged site of anti-hegemonic struggles tends to leave that very hegemony intact. What therefore needs to be addressed if the critique of the white university is not to stray into comforting tokenism is how exposure to capitalist and disciplinary forces has not so much coincided with the whiteness of the university as it has become instrumental to it. A main contention of this chapter is indeed that, in its production of neoliberal subjectivity, the contemporary university has managed to appropriate identity politics and the unsubstantiated celebration of diversity in order to obscure issues of raced, gendered, and classed difference and thereby leaves the hegemonic invisibility of whiteness untouched. This distorted version of identity politics that accommodates a segregated understanding of particular antagonisms in order to leave class antagonism unaffected has been addressed by Žižek as elemental to the continued survival of global capitalism (2008: 271). It is through neoliberalism's "silent" revolution (Hall, 2011), which has greatly affected the contemporary university, that this misappropriation of identity politics and struggles for recognition can be understood as a restorative project of class power. Through this appropriation, discourses of diversity that do not directly challenge global capitalism are rearticulated within the latter as a mechanism that further strengthens raced and gendered exploitation.

So how has neoliberalization impacted the contemporary university? In order to answer this question, a little further dwelling is required on neoliberalism as an economic, political, and moral project of subjectification and subjugation. Advocating the maximization of entrepreneurial freedom through an emphasis on radical marketization and market deregulation, protection of property rights, free trade, and liberal individuality, neoliberalism ultimately revolves around "attempts to restore or reconstruct upper-class power" (Harvey, 2007: 28). As these policies never stimulated the worldwide growth they promised and disproportionally benefitted the upper-class while exacerbating living and working conditions for the labouring classes, neoliberalism has been successful at restoring class power. Specifically then, the neoliberal revolution was not as much a shift of ideologico-economic paradigms, as it radically changed the very workings of these paradigms as such. This is due to the simple fact that neoliberalism is not just a project of managing governments or economies but also one of managing or reconstituting the human subject. Through its fundamental commitment of dismantling the welfare state, it also sought to produce a subjectivity that not merely condoned this project but also actively supported and even executed it. For Foucault (2008: 226), this subject is the entrepreneur of the self or *homo economicus*, "being for himself his own capital, being for himself his own producer, being for himself the source of his earnings". As neoliberalism restructures every aspect of life along economic lines, neoliberal subjectivity becomes its own enterprise: every action is charted along a cost/benefit analysis that rewards profit and penalizes expenditure. Under neoliberalism, the human subject is no longer a worker or labourer but human capital: an abstracted capacity with an ever-potential return on investment.

Although explicitly geared towards an individually liberated society of governmental deregulation, it has become clear that neoliberalism's fetishization of competitiveness necessitates mechanisms of perpetual control that are implemented to secure the further mutation and survival of capitalism. As every aspect of life becomes commodified, so is it transformed into a mechanism of control: "Man is no longer man enclosed, but man in debt" (Deleuze, 1992: 6). Indeed it is debt, as a mechanism that incessantly prolongs the neoliberal programme, which infuses an element of morality into neoliberal ideology. Similar to the German word *Schuld* (which refers to both debt and guilt), indebtedness has become elemental to the neoliberal subject's internalization of the enterprise of the self as an existential responsibility. Through debt as a mechanism of socialization, subjectification/subjugation takes place, as it produces an indebted subject that is perpetually responsible (i.e. guilty) for the accumulated debt. When the entrepreneur of the self makes erroneous investments, social rights (insurance, healthcare, unemployment benefits) are transformed into privatized debt that demands repayment. The controlling function of debt is then a subjugating mechanism for stifling action and possibilities under the guise of personal guilt, which further entrenches the subject into the neoliberal condition that is indebtedness for life: when everyday life is commodified into a profitable market for private actors that replaces the dismantled welfare state, debt effectively becomes an instrument for disrupting working class power into an atomized precariat of entrepreneurs of the self that are conditioned to live in endless debt/guilt.

When we then look at the effects that neoliberalism has had on higher education, it is first and foremost important to clarify, as it was already indicated in the previous section, that the university "has always been an institution supporting authorities, if not altogether an institution for exercising and accommodating subjugation" (Raunig, 2013: 24–25). As an institution that is fully integrated into the developments described over the previous paragraphs, educational policies in the UK have been subject to neoliberal restructuring since the government of Margaret Thatcher (Edwards, 1989). Through neoliberalism's discursive toolbox of increased corporatization, fiscal responsibility (i.e. guilt), fetishized competitiveness, and dogmatic marketization, a series of processes are unleashed that not only produce the privatization and commodification of education and research or the consumerization of students but also the precarization of academic workers. To clarify this, it is necessary to reconceptualize the neoliberal university as a platform capitalist institution (Hall, 2016: 49). Increasingly widespread through the success of companies such as Facebook, Google, Amazon, and, especially, Airbnb and Uber, platform capitalism concerns a predominant business model that is premised upon bringing different actors together (suppliers, customers, workers, advertisers) on a platform that profits from the accumulation of vast amounts of data as well as the renting or leasing of its infrastructure (Srnicek, 2017). In higher education, this post-welfarist model has been adopted by the university, which rents or leases its services and facilities like a platform or "licensed storefront for brand-name corporations – selling off space, buildings, and endowed chairs to rich, corporate donors" (Giroux, 2009: 674). Corporate culture has greatly affected the university where customer satisfaction has hegemonized the neoliberal ethos.

In the UK, the commodification of education has developed through the introduction of tuition fees in 1998 and, especially, the raising of the tuition fee cap to £9000 a year. Through the UK government's Student Loan Company, for instance, students are gradually bogged down in debt – something that adds to entrepreneurial subjectification and entrenchment into indebtedness for life. Yet it also has a disciplinary function of class power restoration that forces poorer students out of access to tertiary education and reconfigures education itself as a tool to earn money to repay the debt (Cowden and Singh, 2013: 33). This then incentivizes students to regard their education as a return on investment that enhances the profitability of the micro-enterprise that they are (Patrick, 2013). In other words, learning becomes vocational training tailored to the needs of the market. Yet this also involves an endless cycle of learning, as the self is in constant need of self-improvement and further training – a condition that makes "students permanently start over from the beginning and repeatedly call previously acquired certificates into question" (Raunig, 2013: 32).

However, one of the most detrimental effects of the platform capitalist university has been the precarization of academic workers (Standing, 2011). The institution namely wants "to be able to draw from a pool of part-time, hourly-paid, zero-hours and no-contract workers who are available 'on tap', often at extremely short notice" (Hall, 2016: 18). In the UK, it is, for instance, estimated that when "the use of atypical staff is factored in, 54% of all academic staff and 49% of all academic teaching staff are on insecure contracts" (UCU, 2016: 1). It is for these reasons not entirely surprising that 60% of part-time teachers are unable to improve their professional development or consider a future within their profession (Leigh, 2014). Neoliberalism once again restores class power, as the precarization and exploitation of academic workers who engage in mass teaching coincides with a privileged elite of so-called researchers of excellence that consolidate their privilege through a series of gatekeeping mechanisms (Raunig, 2013: 34). Discourses of excellence therefore control academic workers by producing a "domain of professional self-marketeers" (Graeber, 2015: 135) and this is effectively where one of neoliberal's most powerful levers operates: control mechanisms of subjectification/subjugation. Performance-based funding systems like, for example, the Research Excellence Framework and the Teaching Excellence Framework fetishize competitiveness as a tool for fostering excellence yet they also operate to consolidate the university's close ties to governmental and corporate power as well as to ideologically facilitate patterns of inclusion and exclusion on a global scale (Watermeyer and Olssen, 2016; Raaper, 2017).

They are arguably accompanied by increased levels of surveillance (Page, 2017), yet they also need to be understood against the background of "a set of mechanisms of *perpetual pedagogical control* in which *the market will become a regulator of pedagogical possibilities*" (Heaney and Mackenzie, 2017). Perpetual control and incessant subjectification/subjugation are mediated through an ideology of excellence and competitiveness that is manifested through surveillance, assessments, evaluations, rankings, and so on. Yet what effectively operates through these normative frameworks are modes of exclusion and inclusion, which function like consumer product ratings that systematically associate excellence and "world-class" status with a position of privilege for elite institutions, the hegemony of English in both research and teaching, a disadvantaged position of institutions in the Global South, as well as a more exacerbated racialized and gendered privilege that accompanies the white university (Amsler and Bolsmann, 2012; Brøgger, 2016; Saunders and Ramírez, 2017; Heaney and Mackenzie, 2017).

Through the contemporary university as a platform that hosts gatekeeping mechanisms of perpetual control and surveillance, gendered, racialized, and classed patterns of inclusion and exclusion are reproduced that effectively condition participation on an acceptance of white, masculine, and middle-class values. This not only refers to how platform capitalism itself reproduces gendered and racialized biases through its restoration of upper-class power (Srnicek, 2017: 77), but also a broader appropriation of hollow discourses about diversity that serve to sanitize issues of difference. A platform of subjectification/subjugation, then, ultimately functions on the obfuscated positionality of whiteness as a hegemonic category that distributes racialized hierarchy.

Through the cracks in the white university: The pluriversity-to-come

This chapter has argued that the obscured positionality of the white university is reproduced through the gatekeeping mechanisms of neoliberal subjectification/subjugation. In a brief concluding section, two questions need to be answered: what effectively underpins this entire dynamic and what strategies are there for those who seek to resist the white university and consider alternatives to it? Regarding the first question, it needs to be noted that the Eurocentric identification with universality "lies in the confusion between abstract universality and the concrete world hegemony derived from Europe's position as centre" (Dussel, 2000: 471). Although both the foundation of the Eurocentric identification with universality as well as the concealed whiteness of humanity were anchored within the history of European colonialism, this imperial positionality gradually evaporated into what Hamid Dabashi (2009: 213) describes as "*epistemic endosmosis*" – i.e. "a mode of knowledge that is devoid of agential subjectness, which is the modus operandi of an empire without hegemony". Although it is essential to retrace the white university's epistemic conjuncture within the history of European colonialism, the evaporated *ego cogito* needs to be reprovincialized if one is to effectively unsettle the invisibility of whiteness in the contemporary university. Indeed, by actively re-positioning or re-situating the European perspective, whiteness is stripped from its invisible transcendence. From a decolonial perspective then, this embodied standpoint of knowledge production intersects with different disenfranchised modes of knowledge production, which, as mentioned earlier, have either been appropriated into the Eurocentric canon or gentrified into non-existence. As Robbie Shilliam (2015: 11) argues, the resistance against "racial inequality requires the cultivation of spiritual, philosophical and political standpoints that reach across these lines to rebind the various descendant communities who have and continue to suffer from such exploitation and dispossession". Although configuring such a cultivation into extensive detail would exceed the scope of this chapter, it is nonetheless important to note that unsettling humanity-as-whiteness involves a practice of recollecting standpoints of knowledge cultivation that disavow whiteness as a hegemonic category and can, for these reasons, pose an effective alternative against it (Collins, 1990).

This brings us to the second and remaining question: what strategies can be deployed to unsettle the white university in the daily practice of academic work? Such strategies concern both dislodging the contemporary university's integration within capitalist circuits of (knowledge) production and subjectification/subjugation as well as displacing the whiteness of the identity that is produced through the university's institutional organization and practices. As such, the two questions seem to be interrelated, as the development of alternative modes of knowledge cultivation necessitates the conceptualization of particular strategies and practices that can effectuate them. More so, as these practices seek to disrupt the hegemony of the raced, classed, and gendered subjectivities that are constitutive and themselves constituted through the neoliberal restoration of upper-class power, the latter is once again exposed as an intensification of white invisibility. What is therefore needed is "the creation of emotional, embodied, and intellectual pedagogies of possibility" (Motta, 2013: 86). This not only entails reconstituting stifled possibilities that facilitate alternative practices of learning, but also a resistance against the perpetual control that subjects the academic precariat and commodifies students as consumers. There is arguably a broad string of practices that reflects this, ranging from building robust and effective unions to grafting critiques of the white university into teaching and research practices. Yet, as these examples risk being subsumed into the neoliberal discourse of self-management, more fundamental work would involve "the building of organisational power in autonomous, directly democratic workplace associations" (Kamola and Meyerhoff, 2009: 18). In other words, against the atomized entrepreneur of the self, counter-organizing can take place in which the fragmentation of the neoliberal subject is recomposed "around the project of class struggle and the production of anti-capitalist commons" (ibid.: 20).

It goes without saying that these struggles take place on many fronts and materialize in different ways depending on different research communities and teaching contexts. As such, it does not merely concern a struggle *against* the university, as it also concerns one *within* it as well as *beyond* it. Indeed, it is a "struggle for autonomous free spaces *in* the university and simultaneously self-organisation and *auto-formazione* beyond existing institutions" (Raunig, 2013: 48–49). This heterogenous positionality has been referred to by Harney and Moten (2013) as a struggle for and of the undercommons. The latter refers to "the underground of the university" (ibid.: 28), informal spaces that constitute formal spaces yet can never be contained by them: dining halls and smoking lounges, extra-curricular workshops and trans-academic conferences. These spaces are formative to academic life yet they elude being institutionalized into it. It is within and through the undercommons that "refugees, fugitives, renegades, and castaways" (Harney and Moten, 2013: 29) can be harboured and intersect with professionalized academic labour. It is within this crack – as "the perfectly ordinary creation of a space or moment in which we assert a different type of doing" (Holloway, 2010: 21) – that the dependency of the university on the undercommons can be exploited for counter-organization. These cracks then give space for an active refusal to repay debt and feel guilt for indebtedness. It is through the undercommons then, that the heterogeneous positionality can be conceptualized as one that is not a modified view from nowhere that re-centralises the Eurocentric university but an intersectional standpoint that draws on the different perspectives that are subjugated through "racial neoliberalism" (Maldonado-Torres, 2016: 39). Indeed, what enshrines both the critiques of the neoliberalized and the westernized university is a challenge to "work in, through and outside the sites of knowledge production – with local or global social movements - participate in the slow careful process of decolonising the westernised university from within and without" (Velázquez, 2016: ix). By exploiting the cracks in the white university and experimenting with new practices of doing and self-organizing in solidarity with those working without or outside, a pluriversity-to-come can be prefigured as a daily praxis.

References

Amsler, S.S. and Bolsmann, C. (2012) University rankings as social exclusion. *British Journal of Sociology of Education* 33(2): 283–301.

Bartoli, E., Michael, A., Bentley-Edwards, K.L., Stevenson, H.C., Shor, R.E. and McClain, S.E. (2016) Training for colour-blindness: White racial socialisation. *Whiteness and Education* 1(2): 125–136.

Brøgger, K. (2016) The rule of mimetic desire in higher education: Governing through naming, shaming and faming. *British Journal of Sociology of Education* 37(1): 72–91.

Chomsky, N. (1993) *Year 501: The Conquest Continues*. London: Verso.

Clegg, S., Parr, S. and Wan, S. (2003) Racialising discourse in higher education. *Teaching in Higher Education* 8(2): 155–168.

Collins, P.C. (1990) *Black Feminist Thought: Knowledge, Consciousness, and the Politics of Empowerment*. New York: Routledge.

Connell, R. (2007) *Southern Theory: The Global Dynamics of Knowledge in Social Science*. Cambridge: Polity Press.

Cowden, S. and Singh, G. (2013) The poverty of student life: The politics of debt. In S. Cowden and G. Singh (eds) *Acts of Knowing: Critical Pedagogy In, Against and Beyond the University*. London: Bloomsbury, pp. 15–40.

Crozier, G., Burke, P.J. and Archer, L. (2016) Peer relations in higher education: Raced, classed and gendered Constructions of Othering. *Whiteness and Education* 1(1): 39–53.

Dabashi, H. (2009) *Post-Orientalism: Knowledge and Power in Time of Terror.* London: Transaction Publishers.

Deleuze, G. (1992) Postscript on the societies of control. *October* 59: 3–7.

De Sousa Santos, B. (2014) *Epistemologies of the South: Justice against Epistemicide.* Boulder: Paradigm Publishers.

Dussel, E. (2000) Europe, modernity, and Eurocentrism. *Neplanta: Views from South* 1(3): 465–478.

Dussel, E. (2013 [1998]) *Ethics of Liberation: In the Age of Globalisation and Exclusion.* Durham: Duke University Press.

Edwards, R. (1989) Margaret Thatcher, Thatcherism and education. *McGill Journal of Education* 24(2): 203–214.

Foucault, M. (1991 [1975]) *Discipline and Punish: The Birth of the Prison.* London: Penguin.

Foucault, M. (2008) *The Birth of Biopolitics: Lectures at the Collège de France, 1978–1979.* New York: Palgrave Macmillan.

Frankenberg, R. (1993) *White Women, Race Matters: The Social Construction of Whiteness.* London: Routledge.

Gillborn, D. (2005) Education policy as an act of white Supremacy: Whiteness, critical race theory and education reform. *Journal of Educational Policy* 20(4): 485–505.

Giroux, H.A. (2009) Democracy's nemesis: The rise of the corporate university. *Cultural Studies – Critical Methodologies* 9(5): 669–695.

Graeber, D. (2015) *The Utopia of Rules: On Technology, Stupidity, and the Secret Joys of Bureaucracy.* New York: Melville House.

Grosfoguel, R. (2009) A decolonial approach to political-economy: Transmodernity, border thinking and global coloniality. *Kult* 6: 10–38.

Grosfoguel, R. (2013) The structure of knowledge in westernized universities: Epistemic racism/sexism and the four genocides/epistemicides of the long 16th century. *Human Architecture: Journal of the Sociology of Self-Knowledge* 11(1): 73–90.

Hall, G. (2016) *The Uberification of the University.* Minneapolis: University of Minnesota Press.

Hall, S. (1996) The West and the Rest: Discourse and power. In S. Hall, D. Held, D. Hubert and K. Thompson (eds) *Modernity: An Introduction to Modern Societies.* Oxford: Wiley-Blackwell, pp. 185–227.

Hall, S. (2011) The neo-liberal revolution. *Cultural Studies* 25(6): 705–728.

Harney, S. and Moten, F. (2013) *The Undercommons: Fugitive Planning and Black Study.* New York: Autonomedia.

Harvey, D. (2007) Neoliberalism as creative destruction. *The ANNALS of the American Academy of Political and Social Science* 610(1): 22–44.

Heaney, C. and Mackenzie, H. (2017) The Teaching Excellence Framework: Perpetual pedagogical control in postwelfare capitalism. *Compass: Journal of Learning and Teaching* 10(2).

Heleta, S. (2016) Decolonisation of higher education: Dismantling epistemic violence and Eurocentrism in South Africa. *Transformation in Higher Education* 1(1): 1–8.

Higher Education Funding Council for England (2005) Equal opportunities and diversity for staff in higher education. Available at: www.hefce.ac.uk/pubs/hefce/2005/05_19/05.19.pdf

Holloway, J. (2010) *Crack Capitalism.* New York: Pluto Press.

Holt, M. and Griffin, C. (2005) Students versus locals: Young adults' constructions of the working-class Other. *British Journal of Social Psychology* 44(2): 241–267.

Kamola, I. and Meyerhoff, E. (2009) Creating commons: Divided governance, Participatory management, and struggles against enclosure in the university. *Polygraph* 21: 5–27.

Kaplan, M. (1995) Panopticon in Poona: An essay on Foucault and colonialism. *Cultural Anthropology* 10(1): 85–98.

Law, I. (2017) Building the anti-racist university: Action and new agendas. *Race Ethnicity and Education* 20(3): 332–343.

Leigh, J. (2014) "I still feel isolated and disposable": Perceptions of professional development for part-time teachers in HE. *Journal of Perspectives in Applied Academic Practice* 2(2): 10–16.

Maldonado-Torres, N. (2016) The crisis of the university in the context of neoapartheid. In R. Grosfoguel, R. Hernández and E.R. Velázquez (eds) *Decolonising the Westernised University: Interventions in Philosophy of Education from Within and Without*. London: Lexington Books, pp. 39–52.

Martineau, H. (ed) (2000) *The Positive Philosophy of Auguste Comte. Volume 1*. Kitchener: Batoche Books.

Mignolo, W.D. (2011) *The Darker Side of Western Modernity: Global Futures, Decolonial Options*. Durham: Duke University Press.

Modood, T. and Shiner, M. (1994) *Ethnic Minorities and Higher Education*. London: Policy Studies Institute.

Motta, S. (2013) Pedagogies of possibility: In, against and beyond the imperial patriarchal subjectivities of higher education. In S. Cowden and G. Singh (eds) *Acts of Knowing: Critical Pedagogy In, Against and Beyond the University*. London: Bloomsbury, pp. 85–123.

Page, D. (2017) The surveillance of teachers and the simulation of teaching. *Journal of Education Policy* 32(1): 1–13.

Paraskeva, J.M. (2016) *Curriculum Epistemicide: Towards an Itinerant Curriculum Theory*. New York: Routledge.

Patrick, F. (2013) Neoliberalism, the knowledge economy, and the learner: Challenging the inevitability of the commodified self as an outcome of education. *ISRN Education* 1–8.

Pilkington, A. (2013) The interacting dynamics of institutional racism in higher education. *Race Ethnicity and Education* 16(2): 225–245.

Puwar, N. (2004) Fish in and out of water: A theoretical framework for race and the space of academia. In I. Law, D. Phillips and L. Turney (eds) *Institutional Racism in Higher Education*. Stoke on Trent: Trentham Books, pp. 49–58.

Quijano, A. (2000) Coloniality of power, Eurocentrism, and Latin America. *Nepantla: Views from South* 1(3): 533–580.

Raaper, R. (2017) Tracing assessment policy discourses in neoliberalised higher education settings. *Journal of Education Policy* 32(3): 322–339.

Raunig, G. (2013) *Factories of Knowledge, Industries of Creativity*. Los Angeles: Semiotext(e).

Said, E.W. (2003 [1978]) *Orientalism*. New York: Penguin.

Saunders, D.B. and Ramírez, G.B. (2017) Against 'Teaching Excellence': Ideology, commodification, and enabling the neoliberalisation of postsecondary education. *Teaching in Higher Education* 1–12.

Shilliam, R. (2015) *The Black Pacific: Anti-Colonial Struggles and Oceanic Connections*. London: Bloomsbury.

Smith, L.T. (2012) *Decolonizing Methodologies: Research and Indigenous Peoples*. London: Zed Books.

Srnicek, N. (2017) *Platform Capitalism*. Cambridge: Polity Press.

Standing, G. (2011) *The Precariat: The New Dangerous Class*. New York: Bloomsbury.

Tate, S.A. and Bagguley, P. (2017) Building the anti-racist university: Next steps. *Race, Ethnicity and Education* 20(3): 289–299.

Todorov, T. (1984) *The Conquest of America: The Question of the Other*. New York: HarperPerennial.

University and College Union (2016) Precarious work in higher education: A snapshot of insecure contracts and institutional attitudes. Available at: www.ucu.org.uk/media/7995/ Precarious-work-in-higher-education-a-snapshot-of-insecure-contracts-and-institutional-attitudes-Apr-16/pdf/ucu_precariouscontract_hereport_apr16.pdf

Velázquez, E.R. (2016) Introduction. In R. Grosfoguel, R. Hernández and E.R. Velázquez (eds) *Decolonising the Westernised University: Interventions in Philosophy of Education from Within and Without*. London: Lexington Books, pp. ix–xvi.

Watermeyer, R. and Olssen, M. (2016) 'Excellence' and exclusion: The individual costs of institutional competitiveness. *Minerva* 54: 201–218.

Why is My Curriculum White? Collective (2015) 8 reasons the curriculum is white, *Novara Media*. Available at: http://novaramedia.com/2015/03/23/8-reasons-the-curriculum-is-white/.

Willinsky, J. (1998) *Learning to Divide the World: Education at Empire's End*. Minneapolis: University of Minnesota Press.

Žižek, S. (1999) *The Ticklish Subject: The Absent Centre of Political Ontology*. London: Verso.

4

CAN THE MASTER'S TOOLS DISMANTLE THE MASTER'S LODGE?

Negotiating postcoloniality in the neoliberal university

Lili Schwoerer

Introduction

On the 16th of January 2016, a majority of students at the Oxford University Union voted in favour of removing a statue of Victorian imperialist Cecil Rhodes from the top of Oriel College (BBC, 2016).[1] This decision was preceded by months of student-led campaigning, inspired by calls to "decolonize" education in South Africa. Beginning at the University of Cape Town, activists participating in what has come to be known as the "XMustFall" movement aim to free Higher Education of its colonial remnants and neo-imperialist implications. Despite widespread public attention for this campaign, Oriel College announced two weeks after the vote that the statue would remain in place. Newspapers reported that wealthy donors, in anger about the "shame and embarrassment" caused by the College's response to the campaign, had threatened to withdraw donations amounting to £100m, thus leading Oriel's board of governors to act against the student vote (Mortimer, 2016).

The message of the decision seemed all too clear: money talks in today's university. The threat of losing donors' support can consequently overrule decisions made by the academic body, no matter how democratic they might be. The incident corroborates recent scholarship on neoliberal trends in UK Higher Education (henceforth: HE) – educational scholars widely agree that the increasing "marketization" of HE limits institutional autonomy, and aligns universities' values with those of the market (e.g. Canaan and Shumar, 2008; McGettigan, 2013; Slaughter and Rhoades, 2004).

In addition, the events at Oxford highlight a close connection between this neoliberal "marketization" of UK HE and possibilities for anti-colonial critiques and interventions. Scholars and activists have long argued that the Western university has since its inception served to reflect and reproduce racist, colonial structures of power-knowledge (e.g. Ashcroft, Griffiths and Tiffin, 1995; Mihesuah and Wilson,

2004; Said, 1978; Smith, 1999). However, these critiques have only rarely been taken into account by those critiquing neoliberal developments in HE. The events of January 2016 underline the urgency of exploring the connections between neoliberal trends in UK HE and the ongoing reproduction of colonial hierarchies of knowledge.

This chapter attempts to begin such an exploration.[2] Drawing on research at my home institution and the department in which I was based in 2018 at the time of research – the Sociology department at Cambridge University – I will sketch out some of the ways in which the relationship between colonial discourses and neoliberalism manifested itself in this context. Interviews with students and staff explore how those embedded within the contemporary university sought to unsettle eurocentrism in Sociology teaching, and the possibilities for anti-colonial agency that they identify as arising amidst the entanglement of (post-)coloniality and neoliberal governmentality. It will become clear that navigating these complex interconnections can often consist of striving to gain legitimacy while simultaneously seeking to challenge the basis on which legitimacy is granted. Students and staff navigated the contentious ground between being marginalized or ignored on one hand, and being read in the language of neoliberal multicultural discourse on the other, when taking action to unsettle eurocentrism in Sociology.

Conceptualizing (post-)coloniality in contemporary HE

Educational scholars widely agree that British universities have experienced intensive restructuring measures during the last thirty years. The introduction of university fees and scrapping of maintenance grants, 'Research Assessment Exercises' and the establishment of external evaluations through bodies such as the Quality Assurance Agency (QAA), paralleled by a rise in New Public Management strategies in university governance, many argue contribute to an increasing influence of market forces on university governance, and thus a move away from the self-governing nature of the sector (Ball, 2012; Canaan and Shumar, 2008; Collini, 2017; McGettigan, 2013; Slaughter and Rhodes, 2004). Ever-increasing "audit" aims to improve the profitability of research and teaching, thus tying the survival of academic disciplines to the achievement of externally defined goals and benchmarks (Davies, 2005; Strathern, 2000). It is argued that these neoliberal trends manifest themselves on a multiplicity of levels within the universities, "diminish(ing) the relationships, ideas and subjectivities that maintain critical spaces external to pervasive market relationships" (Shahjahan, 2015: 491).

While it is undoubtedly useful to draw out the multiple levels on which neoliberal capitalism affects academic life, bemoaning the increasing dominance of "audit" and the subsequent loss of spaces for critical knowledge production tends to paint a picture of a linear development (or downfall) from a critical, self-governed public university to an externally regulated private one that continually decreases the space for the production of non-commodifiable scholarship (Gerrard, 2015). Portraying the rise in "audit" and managerialism as a neoliberal attack on the "golden era" of autonomous pre-neoliberal education then easily glosses over the fact that UK HE has never been a disinterested quest for pure knowledge, but instead has since its inception been closely tied to imperial interests, shaping the criteria for what constitutes legitimate academic knowledge and who is able to produce it, as well as in terms of more direct, material implications of the university with colonization and imperialism (see for example Coloma, 2010; Kuokkanen, 2007; Mihesuah and Wilson, 2004; Morgensen, 2012; Ngugi wa Thiong'o, 1986; Said, 1978; Smith, 1999; Wilder, 2013; Wynter, 1994). This oversight is endemic to much scholarship on the "neoliberal university", despite the fact that postcolonial and other anti-colonial scholars have for a long time extensively problematized the manifestations of colonial power relations in knowledge production, and consequently explored possibilities for dismantling them.

UK sociologists, drawing on such scholarship, increasingly concern themselves with the racialized and gendered legacy of their own discipline, and knowledge production in HE more generally (Ali, 2007; Bhambra, 2015; and Back and Tate, 2015; Connell, 2007; Gutiérrez Rodríguez, Boatca and Costa, 2010). The ways in which intersecting power structures of race and gender shape experiences in UK HE are also widely investigated (Ali, Benjamin and Mauthner, 2004; Andrews and Palmer, 2016; Bhopal, 2014, 2015; Law Phillips and Turney, 2004; Mirza, 2008; Simmonds, 1992). While not focusing specifically on the relationship between colonial discourses and recent neoliberal shifts in HE policy, this important set of literature invites us to look closely at their interconnections, and how they might work in conjunction to shape epistemic hierarchies in contemporary HE.

This volume's theoretical underpinning, the "coloniality/modernity" paradigm, can also provide us with useful insights for such an investigation. Aníbal Quijano's (2000, 2007) "coloniality of power" is helpful for my analysis in emphasizing the fundamentally intertwined nature of global structures of subjugation with eurocentric, colonial systems of knowledge. María Lugones (2007) importantly highlights the necessity of investigating the coloniality of categories of gender and sexuality. We are thus invited to trace the connections between colonial knowledge subjugation, historically and ongoing, and contemporary curricula and scientific canons. It is, however, essential to remember that such an analysis must be sensitive to the specificities of histories of colonization, and the ways in which these shape contemporary relations between colonizer and colonized. Tuck and Yang (2012: 17) powerfully remind us that "Decolonization is not a metaphor": subsuming all anti-racist and social justice work under the banner of "decolonization" easily falls prone to erasing the specificities of ongoing structures of settler colonialism, and struggles against them. Using "decolonization" as a metonym for social justice can constitute a "move to innocence", enabling settler scholars to deny their complicity in settler colonial oppression. Tuck and Yang write on occupied indigenous land, but their argument remains relevant in the UK context: as we seek to unsettle eurocentrism in the UK university, we must strive to understand, name and dismantle our own implications in reproducing epistemic hierarchies, rather than using the decolonial paradigm's claim about the all-encompassing nature of coloniality to deny our complicity in reproducing colonial power relations.

Nevertheless, the coloniality/modernity framework can be, alongside other anti-colonial and anti-racist epistemic critiques, valuable in highlighting the connection between knowledge production in universities, and intersecting racialized and gendered hierarchies in wider society. Drawing on such critiques, while closely investigating the ways in which sexist and racist epistemic hierarchies are reproduced on a micro-level, will then perhaps bring us slightly closer to understanding the way in which the "complexly mutating entity" (Mbembe, 2016: 8) that is the contemporary university operates.

Methodology and sample

I am participating in what I critique. The discipline of Sociology has shaped my thinking and my academic work, and thus has led me to voice the arguments presented in this chapter in the first place. My critique of the neoliberal academy is made possible by my embeddedness within it – and by my profiting from what it has provided, and continues to provide me with. As a white woman, I benefit from the racist structures of power-knowledge that this chapter is critiquing, and my academic work relies, at least to an extent, upon the reproduction of sexist and racist discourses. This chapter, then, also attempts to begin an exploration of such complicities, and of the possibilities of critiquing the neoliberal university from within it.

My research focused specifically on sexist and racist discourses and their intersections, thereby sidelining the many other ways in which colonial relationships of power-knowledge manifest themselves, in UK HE and beyond. It is, however, essential to acknowledge that these discourses cannot be understood separate from each other, and that they interlock with a multiplicity of other oppressive structures (including, but not limited to, sexuality and class) at all times (Combahee River Collective, 2016). I focused on racism and sexism in an attempt to balance time constraints with analytical precision. I sought to cut through some of the layers of association that accompany the word "colonial", perhaps also causing some productive discomfort in those whose curricula are named as that – sexist, and racist.

The process of researching this study was always mediated through my positionality as a white woman, and so are the findings presented here. The ways in which my white privilege played out while researching told its own story about how racial and gendered power relations manifest in HE. Whiteness hides the specificity of its subject position and accords authority to the speaker; it is reasonable to assume that this affected the ways in which my participants understood and responded to the project, as well as the wider reception of my work. Feminist methodology emphasizes that interviews are always saturated in power, and these power relations are never symmetrical (Nagar and Swarr, 2010). I am necessarily positioned differently from my respondents due to the fact that I pose the questions, and ultimately select how to present the responses. The power relations between my respondents and me differed in each interview, but the ways in which they were mediated by race and gender were central, affecting what was talked about and what was left out, and the ways in which stories were told. As a white woman, I was situated as an outsider when my respondents discussed issues of racism. On the other hand, some aspects of narratives about sexist exclusion resonated with my own experiences as a woman in academia. Such feelings of recognition were always partial, mediated by our specific positionalities.

Finally, it is also important to be mindful of the stories that weren't told – the call for interviews, dispersed through an undergraduate e-mail list and snowball sampling, builds barriers that exclude those who do not have the time, energy and security to respond.

Cambridge should not be read as a representative case for an audited "neoliberal university", but rather stands out as an interesting case both due to its historical role in the British Empire and its importance in shaping what is considered legitimate knowledge in wider society today (Symonds, 1993; Wilder, 2013). The Sociology department at Cambridge is comparatively small, and of fifty-two academic posts that existed in the academic year 2015–2016, sixteen taught undergraduate students. The department consistently ranks highly in national and international league tables on teaching and research "excellence" (THE, 2015; Top Universities, 2016).

A brief content analysis of the 2015/2016 undergraduate curriculum reveals that men, authors racialized as white and authors located in Europe and North America were heavily represented in reading lists and exam essay questions. It is important to note the problems of this preliminary analysis. Perhaps most importantly, it equates (visible) identity markers with subject positions. Further, by drawing attention to the number of women writers, and writers of colour in undergraduate syllabi, I am not trying to claim the existence of an essentialized knowledge stemming from the "black" or "female" standpoint that is inaccessible to those not belonging to that particular identity group. However, accepting that epistemic location *does* matter with regards to knowledge production, I offer a starting point for further, more nuanced analysis. For now we can say that in the majority of courses, topics concerning themselves with geographical areas outside of Europe and North America were largely talked about, rather than speaking for themselves. This was particularly prevalent in courses which concern themselves with "theory" or with large-scale, arguably universal phenomena such as "modernity", "globalization" and "capitalism". Some courses on the other hand stand out for their diversity of authors and locations that are written about and, at times, from. The modules on "race" and "racism" and feminism stand out for diversity of speaker positions, and the course in "advanced social theory" moves away from the conventional, white cisgender male cannon to look at places beyond Europe and the USA as sites of theory production. We must also note that many of these courses were optional, third-year modules. It could be argued that these interventions therefore stood in a non-threatening relationship to the white, male "building blocks" that constitute the discipline, are pushed to the margins and therefore legitimately ignored. Interviewing students and staff allowed me to explore this dynamic further.

Navigating legitimacy

Through semi-structured interviews with eight students and seven teaching staff in Spring 2016, I explored the ways in which sexist and racist discourses in Sociology operate, and their practices of resistance against them. I recruited student interview participants through the department's undergraduate e-mail list, and gained some additional respondents via snowball sampling. I chose to interview only under-graduate students, and only those for whom Sociology had been a substantive part of their degree. I e-mailed all lecturers in the department individually. This likely produced a sample that was skewed towards those interested in curriculum reform, (post-)coloniality or critiques of HE, but resulted in in-depth, collaborative discussions on themes such as the history and socio-political role of the social sciences, individuals' relationships to their subjects, their experiences teaching and learning and the relationship of these to their subject position, as well as their thoughts on the "decolonization" movement in UK HE.

Overall, it became clear that both students and teaching staff experience ways in which canonical standards are shaped by neoliberal university reforms and "audit" culture. Mechanisms of "audit" often work in conjunction with "older" hierarchies, manifesting themselves through ritual and symbolism, when structuring what con-stitutes "legitimate" sociological knowledge along gendered and racial lines, amongst others. My respondents tell stories of how epistemic norms form barriers that are often subtle and hard to pin down and articulate. Knowledge that seeks to challenge such norms, as well as the bodies producing it, often finds itself at the margins of the hegemonic disciplinary discourse. Students and staff nevertheless make inter-ventions that subvert and contest racist and sexist epistemic hierarchies, while often at risk of being either incorporated into logics of neoliberal "inclusion", or alter-natively marginalized. My interviews underline how this tension does not prevent creative, critical engagement with racialized and gendered epistemic hierarchies.

Rules of legitimacy: Auditing the sociological

In my interviews, students and staff frequently identify ways in which canonical rules, subtly defining what counts as a sociological *way of doing things* (Connell, 1997), are influenced and shaped by mechanisms of external evaluation. Some, however, also suggest that sociologists might be able to challenge discourses about the subject's character and purpose by drawing on existing scripts about the pro-gressive nature of the discipline. The tension between these different conceptions of the discipline is reflected in curricula, and is thus worth discussing in some detail.

When asked about the limitations they face when attempting to challenge euro-centrism in their own teaching, Dr Jones, a lecturer in the department, answers:

> Just the canon [...] Western science is Western science. When we look back on the legacy of our craft, it's a primarily eurocentric legacy. Which means it's hard, when we do research we have to address questions that have

some endurance in the canon. That have some recognized legitimacy and importance.

(Dr Jones, lecturer)

Challenges to norms and conventions must necessarily be rendered intelligible. In the case of contemporary Sociology this means that it must in some sense relate to a set of largely eurocentric texts and also must work within a defined set of epistemological and methodological approaches (Hunter, 2002). How set, then, are these rules? How can we imagine possibilities for their subversion, or contestation, within Sociology at Cambridge? And how are these possibilities shaped by neoliberal HE reforms?

When looking closely at how the canon is produced at a micro-level, we can recognize that the perception of *what Sociology is* contributes significantly to *what Sociology can be*. Despite the relatively recent institutionalization of the alleged "founding fathers" as central to the study of Sociology (Connell, 1997), several of my participants perceive their location at the "core" of the discipline to be as a major barrier to challenging epistemic hierarchies in curricula. This is, for instance, illustrated by Jie, who claims that there is need to change

> [...] the way that students think about it, the way that – "This is just what Sociology is" and then we sometimes study other people.
>
> *(Jie, first year HSPS student)*

My interviews, however, suggest that the script of Sociology as being "just old white men" (Jie) is not the only available one. Many of my respondents underline the progressiveness, interdisciplinarity and radical history of the "promiscuous social science" (Dr Randall). The fact that the discourse of Sociology as being inherently emancipatory exists, then, might potentially allow for challenging the canonization of eurocentric literature, methodology and epistemology to a larger extent than would be possible otherwise. The script of Sociology as "just old white men" stands next to the script of Sociology as concerned with "debates about gender, race, class and diversity" (Dr Goodwin). The latter might provide a field of possibilities for destabilizing the former.

However, it seems like discourses of Sociology as a counter-hegemonic discipline are shrinking rather than expanding in the context of neoliberal university reforms (Back and Tate, 2015). Indeed, many of my respondents agree that Sociology loses its "radicalism" (Dr Randall) as it progressively moves towards producing outputs that are aligned with market values. This would suggest that possibilities to challenge canonical norms are progressively diminishing in a culture of "audit" and managerialism. Corroborating critical educational scholarship's claims about the subtlety of "audit culture", Dr Birch says that the Research Excellence Framework influences the Sociology curriculum by discursively constructing knowledge that runs counter to market logics as less valuable:

> I think that filters down into the curriculum, what people think is core knowledge that students must get and then what's sort of marginal stuff that's just easy.

You know, what's easy and what's more difficult. Or rigorous. What's opinion. You know. Things like that, that kind of division. Things get coded that way.

(Dr Birch, lecturer)

Coding anti-hegemonic knowledges, such as postcolonial critiques, as "less rigorous" enables them to be sidelined, and to not be considered as a significant challenge to the canonical "building blocks" of the discipline. Mechanisms such as the Research Excellence Framework, in their close connection to marketable outputs, subtly shape what constitutes "just opinion", as opposed to "rigorous research". At Cambridge, the influence of such external regulators is often subtle, concerning research outputs and discourses of "legitimacy" rather than directly shaping curricula. However, the shape that mechanisms of "audit" take, my interviews suggest, differs greatly amongst universities.

When asked about what factors influence what she includes in her Cambridge syllabi, Dr Birch claims that she decides this based on her own preferences:

I think we have a lot of freedom to put what we want to [...] I think I could put anything I wanted there and nobody is there to look over me and say "she shouldn't be teaching this".

(Dr Birch, lecturer)

Dr Young compares this with his experience of teaching at his previous university, a post-1992 institution:[3]

I think probably in Cambridge you do have much more freedom to do what you want [...] [I'm] arriving here, looking for the guidelines, looking at what one's supposed to put in, and there's nothing there, and I'm thinking "here do I find all this out, and perhaps you don't, you just do what you want". [At] other universities, there are many more criteria, guidelines, ways of designing things that you have to put in, content that you have to put in.

(Dr Young, lecturer)

This relative freedom to teach is reaffirmed by the university's ranking in league tables, if not a direct result thereof. The ubiquity of "audit" is thus linked to an increasing bifurcation between post-1992 universities and Russell Group universities, Oxbridge in particular; while the latter still remain (relatively) free to choose what knowledge they produce, the former progressively align themselves with market goals, and provide knowledge that is relevant to policy and industry (Jones and Thomas, 2005). At Cambridge, according to Dr Randall:

[...] the way that the classroom and the curriculum is structured is anarchic in a certain sense because they don't really tell you what to do with your syllabus, but it's also very structured. Structured in a hierarchical way. What people know. You got to know stuff to put it on the syllabus.

(Dr Randall, lecturer)

Staff at Cambridge teach what they know, what they have been taught themselves and what they are used to teaching. A syllabus, once it is in place, tends to stay in place if no active interventions are made. The relative freedom of Cambridge lecturers to teach what they want heightens the importance of who teaches. Progressive lecturers then have relative freedom to challenge the limits of legitimacy by bringing in critical readings and diversifying the curriculum. As Dr Birch puts it, at Cambridge there is "room for a bit more [...] playing with the system'. This highlights the importance of hiring practices – of who gets to teach and who gets to study in the institution, an analysis of which is beyond the scope of this chapter. Instead, I will focus on what happens when those who are willing to contest colonial discourses in Sociology teaching do enter.

Bodies

> One of my biggest accomplishments in Michaelmas [first term of the academic year] was raising my hand and asking a question in a politics lecture. Because it was just so scary, and so intimidating, and I felt like everyone was just scrutinizing what I was saying. And I've never been more sensitive of the fact that I look different than in a lecture, because I look around, in the gender and intersectionality lecture, she talked about race and gender and I looked around and it's like, me, and then maybe 10 people from like 180 people in a lecture theatre [...]
>
> *(Mei, first year HSPS student)*

Mei, discussing her experiences as a woman of colour in the white-dominated environment of Cambridge, points towards a theme that I encountered frequently in my interviews with students, and at times with teaching staff: a feeling of not-quite-belonging, of sticking out due to their skin colour, gender, sexual identity and/or class background. I heard many stories of how the often unspoken rules and rituals students were confronted with in their courses and in the larger Cambridge University environment formed barriers that are "slippery to recognize and name" (Puwar, 2004: 23), and that draw an imaginary line between those who are "politically, historically and conceptually" marked as belonging, and those who are not (Puwar, 2004: 8). These barriers, enacted through rituals and embodied norms, are structured around gendered and racial lines, amongst others. They shape possibilities for resistance and critique, thus serving to reproduce white maleness as the academic norm – and by extension affect what is taught and the possibilities for contesting it.

Jess, a student in their second year, aptly describes to me how such invisibilized boundaries manifest themselves in a Sociology lecture:

> There's never the space to be like "oh actually, I have experienced this", or like "actually, I don't experience what you're talking about because the 'we' that you're using just doesn't include me at all". A lot of time lecturers use

an inclusive "we" and you are sitting there like "this doesn't include me" or like, my friend who's sitting there is like "that's not me". So I think there is actually not that much space, yeah, for people to be like – this doesn't relate to me, because of the way that lecturers frame it as being like an "us" thing.

(Jess, second year HSPS student)

Jess feels unable to challenge the lecturer's implicit assumptions about who constitutes the body of students they are speaking to. In the instance they describe, Jess recognizes that disembodied knowledge indeed has a body, but that this body does not include them, or their friends. However, the authority with which the lecturer takes the "we" as a given provides little space for challenging its foundation.

Nirmal Puwar (2004) offers an illuminating study on the effects and meanings of gendered racialized "bodies out of place" in institutional settings, including academia in the UK. With the move towards policies of multicultural inclusion, bodies can enter institutional contexts that were previously denied to them. However, Puwar argues that upon entering, these bodies become clearly marked as somatically non-normative, as "space invaders" who, in their "out-of-placeness", simultaneously disrupt these spaces and serve to reaffirm their normative whiteness. Jess' and Mei's quotes tell stories of being "space invaders" who need to change and adjust themselves in order to fit into the "we" that is the imagined, normative community of Cambridge students – or alternatively, be excluded from it. The normative body of the Cambridge student is discursively re-enacted through rituals, symbols and speech acts such as the ones described by Jess. Other students elucidate how subtle mechanisms of control police, and thereby reproduce it:

[We are] monitoring ourselves and each other. When you walk through a Cambridge quad you are very aware of where and how you look walking through the quad, what that looks like and how you fit into this particular image.

(Heather, first year HSPS student)

Authors have drawn on Bourdieu's (1984: 467) concept of habitus – "internalized embodied schemes" that are acquired throughout the life course when navigating social fields – to theorize a distinct academic habitus. This academic habitus marks the person who holds it as possessing attributes of intellectual worth (see Reay, 2004; Reay, David and Ball, 2005). The rules that constitute the academic habitus are clearly gendered and racialized (Puwar, 2004). While bodies that are constructed as not belonging will not cease to be "out of place", those "space invaders" who manage to embody the correct habitus are preferred by the standards of the institution, and find its rules easier to navigate (Puwar, 2004). Indeed, my respondents frequently describe the process of adjusting to the academic habitus, learning how to do things in the "right" way; what it means to be writing, or speaking in legitimate language, or to be drawing on legitimate sources in an essay. Several respondents express finding themselves in a dilemma: they wish to challenge the normative way of doing things in Sociology, for example by drawing on alternative sources in their coursework, or using creative methodological approaches, but find that they simply do not have the time or energy to do so.

> Most of the time you basically have enough time to look at the core readings, and write an essay from that. You don't actually have time to go and find lots of different sources [...] unless you're doing pretty well at that exact moment in time you're not gonna be able to do that. Like, there are some people who are just going to be way less able to do it all.
>
> *(Jess, second year HSPS student)*

Learning to be (to some extent) part of the normative body is often easier than "sticking out": navigating the process of being a "space invader" requires emotional, as well as physical labour (Hochschild, 1983; Puwar, 2004). "Space invaders" find themselves in a position of precarity: since their body is not assumed to be a producer of knowledge, they have to constantly prove themselves, "make a concerted effort to make themselves visible as proficient and competent, in a place where they are largely invisible as automatically capable" (Puwar, 2004: 60). This can then manifest itself in a "burden of representation"; since women and racialized minorities are constructed as representing a specific identity group, their success in institutional settings tends to be read as representative of that group (1994: 62).

Mei points towards this dynamic when reflecting on the possibilities she has for critically engaging with what she is taught in her degree:

> I talked to some other people – other people of colour – and I also feel like there's an implicit kind of thing where – and even being a woman too [...] there's always this expectation or this pressure on you to be successful because, or not because of, so, but in spite of being something else? Which is really bad as a mentality because I need to like, prove myself as an Asian person, prove myself as a woman. And I shouldn't have to do that; I should just focus on proving myself as a student. But there is this pressure that certain types of people don't really have to think about. They can just succeed on their own terms.
>
> *(Mei, second year HSPS student)*

The emotional labour involved in being a "space invader" is added to constant time pressure and high academic expectations, thus structuring possibilities to produce counter-hegemonic knowledge along racial and gendered lines. Additionally, being considered not as "succeeding on [your] own terms", but as always attached to a racialized and gendered body influences the way in which a person's critiques are being read, and contributes to defining their legitimacy. The white cisgender man, on the other hand, can continue to speak as the universalized human – a point reaffirmed by anti-colonial scholarship.

We must, however, recognize that these obstacles do not prevent such interventions being made. On the contrary, we can find a multitude of practices of resistance against colonial discourses in staff and student's daily interactions with institutional norms. I spend the remainder of the chapter discussing some of these interventions.

Marginalization, assimilation, legibility

> There are very specific voices on the topic of gender and intersectionality. On the intersectionality questions, there are lots of women of colour, black women, and on the questions about gender, there are writings by women and writings by transgender people and that's very cool. I've never seen that in an academic context. But outside of that particular context, it's very much like the normal white man.
>
> *(Jess, second year HSPS student)*

Jess points towards a dynamic that students and teaching staff frequently identified during my interviews: while some courses are highly diverse in terms of authors and syllabus content, these are often optional third year courses. The largely white, largely cisgender male classical canon remains the "core". Academics who aim to diversify the curriculum feel like they usually only have the option to "tinker at the edges" (Dr Birch), making counter-hegemonic knowledges a supplementary 'Other', which stands in an uncomfortable relationship to the canon, whose legitimacy it often critiques and whose basic propositions it frequently challenges (Connell, 2007). This is aptly summarized by Jie, when she describes her course as follows:

> The first term was all just like old white men and other old white men writing about those three canonical authors [...] The dynamic is really strange because then in the later section of the course we study all of those amazing social theories that do not come from these three.
>
> *(Jie, first year HSPS student)*

Some staff identify these processes of marginalization as neutralizing moves. They argue that allowing marginal spaces for anti-racist and anti-sexist knowledge to exist can serve as a means to preserve hegemonic discourses at the "centre":

> I think that gender, and people who teach gender, can be treated sort of like "oh you just do your stuff and get on with it" and you know, they can do their gender thing. [...] It can give you a certain amount of freedom, but it can also re-entrench that idea that you don't really need to worry about that topic because it's not that important anyway.
>
> *(Dr Goodwin, lecturer)*

Sara Ahmed, in her 2012 study of diversity work in UK universities, highlights how using the language of diversity can indeed gloss over structural dynamics of institutional racism rather than disrupting them. "If diversity becomes something that is added to organizations, like colour," she claims, "then it confirms the whiteness of what is already in place" (Ahmed, 2012: 33). Placing "diverse" bodies in institutional settings enables the institution to perform an image of itself as being "good at diversity" (Ahmed, 2012: 84), as appropriately reflecting multicultural society. Meanwhile, the power structure between the one who includes and the one who is being included often remains untouched – "this very structural position of being the guest, or the stranger, the one who receives hospitality, allows an act of inclusion to maintain the form of exclusion" (Ahmed, 2012: 45). This argument can be extended to the inclusion not only of bodies, but also of knowledge created by them. Rauna Kuokkanen (2007) discusses processes of integration of indigenous epistemologies into the settler academy when claiming that the institution, while technically allowing for the inclusion of knowledges that do not fit, requires them to be "fittable". In order to be included, Indigenous epistemologies must have the ability to be placed next to hegemonic approaches as a non-threatening exception, to be read in a way that keeps hegemonic discourses safely in their place. In order for a counter-hegemonic critique to be unthreatening to the hegemonic discourse, it is constructed as something external, or "Other" to it. Knowledges produced by those with gendered and racialized bodies tend to be constructed as particular to their specific subject position, instead of being regarded as universal (Puwar, 2004). However, as bell hooks powerfully reminds us, such space at the margins can hold potential for resistance. The consciousness fostered "as part of the whole but outside the main body [...] look(ing) both from the outside in and from the inside out" (hooks, 2010: ix) affirms "the necessity of resistance that is sustained by remembrance of the past, giving us ways to speak and de-colonize our minds, our very beings" (hooks, 1990: 341).

hooks' discussion focuses on her epistemic standpoint as a black woman, a specific location that should not be conflated with other spaces in which anti-hegemonic knowledge is produced. However, she inspires us to recognize the subversive potential of marginal disciplines stemming from, and rooted in, anti-racist and feminist social movements. Such marginal knowledges, brought into the "centre" of the UK university, can unsettle the naturalized dominance of eurocentric curricula. Dr Goodwin talks about her experiences teaching and researching gender when highlighting its importance for disrupting hegemonic knowledges:

> You can do these innovative and challenging and radical things within it and then some of that will hopefully filter through into the main discipline of which you are part and which you are allied with. And then people will start to see that, oh yeah, so maybe we need to think about women more. Or maybe we need to think about gender [...] – so it's a process of integration I think which necessarily has to have this kind of step in between of being not integrated because it's precisely about challenging what you might become integrated with.

According to Dr Goodwin, the space on the margins of the discipline is useful when inhabited temporarily, as a tactic that ultimately aims to challenge the foundations of the "core". Puwar (1994: 73) argues that

> the restricted grounds from which women of colour within academia are enabled to speak can become especially apparent when they go outside the remit of 'benevolent multiculturalism' and write about mainstream academic subjects that occupy a central place in the academic hierarchy of knowledge.

The necessity of reintegrating knowledge produced at the "margins" is underlined by my respondents when they illustrate the radical potential of contextualizing the classical canon in a way that problematizes it "from different spaces" (Dr Randall). Moving away from "additive" approaches to diversifying Sociology teaching by placing authors of colour, women and authors writing from the "South" within the centre disrupts hegemonic conceptions of who can produce knowledge about "legitimate" topics such as political economy. Dr Birch draws on an example of feminist economists who are located in a country outside of the UK when arguing that

> While sometimes things get coded as not rigorous or easy and so forth, there's the alternative coding of like, anything to do with Marx, Weber, Durkheim or, you know, political economy is like "male". Why are we using binaries like that? I wonder whether it's useful. [...] There are lots and lots of feminist economists [...], who are using like, the tools of the master but they are using it in ways that are quite radical.
>
> *(Dr Birch, lecturer)*

Asserting one's presence in a space where one, and the knowledge one produces, do not traditionally belong means having to navigate the dangers of being read as a "box-ticking exception" (Dr Goodwin) and being confined to the margins. However, it nevertheless holds the potential to open up space for challenging the sexist and racist assumptions of hegemonic discourses, and the deeply ingrained conceptions of who can produce knowledge that are intertwined with it.

Conclusion

Neoliberal HE reforms progressively transform UK education from a social good into an investment, and a commodity that can be bought and sold. Spaces for critical knowledge production are increasingly closed off – critical humanities and social sciences run counter to the logic of the market. The neoliberal era also continued the transformation of older colonial structures of domination, which were historically supported by, and reflected in, academic endeavours into (at times) subtler neo-colonial dependencies. These continue to be underpinned by racist and sexist discourses, which construct those located in the global "South" and people of colour, amongst other attributes, as intellectually inferior and subjects to be "known" instead of producers of knowledge themselves.

My interviews speak of complex forms that resistance against such epistemic hierarchies takes in the neoliberal academy, and in Cambridge specifically. Cambridge, as an institution that is specific due to both its historical significance in conceptualizing, administering and inscribing Empire and its global significance today, provides an elucidating, if not fully typical case to explore such resistances. My small-scale exploratory study conducted in 2016 suggests that the process of unsettling eurocentrism in Sociology here is not straightforward, and can be ambivalent and self-contradictory; that interventions often remain in tension between co-optation by hegemonic discourses on one hand and illegibility on the other. These tensions are at times used creatively for advancing such forms of resistance. My interviewees also tell stories about how the multiplicity of invisible barriers, built into the "way of doing things" in an elite university, structure possibilities for counter-hegemonic agency, amongst others, along racial and gendered lines. Students and teaching staff at the University of Cambridge navigate these barriers when contesting hegemonic notions of what constitutes legitimate knowledge in Sociology, and who is able to produce it.

Notes

1 The title of this chapter is, of course, in reference to and admiration of Audre Lorde (1983).
2 This research was conducted in the year 2015/16, and can therefore only provide a snapshot of students' and staff's opinions and efforts from this given time. Since writing, decolonization at Cambridge has grown to be a wide-ranging active movement, and inspirational decolonizing changes have been, and continue to be made in and outside the Sociology Department.
3 Former Polytechnics who were granted university status with the Further and Higher Education Act 1992.

References

Ahmed, S. (2012) *On Being Included*. Durham: Duke University Press.
Ali, S. (ed). (2007) Feminism and postcolonialism: Knowledge/politics [Special issue]. *Ethnic and Racial Studies* 30(1): 191–212.
Ali, S., Benjamin, S. and Mauthner, M. (2004) *The Politics of Gender and Education*. Basingstoke: Palgrave Macmillan.
Andrews, K. and Palmer, L. (2016) *Blackness in Britain*. London: Routledge.
Ashcroft, B., Griffiths, G. and Tiffin, H. (1995) *The Post-Colonial Studies Reader*. London: Routledge.
Back, L. and Tate, M. (2015) For a sociological reconstruction: W.E.B. Du Bois, Stuart Hall and segregated sociology. *Sociological Research Online* 20(3): 1–12.
Ball, S.J. (2012) Performativity, commodification and commitment: An i-spy guide to the neoliberal university. *British Journal of Educational Studies* 60(1): 17–28.
BBC News (2016) Oxford union votes to remove Cecil Rhodes statue. *BBC News* 20 January. Available at: www.bbc.co.uk/news/uk-england-35359530 (Accessed 15 May 2016).
Bhambra, G. (2015) *Connected Sociologies*. London: Bloomsbury.

Bhopal, K. (2014) The experiences of BME academics in higher education: Aspirations in the face of inequality. *Leadership Foundation for HE Stimulus Papers*.

Bhopal, K. (2015) Race, identity and support in initial teacher training. *British Journal of Educational Studies* 63(2): 197–211.

Bourdieu, P. (1984) *Distinction*. Cambridge, MA: Harvard University Press.

Canaan, J. and Shumar, W. (2008) *Structure and Agency in the Neoliberal University*. New York: Routledge.

Collini, S. (2017) *Speaking of Universities*. London: Verso.

Coloma, R. (2010) *Postcolonial Challenges in Education*. New York: Peter Lang.

Combahee River Collective (2016) A black feminist statement. In S. Mann and A. Patterson (eds) *Reading Feminist Theory: From Modernity to Postmodernity*, 1st ed. Oxford: Oxford University Press, pp. 247–252.

Connell, R. (1997) Why is classical theory classical? *American Journal of Sociology* 102(6): 1511–1557.

Connell, R. (2007) *Southern Theory*. Cambridge: Polity.

Davies, B. (2005) The (im)possibility of intellectual work in neoliberal regimes. *Discourse* 26(1): 1–14.

Gerrard, J. (2015) Public education in neoliberal times: Memory and desire. *Journal of Education Policy* 30(6): 855–868.

Gutiérrez Rodríguez, E., Boatca, M. and Costa, S. (2010) *Decolonizing European Sociology*. Farnham: Ashgate.

Hochschild, A. (1983) *The Managed Heart*. Berkeley: University of California Press.

hooks, b. (1990) Marginality as a site of resistance. In R. Ferguson, M. Gever, T. Minh-ha and C. West (eds) *Out There: Marginalization and Contemporary Cultures*. New York, NY: MIT Press, pp. 341–343.

hooks, b. (2010) *Feminist Theory: From Margin to Center*. Cambridge, MA: South End Press.

Hunter, M. (2002) Rethinking epistemology, methodology, and racism: Or, is White sociology really dead? *Race and Society* 5(2): 119–138.

Jones, R. and Thomas, L. (2005) The 2003 UK government higher education white paper: A critical assessment of its implications for the access and widening participation agenda. *Journal of Education Policy* 20(5): 615–630.

Kuokkanen, R. (2007) *Reshaping the University*. Vancouver: UBC Press.

Law, I., Phillips, D. and Turney, L. (2004) (eds) *Institutional Racism in Higher Education*. Stoke-on-Trent: Trentham Books.

Lorde, A. (1983) The master's tools will never dismantle the master's house. In C. Moraga and G. Anzaldúa (eds) *This Bridge Called My Back: Writings by Radical Women of Color*. New York: Kitchen Table Press, pp. 94–101.

Lugones, M. (2007) Heterosexualism and the colonial/modern gender system. *Hypatia* 22(1): 186–219.

Mbembe, A. (2016) Decolonizing knowledge and the question of the archive. Available at: http://wiser.wits.ac.za/content/achille-mbembe-decolonizing-knowledge-and-question-archive-12054.

McGettigan, A. (2013) *The Great University Gamble: Money, Markets and the Future of Higher Education*. London: Pluto Press.

Mihesuah, D. and Wilson, A. (2004) *Indigenizing the Academy*. Lincoln: University of Nebraska Press.

Mirza, H. (2008) *Race, Gender and Educational Desire: Why Black Women Succeed and Fail*. London: Routledge.

Morgensen, S. (2012) Destabilizing the settler academy: The decolonial effects of indigenous methodologies. *American Quarterly* 64(4): 805–808.

Mortimer, C. (2016) Cecil Rhodes' statue will not fall, Oxford college confirms. *The Independent* 28 January. Available at: www.independent.co.uk/news/uk/home-news/cecil-rhodes-statue-will-stay-at-oxford-despite-student-campaign-oriel-college-says-a6840651.html (Accessed 15 May 2016).

Nagar, R. and Swarr, A. (2010) *Critical Transnational Feminist Praxis*. Albany: State University of New York Press.

Ngugi wa Thiong'o (1986) *Decolonising the Mind*. London: J. Currey.

Puwar, N. (2004) *Space Invaders*. Oxford: Berg.

Quijano, A. (2000) Coloniality of power, eurocentrism, and Latin America. *Nepantla: Views from South* 1(3): 533–580.

Quijano, A. (2007) Coloniality and modernity/rationality. *Cultural Studies* 21(2): 168–178.

Reay, D. (2004) Cultural capitalists and academic habitus: Classed and gendered labour in UK higher education. *Women's Studies International Forum* 27(1): 31–39.

Reay, D., David, M. and Ball, S. (2005) *Degrees of Choice*. Stoke-on-Trent: Trentham.

Said, E. (1978) *Orientalism*. New York: Vintage Books.

Shahjahan, R. (2015) Being 'lazy' and slowing down: Toward decolonizing time, our body, and pedagogy. *Educational Philosophy and Theory* 47(5): 488–501.

Simmonds, F. (1992) Difference, power and knowledge: Black women in academia. In H. Hinds, A. Phoenix and J. Stacey (eds) *Working Out: New Directions for Women's Studies*. London: Routledge, pp. 51–60.

Slaughter, S. and Rhoades, G. (2004) *Academic Capitalism and the New Economy*. Baltimore: Johns Hopkins University Press.

Smith, L.T. (1999) *Decolonizing Methodologies*. London: Zed Books.

Strathern, M. (2000) *Audit Cultures: Anthropological Studies in Accountability, Ethics and the Academy*. London: Routledge.

Symonds, R. (1993) *Oxford and Empire*. Oxford: Clarendon.

Times Higher Education (THE) (2015) Subject ranking 2014–15: Social sciences. Available at: www.timeshighereducation.com/world-university-rankings/2015/subject-ranking/social-sciences#!/page/0/length/25/sort_by/rank_label/sort_order/asc/cols/rank_only (Accessed 9 June 2016).

Top Universities (2016) QS world university rankings by subject 2016 – Sociology. Available at: www.topuniversities.com/university-rankings/university-subject-rankings/2016/sociology#sorting=rank+region=+country=+faculty=+stars=false+search= (Accessed 9 June 2016).

Tuck, E. and Wayne Yang, K. (2012) Decolonization is not a metaphor. *Decolonization: Indigeneity, Education and Society* 1(1): 1–40.

Wilder, C. (2013) *Ebony and Ivy: Race, Slavery and the Troubled History of America's Universities*. New York: Bloomsbury Press.

Wynter, S. (1994) "No humans involved": An open letter to my colleagues. *Forum N.H.I.: Knowledge for the 21st Century* 1(1): 42–73.

5

BLACK STUDIES IN THE WESTERNIZED UNIVERSITY

The interdisciplines and the elision of political economy

Charisse Burden-Stelly

Introduction

When the Black studies[1] movement was inaugurated in the late 1960s, it represented the intellectual expression of political Pan-Africanism in United States universities. It was formed to fundamentally challenge the statist, imperialist, racist, and Eurocentric underpinnings of the traditional disciplines in westernized universities by centring community development, African and African descendant struggles for liberation and self-determination, and the importance of internationalism to the larger project of Black freedom. As early as 1900, Pan-Africanism and radical Black internationalism rendered legible the ideological convergences and structural imperatives of decolonization in Africa and the Caribbean and the quest for Black self-determination and liberation in the United States. However, the institutionalization of Black studies in the westernized university engendered a turn away from these early commitments. With American studies and area studies setting the precedent, the struggle over redistribution was replaced by a struggle over representation (Ferguson, 2012). The mobilization of culture became the means of securing recognition and reward. History and literature became the two areas through which Blackness was studied, defined, and codified, and cultural production and formation became the focus of anthropological and sociological studies of the Black condition. Political economic and structural approaches became relegated to "culturalism" as Black studies became institutionalized and legitimated. Culturalism can be understood as the regime of meaning making in which Blackness is culturally specified and abstracted from material, political economic, and structural conditions of dispossession through statist technologies of antiradicalism (Burden-Stelly, 2016).

By the late 1980s, when Black studies had become more or less fully incorporated into the westernized university, there was a noted and distinct absence of political economic and material critiques of racialized dispossession in its formulation, and an overwhelming overrepresentation of literary and cultural studies (Mason and Gīthīnji, 2008). The organization of Black studies around the analytic of "diaspora", starting in the late 1980s and early 1990s, acted to reify the abstraction of Blackness from material and structural realities through the transnationalization of the study of Black sociality and culture. In other words, given the turn to cultural specifications in Black studies, African diaspora studies became institutionalized as another culturalist project. Despite the reality that the shift in global accumulation – that is, the transformation of capital from its liberal to corporate form (Robotham, 2005) – precipitated by the 1973 world recession and oil crisis (Arrighi, 1991) had homologous, albeit geospatially specific, effects on all African descendant peoples, African diaspora studies was nonetheless "content to confine itself to the surface of social, cultural, and economic life … while avoiding 'the hidden abode' of production" (Robotham, 2005: 21). Because political economy has essentially become anathema to Black Studies, it has remained a largely academic enterprise divorced from the increasingly immiserated material conditions of Black people. The cultural politics of recognition continue to dominate the conceptual and theoretical frameworks of Black studies, to the detriment of an approach focused on the practical necessity of radical economic praxis to apprehending the actually existing realities of Black communities. Largely missing from Black studies curricula are political economic histories that can adequately explain the parallels in global Black economic conditions, and the gaps between "the West and the rest" in income, wealth, and the ownership of factors of production. Such exclusions hamper the decolonial potential of Black studies.

Julianne Malveaux (2008: 785) contends that "traditional and theoretical disciplinary formulations" of Black studies have been largely influenced by the exigencies of accommodation to the westernized university. Such imperatives are best served by the field's culturalist approach. This chapter argues that once Black studies became the object of academic disciplining and control through management by the westernized university, it became abstracted from its activist, community-oriented, and militant origins. Specifically, the process of westernization occurred as Black studies began to take on the epistemological and structural form and function of American studies and area studies. The latter were the founding interdisciplinary projects that emerged in the Cold War university, purged of Marxism by McCarthyism (Pfister, 1991). These provided the grammar, form, and function for subsequent "interdisciplines", which, like their predecessors, elided class and political economic analysis.

American studies, area studies, and the inauguration of the "interdiscipline"

Roderick Ferguson (2012) analyses the role of the westernized university in harnessing minority difference to the reconstitution of capital and to the needs of the U.S. State. He does so in the context of demands for inclusion by minoritized subjects on the one hand, and of the global entrenchment of the U.S.-led political and economic order on the other. He argues that, by the late 1960s, the westernized university had become the "training ground" for how the U.S. state and capital ought to contend with the meaning, representation, and accommodation of minority difference (Ferguson, 2012). It was concerned with developing ways to "combin[e] tolerance for diversity with the imperatives of world order" (Gleason, 1984: 356). With the proliferation of domestic and global minorities in the westernized university, disciplines that focused on specific races, ethnicities, global areas, and identities increasingly became sites of surveillance (Schueller, 2007).

Ferguson (2012) locates the rise of the interdisciplines in the institutionalization of ethnic and gender studies. He argues that with the ascent of U.S. empire came the concomitant need to regulate and manage groups of "others" domestically, and nations of "others" internationally. This demanded a shift in the westernized university away from specialized disciplines to interdisciplinary forms of knowledge. However, Ferguson's analysis overlooks that the narratives of freedom and democracy necessitated by the spread of the ideology of embedded liberalism rested upon the spread of knowledge about a distinctive American culture, civilization, and society that differentiated itself from Europe and Asia. It was this imperative that led to the emergence of one of the first major interdisciplinary fields: American Studies. This was accompanied by the development of area studies – the other foundational interdiscipline – which aimed to acquire and produce knowledge about other parts of the world with which the U.S., as the dominant empire, would have to contend (Schueller, 2007). Liberated from the restrictive methods, canons, and approaches of dominant disciplines, the foundational interdisciplines were able to deploy the power of the state and the westernized university to position difference as their object of critique and engagement (Ferguson, 2012). Acting in the service of U.S. state and empire, these interdisciplines provided the acceptable institutional model for fields of study focused specifically on minority difference as it related to history, culture, and society. They allowed the westernized university to strategically accommodate the demands for Civil Rights, "relevant education," and equality. American studies provided the grammar, and area studies structured the relationship between the state, the westernized university, and the study of difference. With American studies and area studies as the intellectual arms of the U.S. state, "anti-fascism, anti-communism, and anti-imperialism coexisted in an uneasy partnership in the governing vision of U.S. post-World War II promise: universal nationhood and liberal, capitalist democracy" (Singh, 1998: 488). Such coupling had consequences for the development of Black studies. Contrary to the desire of Black radicals and nationalists to use institutions of higher education to inculcate their vision of the state and society, Black studies was institutionalized and legitimated to a large degree in the image of the foundational interdisciplines.

The westernized university became a site of "reactive crisis management" (Habermas, 1975: 60) that both accommodated and disciplined challenges to Eurocentrism, Euro-American coloniality, and white supremacy. While it incorporated new cultures, knowledges, and bodies, it did so in a way that produced new means of exclusion in the service of U.S. empire. The state –by way of the university – was able to reconstitute its pedagogy through the "distribution [of] recognition and legitimacy" (Ferguson, 2012: 22). It transformed its power in order to "incorporate formerly marginalized and excluded subjects and societies, an ability signified through the extension of recognition and sovereignty for people who spent much of their histories under the colonial yoke", and an extension of recognition and rights for people who had hitherto been marginalized from citizenship through processes of racial exclusion. In effect, "the modern idea of empire and the modern idea of difference" (Ferguson, 2012: 24) coalesced into an ideological project that entangled "the management of the international … with the management of diversity" (ibid.). The "interdisciplines", especially area studies, American studies, and, later, Black studies, were essential to the efforts by the United States to refashion itself as anticolonial and antiracist in order to legitimate itself as the world leader and to absorb heterogeneity.

The westernized university plays a fundamental role in the maintenance of the state as the institution by which national values are cultivated, stored, and reproduced. Further, it serves a political function, operating in relationship to the state, which is dedicated to maintaining the interests of empire "over and against various communities which exist in this country that are committed to radical change" (McWorter, 1969: 63). With the rise of the United States after World War II came the necessity to develop an epistemic practice that demanded consolidation of knowledges from different fields of study as interdisciplines. This was because,

> The interdisciplines were an ensemble of institutions and techniques that offered positivities to populations and constituencies that had been denied institutional claim to agency. The interdisciplines connoted a new form of biopower organized around the affirmation, recognition, and legitimacy of minoritized life.
>
> *(Ferguson, 2012: 13)*

This was in response to the changing demands of post-World War II and postcolonial formations.

The westernized university became the space in which demands for the reformulation of epistemological and cultural representation converged with the capitalist-imperialist interest in difference. It provided the means by which state and capital developed its "methods of representation and regulation". This is especially true in the context of the 1960s and 1970s, during which "[n]ational liberation, civil rights, and neocolonialism [came to] be understood as part of a larger social context that proclaimed the command of a new mode of power, a mode that was composed of power's new techniques of management, especially around internationalism and minority difference, as well as its insinuation into political agency. This period inaugurated a new intentionality in the university that, as an instrumentality of state and empire, needed to locate, "know", accommodate, and affirm difference so that it could simultaneously legitimate and depoliticize it. American studies and area studies institutionalized the historical imperatives of U.S. power to manage international and sociopolitical difference (ibid.: 25–27) through the inscription of their new knowledge into the pedagogy of U.S. state and empire.

American studies

The position of the United States as the new superpower at the conjuncture of World War II and the Cold War created the conditions for the development of American studies. Nikhil Pal Singh (1998: 488) argues that: "The ideological framework of the cold war ... creat[ed] a new understanding of American universalism that tied together a celebration of America's pluralism and political exceptionalism with the fate of the 'the West' as a whole." Throughout the 1940s, as the United States became more important economically, politically, and militarily in the world-system, there arose a need to narrate the role of American society in Western Civilization. As the U.S. ascended to the status of world power, nation, identity, and culture came to constitute the basis of a powerful state ideology that

> powerfully reinforced existing tendencies toward cultural nationalism, gave great prominence to the ideological dimension of American identity (that is, to the ideas and values for which the nation stand for), and forged a link between the democratic ideology and the idea of culture that became central to the American [s]tudies approach.
>
> *(Gleason, 1984: 345)*

A peculiarly U.S. brand of democracy became identified with the defence and preservation of Western civilization. The democratic ideal became a foundational aspect of Americanness. It was the *ideology* of democracy – not the practice – that the United States came to represent. As an abstract ideal, democracy became the essential component of national culture (Gleason, 1984). The patently undemocratic reality of postwar U.S. society did not interfere with this narrative of democracy as endemic in American culture. "American culture" became abstracted from material and structural realities, including inequality, racism, and capitalist exploitation, and came to be understood as the way in which the American people existed in the world – in other words, as the "American way of life". The collapsing of "democracy" into "America", and "culture" into "way of life", provided the basis for the development and organization of American studies that paved the way for the institutional form of Black studies.

The casting of democracy in cultural rather than structural terms negated an effective critique of the racialized, classed, and gendered materialities of the United States. Thus, the Marxist intellectual tradition was irrelevant as a counter to this narration of the project of democracy insofar as the latter was wholly abstracted from the structural features of U.S. economy and society. In other words, "Popular front and cold war 'Americanism'... created the special conditions that had blocked the development of an American Marxist" tradition (Shank, 1997: 96). The "nation" came to operate metonymically for American "culture" and "civilization". This became reflected in the creation of the first American Civilization departments that presaged American studies. These departments came about because the study of American culture could not be accomplished by a single discipline, and thus required a move to interdisciplinarity. As Gleason (1984: 354) explains, "… understanding the national culture holistically is the task Americanists have always set for themselves … [it is] the implicit (and sometimes explicit) premise of the American [s]Studies approach …" The emphasis on culture was integral to the maintenance of the status of American studies as a discipline; even as American studies began to centralize social science methodology over more humanistic approaches, anthropologically based notions of culture remained foundational to the study of "America". Culturalism in the United States came to inhere in ideological notions of nationalism that tended to be abstracted from historical material realities. The study of culture in this way required little attention to structural power relations. Epistemologically, culturalism came to constitute the ways that Americans understood themselves and their relations to society. Accordingly, such "[s]tudies of culture too far removed from studies of social structure" have left us unable to fully explain the world in which we live (Lipsitz, 1995: 371).

New Black and ethnic studies programmes that entered the academy in the late 1960s and 1970s inherited the interdisciplinary approach from American studies that had become "high pedagogical fashion, but [was] more of a slogan than a serious endeavor" (Marx, 1979: 400). Black studies began to assert forms of cultural politics characterized by Eric Hobsbawm (1996) as essentially assertions of particularistic nationalisms in claims by groups to their right to self-determination. In making these claims, many aspects of American studies came to be reproduced in Black studies. The American studies project was conceived to describe, construct – and later critique – a particular American culture and civilization (i.e. national self-determination) vis-à-vis other "great" civilizations in order to provide a scholarly basis for American empire. Similarly, Black studies employed approaches of American studies, specifically those that mobilized the use of history, literature, and culture to articulate "difference" and cultural particularity in their efforts to challenge white supremacy and Eurocentrism and to assert their right to self-determination.

Area studies

Area studies emerged in the 1950s as the intellectual arm of the Cold War, as "part of the struggle for World hegemony against Communist states" (Paik, 2013: 4), and in response to the emergence of the United States as a global political and economic superpower. Even after a respecification of area studies under a new Higher Education Act in 1968 in the face of increasing criticism of the ethnocentrism and sociopolitical agenda of the field, it continued to be bound up with the imperialist agenda of the American state (ibid.) This has much to do with the fact that it was inaugurated to support the global demands of the U.S. imperial state and its efforts to secure and expand U.S. power (Spivak, 2003). As Mimi White and James Schwoch (2006: 11) assert,

> World War II saw, particularly in the USA ... an incredible state mobiliza-
> tion of cultural analysis, most prominent for the American case ... in the
> restructuring of universities themselves. This led, particularly in the USA, to
> the postwar rise of area studies programs ... During the Cold War era, many
> of these USA area studies programs would become politicized in the service
> of a national security state, most notably through the influence of research
> funding.

In the era of decolonization and international bipolarity, area studies took on the role that anthropology occupied in the era of direct colonial administration (Wallerstein, 1997). The Cold War influenced the types of knowledge that would be sought about strategic areas in the decolonizing world in order to conscript them into the orbit of U.S. empire. Hans Morgenthau confirmed that "[it is not] an acci-dent that the areas around which area studies are centered are generally defined in terms which coincide with the areas of political interest" (cited in ibid.: 207). Built upon the "intellectual, material, and racial pillars" (ibid.) of U.S. national politics, area studies was officially institutionalized in 1958 through the National Defense Education Act (NDEA) and Title VI. A response to the Soviet Union's launching of Sputnik, the Act aimed to address the perceived U.S. educational weakness, and to educate citizens in science, language, and areas studies in order to surpass the Soviet Union in technological capability and international influence (Kuntz, 2003).

Additionally, "to meet the demands of war, scholars of diverse disciplines *were forced* to pool their knowledge [of many areas of the globe which had been inad-equately studied] in frantic attempts to advise administrators and policy makers" (Spivak, 1998: 809). This formalization of the "material and political relationship between area studies and the state" (Schueller, 2007: 44) was bound up in the anti-communist and culturalist ethos of the immediate postwar period. Marxist analysis – which emphasizes class difference, critiques nationalism as an ideology that pro-duces distortions both in scholarship and in reality, and asserts the role of relations of production and political control in specifying "third worldliness" – had no place in area studies because it was patently antithetical to its project and methodol-ogy. Inasmuch as Marxist scholarship highlighted conflict, heterogeneity, resistance, and change (Prakash, 2000), it was fundamentally incompatible with area studies. The possibility of radical scholarship was severely circumscribed, underscoring the inscription of area studies in the anticommunist, Cold War politics of the U.S. state.

As an interdiscipline, area studies was epistemologically grounded in the logics of U.S. empire, locating theory in the Global North and the objects of study – those to be represented – in the Global South (Grosfoguel, 2011). As such, "... areas to be studied/conquered are 'out there,' never within the United States ... [a]rea studies in U.S. academic settings were federally funded and conceived as having a political project in the service of U.S. geopolitical interests ...". Area studies maintained a particular "epistemic privilege" by asserting itself as observer and the Third World as the object to be observed. Mohanty concurs: "The focus on [area studies] implies that it exists outside the U.S. nation-state … [I]ssues are based on spatial/geographical and temporal/historical categories located elsewhere. Distance from 'home' is fundamental to the definition …" (Mignolo and Tlostanova, 2006; Mohanty, 2002: 519–520). Area studies scholars reproduced the investments and ideologies of the U.S. state in the westernized university by representing "others" as culturally relative and nationally specified. The discovery of new histories and cultures, revealed through "reality on the ground", was made possible by essential understandings of the nation in terms of the culture concept. Like American studies, area studies emphasized culture and the nation through their "depiction of vibrant realities [that] fell in line with nationalist celebrations" (Prakash, 2000: 172).

Area studies initiated a political and academic response in the face of decolonization, anticolonialism, and struggles against neocolonialism. Paradigms were established "that worked in the logic of empire: to contain or direct anticolonial movements and, later, to influence African independent states" (Pierre, 2013: 190). Through area studies, the westernized university developed and constructed knowledge about newly decolonized countries of strategic importance that conscripted them into various liberal discourses such as liberal rights, development, and modernization, and ordered them hierarchically based on their level of compliance. It is no coincidence, for instance, that African studies programmes proliferated in the 1960s and 1970s just as African development was ascending as a national policy issue. Area studies combined the social scientific focus on development with humanities and culture-oriented discourses of modernization. Such epistemology (re)produced the United States as model and benefactor, and, as such, rationalized the latter's intervention in strategic areas on behalf of development and modernity. In this way, area studies became an arm of neocolonialism that influenced key foreign policy considerations, including which countries would receive aid and which would be invaded; the level of democracy or authoritarianism that would be accepted; and how Third World populations would be disciplined, managed, and/ or accommodated.

By 1968, area studies had come under heavy scrutiny and critique due to its ethnocentrism, complicity with the Cold War, and its problematic assertions of modernization theory and its "three worlds" division (Wallerstein, 1997). However, the framework had been set for fields of study that both institutionalized state pedagogy and informed the ways in which the state ought to accommodate difference. As Black studies became absorbed into the pedagogy of the state by way of the westernized university, it aided in the reconstitution of capital based on new forms governmentality and exclusion through cultural specifications. The logics of modernization and development that circulated in area studies became transferred to Black studies through their investment in historicism and the privileging of history; through civilization narratives and the privileging of literary and cultural studies; and through cultural explanations of deviance and pathology and the privileging of structural-functionalist sociology. The result was the production of Black studies specialists whose expertise and production of knowledge served the interests of the state and its engagement with notions of difference that became reduced to specifications of cultural condition.

The institutional formation of Black studies

Community activism, Pan-Africanism (Drake, 1984), and student demands for more relevant education were responsible for the introduction of Black studies into the westernized university. Revolutionary cultural nationalism, the movement it supported, its tendency, and its ideology – in other words, its "intellectual spirit" – motivated the demand for Black studies programmes (Cruse, 1969). Despite its community and activist roots, "university administrators were determined to reshape [Black studies] into a purely academic phenomenon" (Drake, 1984: 228). The university was largely successful through its deployment of "networks of power ... [that] work[ed] through and with minority difference and culture ... to redirect originally insurgent formations and deliver them into the normative ideals and protocols of state, capital, and academy" (Ferguson, 2012: 8). As the number of programmes began to shrink considerably – from about 500 in 1971 to about 225 in 1984 – the survival of Black studies programmes came to rest on their ability to successfully perform a purely academic function. With its move away from counterhegemonic struggle and activism, Black studies succumbed to the "institutionalizing ethos", and became part of the "imperial tendency" of the U.S. state to co-opt, manage, and ultimately undermine oppositional tendencies (Colón, 1984; Drake, 1984; Ferguson, 2012). Black studies was to be an intellectually valid, educationally responsible, and socially constructive project, which, like American studies and area studies, demanded little attention to class:

> While overall the movement was positive, particularly in its critique of white supremacy, the movement's blind spot with regard to class, and specifically, working-class issues, subjected the movement to subversion by pro-corporate forces ... the blind spot to class served to increasingly isolate and marginalize the black studies movement ...
>
> *(Fletcher, 1983: 159)*

Relatedly, Manning Marable argued that the majority of Black studies programmes rejected political economy and public policy to focus on arts and humanities, creating an imbalance between literary and cultural studies and structural critique (Gates and Marable, 1983). Thus, the formalization of Black studies instantiated a move away from class and community.

Area studies provided the basis for a new form of governmentality that inhered in liberal inclusion instead of racist foreclosure and conscious and unconscious "ignorance". It set the precedent for racially specific studies to become sites for the management of difference. Area studies specialists became essential to the formulation of policies and practices adopted by the U.S. state in strategic areas of global governance. Black studies scholars, on the other hand, were largely irrelevant to efforts aimed at the implementation of policies that impacted the material realities of Black people. According to St. Clair Drake, Black studies scholars "did not ... have a strong impact in areas not directly related to teaching and research" (Drake, 1984: 236). Black studies did, however, provide an understanding of how Black people should be managed by the state – namely, through discourses of multi-cultural rights that asserted equality in terms of cultural recognition. As Schueller (2007: 52) argues, "This severance of race from rights across a broad sociopolitical spectrum ... made it possible for multiculturalism ... to simply represent politics of cultural recognition without recognition of equal social reward or redistributive justice". Stated differently, redistribution became extricated from cultural recognition in demands for self-determination. The emphasis on abstract representation ultimately allowed the westernized university, the state, and capital to shore up their power, because "the margins" and the "periphery" came to be understood as sites of cultural empowerment and contestation. The goal was to "develop cultural strategies that make a difference" (Hall, 1992: 107). The accommodation of difference, first though area studies, then through Black studies, allowed the U.S. state to acquire knowledge so that it could adequately conscript those occupying the margins into its imperial project. Through its enunciations of culturally coded Blackness, Black studies became self-disciplining as a prerequisite for its institutionalization. Moreover, like area studies, Black studies became an instrumentality for the management of radical and potentially revolutionary movements, thereby helping the state, capital, and empire to rearticulate itself through incorporation, absorption, and regulation.

As was the case with area studies, Black studies was an "attempt to create a systematic body of knowledge and experience based on the history" (Colón, 1984: 268) of a group that had previously been excluded from academic study. Both area studies and Black studies had the impact of inserting into the westernized university, especially predominately white institutions, research and teaching in areas that had been considered outside of the purview of civilization. In order to facilitate imperial expansion in the postwar moment in which the majority of the world was contesting race-based forms of coloniality, the "American" creed had to be expanded beyond the "cultural values of the Anglo-American tradition" so that "others" could be more easily accommodated. Instead of conscripting racialized and colonized subjects into a model that refused to recognize them, through the university the U.S. state inaugurated programmes of research and scholarship that made their experiences, histories, and cultures constitutive of imperial expansion. One such example was the funding of African student exchange programmes throughout the 1960s with monies appropriated under the U.S. Information and Educational Exchange Act of 1948 (commonly known as the Smith-Mundt Act) – passed to institutionalize U.S. cultural and informational programming as a peacetime technology of foreign policy – and from the sale of war surplus materials abroad. These students were closely supervised by the Department of State under the auspices of the Institute of International Education (African Studies Association, 1961: 1–3).

Another example is the intervention of the Ford Foundation in the formation of Black Studies programmes to ensure that they were geared towards solving the problems of racial exclusion and racial integration, but were not tightly bound to Black Power ideologies and did not exacerbate racially inflected structural crises on campuses or in Black communities (Rooks, 2006: 15–30). The accumulation of knowledge became the means by which the state and the university could monitor and influence the direction of potentially revolutionary movements. Black studies became a site of surveillance and discipline to ensure that, eventually, demands for material redistribution would collapse into demands for cultural recognition.

With the focus on difference and cultural politics, the "margins" came to be incorporated into the "centre", and the "periphery" came to be incorporated into the "core" of global capital, while both the "margins" and the "periphery" effectively maintained their subordinate status. As the site of struggle shifted to the *meaning* of difference and to demands for representation and recognition of cultural identities in society, the focus of Black studies necessarily shifted to the politicization of culture, to the development of new identities, and to contesting the cultural hegemony of the "mainstream". With domination culturally specified, demands for change in and reconfiguration of relations of power came to rest on efforts aimed at the deployment of a politics of difference that would place Blackness on an equal footing with white cultural forms. The state was able to accommodate these demands by shifting to a more open, inclusive, and accommodating "disposition of power" (Ferguson, 2012) that was effective without compromising its capitalist and imperialist agenda. While increasingly worse off in material terms given the ascent of neoliberalism, Black people came to be recognized and represented in popular culture and popular discourse, where they could be adequately policed, regulated, and commodified. Through collaboration between the state and the westernized university, culturalized identity politics came to be recognized as the only acceptable articulation of Black struggle and contestation.

American studies provided the grammar for the conflation of Black studies with the interdisciplinary study of culture. Fletcher (1983: 159) asserts, "… black studies in the sixties … focused on the national or 'ethnic' feature of the African American freedom struggle. Culture was prominent, but so too was a nonclass view of African American experience". The cultural specifications of Blackness readily accommodated the transition to African diaspora studies as the 1980s and 1990s inaugurated concerns about globalization and transnationalism. Such specification created the conditions for African diaspora studies to further distance itself from the (Black) left, because, "[a]t a certain level, the Left, through its critique of empire, intersects with the foreign in being cast as unnational" (Schueller, 2007: 54). Diaspora deterritorialized, relativized, and abstracted Blackness in a way that seriously hampered a structural critique of power. Blackness was delinked from the nation–state and its historical specificity and was asserted as a hybrid, transcultural phenomenon, so the location-specific materialities of dispossession and domination that produced and reinscribed Blackness went largely uncontested. As international linkages became culturally specified and divorced from political and economic realities, possibilities offered up by variations of socialism and communism that "reache[d] out to all oppressed colonial subjects ... [and] enabled many different people to identify with other oppressed peoples and to reject patriotism and national identity" were negated (Kelley, 2004: 43). Race became essentialized in globalized relations of representation (Hall, 1996) and the possibilities of international alliances against capitalist oppression were marginalized.

Conclusion

Engagement with global political economy and radical critique is essential to decolonial praxis. Manning Marable (1983) argued that if Black studies was to remain an important interdisciplinary field, it had to continually interpret and understand socioeconomic and global forces that were rapidly restructuring the life chances of Black people throughout the world. However, with the exception of its engagement in anti-apartheid movements, Black studies continued to refuse consideration of the structural and material effects of "global policies" while asserting Black transnationality through a culturally specified diaspora analytic. In the tradition of American studies and area studies, Black studies became conscripted as the academic arm of U.S. imperial multiculturalism by specifying transnational Blackness as a culturalized trope while ignoring critical engagement with the material conditions of racialized abjection. In the final analysis, the integration of the Black into the global axial division of labour is the necessary condition of capitalist accumulation. It is only through structural and antisystemic modes of critique that expose the material conditions of Blackness that decolonial challenge can be mounted by those racialized as Black in a manner that resists reinscription into the statist project of global capital.

Note

1 Black studies is used here as a metonym for the entire (inter)disciplinary project, starting in the 1960s, that came to be known variously as Africana studies, African-American studies, African and African American studies, Africology, Pan-African studies, and Black new world studies. While I acknowledge that there are politics and ideologies associated with the naming of these programmes, departments, and centres, I have chosen to use one iteration for the sake of convenience.

References

African Studies Association (1961) Educational and cultural exchange with Africa: The program of the Department of State. *African Studies Bulletin* 4(2): 1–8.

Arrighi, G. (1991) World income inequalities and the future of socialism. *New Left Review* 189: 39–64.

Burden-Stelly, C. (2016) The modern capitalist state and the Black challenge: Culturalism and the elision of political economy, PhD dissertation. University of California, Berkeley.

Colón, A. (1984) Critical issues in Black studies: A selective analysis. *The Journal of Negro Education* 53(3): 268–277.

Cruse, H. (1969) The integrationist ethic as a basis for scholarly endeavor. In A. Robinson, C. Foster and D. Ogilvie (eds) *Black Studies in the University: A Symposium.* New Haven: Yale University Press, pp. 4–12.

Drake, S. (1984) Black studies and global perspectives: An essay. *Journal of Negro Education* 53(3): 226–242.

Ferguson, R. (2012) *The Reorder of Things: The University and its Pedagogy of Minority Difference.* Minneapolis: The University of Minnesota Press.

Fletcher Jr., B. (1983) Black studies and the question of class. In M. Marable (ed) *Dispatches from the Ebony Tower: Intellectuals Confronting the African American Experience.* New York: Columbia University, pp. 158–161.

Gates, H. and Marable, M. (1983) A debate on activism in Black studies. In M. Marable (ed) *Dispatches from the Ebony Tower: Intellectuals Confronting the African American Experience.* New York: Columbia University, pp. 186–193.

Gleason, P. (1984) World War II and the development of American studies. *American Quarterly* 36(3): 343–358.

Grosfoguel, R. (2011) Decolonizing postcolonial studies and paradigms of political economy: Transmodernity, decolonial thinking and decoloniality. *Transmodernity* 1(1): 1–36.

Habermas, J. (1975) *Legitimation Crisis.* Boston: Beacon Press.

Hall, S. (1992) What is the 'Black' in Black popular culture. *Social Justice* 25(1): 104–114.

Hall, S. (1996) New ethnicities. In D. Morley and K.H. Chen (eds) *Stuart Hall: Critical Dialogues in Cultural Studies.* London: Routledge, pp. 441–449.

Harris, J. (1990) The intellectual and institutional development of African studies. In J. Bobo, C. Hudley and C. Michele (eds) *The Black Studies Reader.* New York: Routledge, pp. 15–20.

Hobsbawm, E. (1996) Identity politics and the left. *New Left Review* 217: 38–47.

Kelley, R. (2004) How the west was one: On the uses and limitations of diaspora. In J. Bobo, C. Hudley and C. Michele (eds) *The Black Studies Reader.* New York: Routledge, pp. 41–45.

Kuntz, P. (2003) The training of an Africana librarian: The Mellon fellowship. *Journal of Education for Library and Information Science* 44(3/4): 316–331.

Lipsitz, G. (1995) The possessive investment in whiteness: Racial social democracy and the 'white' problem in American studies. *American Quarterly* 47(3): 369–387.

Malveaux, J. (2008) Why is economic content missing from African American studies? *Journal of Black Studies* 38(5): 783–794.

Marx, L. (1979) Thoughts on the origin and character of the American studies movement. *American Quarterly* 31(3): 398–401.

Mason, P. and Githinji, M. (2008) Excavating for economics in Africana studies. *Journal of Black Studies* 38(5): 731–757.

McWorter, G. (1969) Deck the ivy racist halls: The case of Black studies. In A. Robinson, C. Foster and D. Ogilvie (eds) *Black Studies in the University: A Symposium.* New Haven: Yale University Press, pp. 55–74.

Mignolo, W. and Tlostanova, M. (2006) Theorizing from the borders: Shifting to geo-and body-politics of knowledge. *European Journal of Social Theory* 9(2): 205–221.

Mohanty, C. (2002) 'Under western eyes' revisited: Feminist solidarity through anticapitalist struggle. *Signs* 28(2): 499–535.

Paik, N. (2013) Education and empire old and new: HR 3077 and the resurgence of the US university. *Cultural Dynamics* 25(1): 3–28.

Pfister, J. (1991) The Americanization of cultural studies. *The Yale Journal of Criticism* 4: 199–229.

Pierre, J. (2013) *The Predicament of Blackness: Postcolonial Ghana and the Politics of Race.* Chicago: University of Chicago Press.

Prakash, G. (2000) Writing post-orientalist histories of the Third World: Perspectives from Indian historiography. In V. Chaturvedi (ed) *Mapping Subaltern Studies and the Postcolonial.* London: Verso, pp. 163–190.

Robotham, D. (2005) *Culture Society, and Economy: Bringing Production Back In.* London: Sage Productions.

Rooks, N. (2006) *White Money, Black Power: The Surprising History of African American Studies and the Crisis of Race in Higher Education.* Boston: Beacon Books.

Schueller, M. (2007) Area studies and multicultural imperialism: The project of decolonizing knowledge. *Social Text* 25(1): 41–62.

Shank, B. (1997) The continuing embarrassment of culture: From the culture concept to cultural studies. *American Studies* 38(2): 95–116.

Singh, N. (1998) Culture/wars: Recoding empire in an age of democracy. *American Quarterly* 50(3): 471–522.

Sklar, R. (1970) American studies and the realities of America. *American Quarterly* 22(2): 597–605.

Spivak, G. (1998) Gender and international studies. *Millennium: Journal of International Studies* 27(4): 809–831.

Spivak, G. (2003) *Death of a Discipline.* New York: Columbia University Press.

Wallerstein, I. (1991) *Geopolitics and Geoculture: Essays on the Changing World-System.* Cambridge: Cambridge University Press.

Wallerstein, I. (1997) The unintended consequences of Cold War area studies. In N. Chomsky, L. Nader, I. Wallerstein and R. Lewontin (eds) *The Cold War and the University: Toward an Intellectual History of the Postwar Years.* New York: The New Press, pp. 195–231.

White, M. and J. Schwoch (2006) (eds) *Questions of Method in Cultural Studies.* Oxford: Blackwell Publishing.

6

BLACK FEMINIST CONTRIBUTIONS TO DECOLONIZING THE CURRICULUM

Francesca Sobande

Introduction

This chapter outlines how Black feminist thought critically contributes to decolonial efforts, both inside and outside of academic institutions. There is consideration of how digital and social media accelerates Black feminist conversations, content, as well as internationally coordinated efforts (Cooper *et al.*, 2017; Hull *et al.*, 2015; Jackson, 2016; Williams, 2015). In addressing these matters, it is not my intention to imply that the compelling contributions of Black feminists are solely a product of twenty-first century politics and online communication channels (Emejulu, 2016). On the contrary, there is recognition of the longstanding concerns and creativity of Black feminists. After all, as Baszile *et al.* (2016: xiii) assert: "Make no mistake about it, Black women have always been theorizing and organizing based on our experiences of the world".

The development of Black feminism has involved building bridges and solidarity between the views and experiences of Black women and women of colour, such as individuals with African and Caribbean backgrounds, as well as those with heritage from other parts of the world, including the Global South (Hill Collins, 2000; Mirza, 1997; Spivak, 1988).

Consequently, this chapter particularly draws upon the seminal work of Black feminists and women of colour, who share a commitment to sustaining intersecting anti-racist and anti-sexist activities (Crenshaw, 2015).

Higher education institutions are now accessed by more "students who formerly had no way to pay for college (class), or students who historically faced discriminatory barriers to enrollment (race, gender, ethnicity or citizenship status, religion)" (Hill Collins and Bilge, 2016: 2). However, despite changes to the makeup of student populations, academia remains steeped in exclusionary practices and processes (Ahmed, 2017), which Black feminism tackles. Many obstacles preventing societally marginalized individuals from entering and thriving in academic settings can be attributed to "the legacy of colonialism" (Bassel and Emejulu, 2017: 89). The ongoing aftermath of empirical colonialist regimes, includes theories of racial science which propagate insidious stereotypes of "Savages – or the colonial other: the Native or Aboriginal peoples, the African, the Indian, the slave – in particular" (Bassel and Emejulu, 2017: 21).

Decolonial action must involve examination of "how we proceed with data collection and analysis – and our modes of inquiry – how we craft our discussion questions – must attend to the pitfalls of erasure" (Means Coleman, 2011: 38). Decolonial efforts are diverse, yet feature concerns with rupturing the imbalance of perspectives, politics, and people represented in academic curricula. Even "the term 'colonization' has come to denote a variety of phenomena in recent feminist and left writings in general" (Mohanty, 2007: 333), of which the concept of "colonized minds" is included. Thus, whilst decolonial action relates to tangible outcomes, including the removal of physical symbols of colonialism, it is also attuned to more abstract ones, such as "decolonization of the mind" (Kilomba, 2010; Mohanty, 2007).

This work focuses on three interlinked components of how Black feminist endeavours fuel decolonial approaches: (1) How Black feminism unpacks normative notions of "knowledge production" and "expertise", (2) The need to (re) contextualize intersectionality and self-reflexivity, as well as (3) Black women's experiences of marginalization (in)side of the academy, and the cultivation of Black feminist community online. This chapter proceeds with discussion of how Black feminist scholarship has been included and excluded from academic spheres. It also incorporates excerpts from in-depth interviews with Black women in Britain, conducted as part of my work as a doctoral researcher.

As a Black feminist in academia, to borrow the words of Liu and Pechenkina (2016: 190), I often "do not quite belong in/with(in) its mainstream narratives", such as those that refute the benefit of researchers reflecting on their lived experience. In the spirit of Black feminism (Cooper et al., 2017; James et al., 2009), which showcases how self-reflection can enrich critical enquiry, the latter stages of this chapter feature more of an autoethnographic tone.

Decolonizing normative notions of "knowledge production" and "expertise"

The crux of agendas to decolonize higher education includes "aims to displace the dominant top-down approach to knowledge production and theory building" (Barongo-Muweke, 2016: 25), such as by focusing on "the empirical worlds and subjective sense making (*consciousness*)" (ibid.) of individuals who are the most subjugated. Movements at universities across the world highlight continued demand for the decolonization of academic institutions, by students and staff who seek to challenge "the ways in which knowledge production – even knowledge about the 'non-West' has been skewed towards the perspectives and modes of articulation of western writers and institutions" (Noxolo, 2009: 56).

Examples of decolonial activity include the University of Cape Town's "Rhodes Must Fall" campaign, which resulted in the removal of a statue of British colonialist Cecil Rhodes (Bosch, 2016). Similarly, at the School of Oriental and African Studies (SOAS), University of London, people participated in action that drew attention to colonial issues which still stain academia (Sabaratnam, 2017). Campaigns and events that posed questions such as "Why is My Curriculum White?" and "Why Isn't My Professor Black?" also gained momentum at University College London (UCL) (Coleman, 2014). Such activity drew further attention to decolonial approaches to academia, which seek to undermine "the power of hegemonic histories that willfully produce collective amnesia" (Noble, 2017: 4). The introduction of an undergraduate degree in Black Studies at Birmingham City University (BCU) (Andrews, 2016) is exemplary of thorough attempts to combat the ways that Black people, as students, staff, as well as the subject of study, are often excluded from higher education.

In addition to addressing issues concerning the accessibility of entry into higher education, decolonial approaches involve aims to disrupt normative perceptions of what constitutes valuable academic enquiry, by "unsettling and reconstituting standard processes of knowledge production" (Bhambra, 2014: 115). Dominant perspectives of what quantifies effective academic work often rest on the belief that such research is an outcome of allegedly objective analytic processes, resulting in "knowledge which is created from outside ourselves, outside our bodies, out of our heads (as it were)" (Simmonds, 1997: 229). Furthermore, "scholarship that does not convey the Eurocentric order of knowledge has been continuously rejected on the grounds that it does not constitute credible science" (Kilomba, 2010: 28).

As Schwoerer (this volume) illustrates, academic curricula are sites that can support the dominance of Eurocentric theory and knowledge produced in the West (Bhambra, 2014; Said, 1978). One of the ways that Black feminist thought challenges impervious power relations is through critiques of the notions of "universality" and "objectivity". The concept of "universal knowledge" is the result of processes by which "a particular defines the universal for the rest of the planet" (Grosfoguel, 2012: 90). The related perception of "objectivity", which is positioned in opposition to that of "subjectivity" (Noxolo, 2009) is based on "two tenets – the belief in the separation of the observers from the observed and the belief in the separation of mind, spirit and matter" (Benjamin, 1997: 2). In short, allegedly universal and objective research is viewed as being the outcome of socio-politically neutral processes.

Black feminism is synonymous with "critiquing the ways of thinking and knowing of the Eurocentric patriarchal paradigm" (Benjamin, 1997: 1), such as by deconstructing hegemonic perceptions of how claims to knowledge should be assessed, as well why and by whom. These efforts confront how knowledge-claims in academia are shaped by a range of context-dependent factors, including geo-political location and the social position(s) of who purports the value of such information. Therefore, Black feminist thought commands consideration of how oppressive forces function ontologically and epistemologically, as well as how political processes of social legitimation bolster knowledge production.

Matters pertaining to knowledge production have been central to "the constitution of colonial identities" (Bassel and Emejulu, 2017: 22) and many hierarchical interactions uphold processes "of knowledge validation" (Kilomba, 2010: 29), which prescriptively outline "what 'true' and 'valid' scholarship is" (ibid.). To suggest that only the work of certain theorists is "political", especially that of individuals whose identities remain scarcely represented in academia, can be a political act itself. The common dearth of core curricula material, by or about Black women, is symptomatic of how "social hierarchies that characterise gender, race and class as intersecting systems of power" (Bassel and Emejulu, 2017: xiii) consistently "reproduce social scripts about who is an expert and who is not" (ibid.).

In line with principles that fortify decolonial academic approaches, which expose "how concepts of knowledge, scholarship and science are intrinsically linked to power and racial authority" (Kilomba, 2010: 27), Black feminist research explicitly deals with questions such as, who is perceived as being able to "teach knowledge? And who cannot? Who is at the centre? And who remains outside, at the margins?" (ibid.) Critical examination of these issues reveals inequalities that feed the disparity between the inclusion and exclusion of certain social groups in academia.

(Re)contextualizing intersectionality and self-reflexivity

This section reflects on, at times, distinctly Black feminist contributions that are (de)contextualized and integrated into academic work, in ways which replace their Black feminist underpinnings with "abstract meta narratives of expert discourses" (Barongo-Muweke, 2016: 25). This contributes to erasure of the unique knowledge and experiences of Black women, which intersectional praxis was intended to address in the first place (Crenshaw, 1989).

The strands that weave the web of colonized academic spaces include entrenched forms of subjugation, such as those pertaining to the "triple oppression of gender, race, and class" (Carby, 1997: 45), as well as other identity characteristics. In order to challenge complex power structures that decolonial efforts are intended to unpick, an intersectional perspective is necessary. It is through accounting for the interlocking nature of oppression (Crenshaw, 2015; Hill Collins and Bilge, 2016) that the different yet related ways that people are marginalized can be articulated and addressed.

In recent years, the term "intersectionality" has acquired a buzz and has even been used in ways suggestive of it being what hooks (2014: 9) may term "a 'hot' commodity to be exchanged in the academic marketplace". When referring to activism, Jackson (2016: 376) asserts that there is an "evolving ideological battle between the margins, where intersectional concerns are centered, and the mainstream, where politics simultaneously make minute accommodations of intersectional frameworks while silencing larger critiques".

Jackson's (2016) claim is also relevant to some of the ways that discussions of intersectionality manoeuvre throughout academic arenas. When conversations about intersectionality avoid reference to Black feminist thought, Black feminists' contributions are (de)contextualized in ways resulting in their surgical removal from the Black feminist grounding that generated them. Such activity suggests that embracement of Black feminist work within academia can involve dilution of the very Black feminist foundations upon which these ideas were developed; hence Nayak's (2014: 40) assertion that "the more Black feminist scholarship is welcomed in the fold of academia, the more vigilant Black feminists need to be about the functions, dangers and consequences of this welcome".

Black feminist approaches enable the formulation of research that moves Black women "from the margins to the center" (James et al., 2009: xiii), whilst scrutinizing the exact structural inequalities that placed them and others in a peripheral position in the first place. The innovative work of Marcus (2015) on the marginalization of Gypsy/Traveller girls in Scotland illustrates how in addition to shedding light on the experiences of Black women intersectional approaches further our understanding of the lives of different individuals who are stigmatized, including in educational contexts.

People "have multiple identities and they use multiple strategies to achieve their aims" (Essed, 1996: 1). Broader teaching of intersectional approaches can enable students to critique how interdependent systems of oppression support each other. It can also encourage students to unpack the power relations that feed the secondary status of knowledge produced by individuals whose identities, ontologies, epistemologies, methodologies, and theoretical frameworks contrast with those most commonly represented in archetypal "canonical" bodies of work. In turn, students can gain greater critical insight into academic writing and structures, including how they connect to omnipresent societal relations.

Black feminism calls "for an epistemology that includes the personal and the subjective as part of academic discourse" (Kilomba, 2010: 31). As well as developing intersectional praxis, Black feminist contributions have historically embraced reflexive methodologies (Hill Collins, 2000), including "writing as autoethnography, poetry, and narrative" (Boylorn, 2006: 651), in addition to other forms of creative practice, which can articulate "one's discomfort as a racially marked body" (Johnson, 2017: 275).

Self-reflexive output, such as "autobiographies, historical diaries, and photographs, makes visible the ways in which regulatory discursive power and privilege are 'performed' or exercised in the everyday material world of the socially constructed black and ethnicized woman" (Mirza, 2015: 2). Such women find themselves traversing predominantly patriarchal and racially white spaces in academia and further afield, within which their hyper-visible embodied presence signifies the common invisibility of Black women and women of colour in these settings (Ahmed, 2007; Gay, 2014). The words of Simmonds (1997: 228) on doing sociology as a Black woman poignantly summarize these feelings: "In the final analysis, I might be an academic, but what I carry is an embodied self that is at odds with expectations of who an academic is".

The work of individuals such as Boylorn (2006), Busia (1993), Kilomba (2010), and Lorde (2002) exemplifies how Black feminist methodologies incorporate creative and unconventional methods, which push against the dominant belief that academic writing should be devoid of expressive introspection. Such writers, scholars, activists, and performers showcase how creative and scholarly work need not be mutually exclusive, whilst demonstrating how the juncture of both aids "strategies of subverting that gaze whose satisfaction lies through reifying us as 'other'" (Busia, 1993: 207). It is through self-reflexive approaches that explore the inter-subjectivity of research(ers) (Noxolo, 2009) and emphasize the "relationship between being and knowing" (Simmonds, 1997: 229) that there is revelation of "the fragility of theory's insistence that we can articulate truths only through a rational and objective epistemology of social reality" (ibid.).

The increased inclusion of Black feminist methodologies in academia can contribute to impactful work such as that of Johnson's (2017: 274), on how "comfort is felt in relation to racially marked bodies", and which challenges the notion that a researcher's lived experience "can be divorced from the academic writing produced from this positioning" (idem). Curricula that fosters such intersectional and self-reflexive perspectives can encourage students to critically and consciously think about their identities in relation to their own research (Arya, 2012), as well as reflect on the identities of others when assessing their work.

"(In)side" of the academy and cultivating community online

On one hand, Black feminist thought has garnered more academic attention in recent decades. On the other, Black feminists still struggle to transcend the disciplinary boundaries of race and gender studies alone, as is indicated by articles such as "What's a Hip Hop feminist doing in teacher education? A journey back to curriculum theory in three acts" (Guillory, 2011). The future of Black feminist scholarship may hold exploration of "new disciplinary, interdisciplinary, and transdisciplinary discourses" (James et al., 2009: xviii), as well as further "acknowledging and exploring the diasporic and cross-cultural dimension of Black women's experiences" (ibid.).

The insightful and rigorous research of North American Black feminists is imperative to the continued galvanization of Black feminist academic discourse. However, North American Black feminist thought has a hegemonic status in comparison with Black feminist work originating in other parts of the world, including Africa and Europe. As such, this chapter argues for the wider inclusion of Black feminist contributions in academic curricula, along with broadening the plurality of perspectives that are included.

Black feminist thought plays a vital role in enabling Black women to articulate and archive their experiences, both inside and outside of academic contexts. The term "(in)side of the academy" encompasses how despite being formally accepted in academic settings, be it through successful student and job applications, Black women may feel as though they simultaneously exist within and to the side of academia. The words of Guillory (2011: 20), when recounting selecting an academic major in curriculum theory, echo the sentiments of the interview participants referenced to in this chapter: "many of the discourses to which I was introduced through required course readings did not include theorizing about Black women's epistemologies".

Participants interviewed as part of my research about the experiences of Black women in Britain as media spectators and producers spoke of the challenges they faced as part of their higher education trajectories. When recalling her years as a social sciences undergraduate student, a participant who is 30 years of age and chose the pseudonym Ralph-Angel, said:

> When I was studying at uni, I felt like [...] looking back now, the educational or academic curriculum doesn't really accommodate [...] they're not teaching anything, you know, that can keep me engaged [...] so I went to a Scottish university, where what is being taught [...] it's not something that I can entirely relate to, so that made me disengaged [...] why do I want to read something that a middle aged white man wrote 300 years ago? How is that relevant to me? Education has so much to do with the way you see the world and the way that the world sees you, you know! For me, I'm also the first person in my family to actually finish high school and go to university [...] so I had to sort of teach myself what's relevant and what's not. Anyone can pick up a book and read something, but it's reading the *right* thing sometimes.
>
> *(Ralph-Angel)*

The words of Ralph-Angel point to the "symbolic and narrative struggle over the defining materiality of her educational experience" (Mirza, 2015: 2), partly due to the lack of scholarship she was exposed to that was by or about Black women.

The absence of perspectives of Black people within academic curricula is a form of "symbolic annihilation" (Gerbner and Gross, 1976), which relates to "how representations, including omissions, cultivate dominant assumptions about how the world works and, as a result, where power resides" (Means Coleman and Yochim, 2008: 2–3). The marginalizing experience that Ralph-Angel references is mirrored by the words of scholars such as Cutts *et al.* (2012: 58), who write of how "the academy rarely welcomed us as citizens", when reflecting on their experiences as Black women studying doctorate degrees.

In the case of Ralph-Angel's narrative, one of the dominant messages conveyed to her over the course of her undergraduate experience was that Black women are rarely deemed capable of imparting academic knowledge upon others. Ralph-Angel's comments signpost how despite gaining entry to academic spaces as students, without access to curricula that relate to their different raced, gendered and classed lived experiences, feelings of inclusion are unlikely to truly manifest for individuals who struggle to connect their realities to those outlined in the material that they encounter at university.

Several of my interview participants commented on having wanted to pursue postgraduate studies but having been put off by the scarcity of Black women role models in academia, as well as their negative experiences of having their work dismissed due to its foregrounding of issues to do with race. Amongst my interview participants, there were also individuals in the midst of conducting postgraduate research, who woefully shared anecdotes about the extensive emotional labour (Mirchandani, 2003) involved in pursuing their work. Their frustrations included concerns about how Black women's academic contributions can be eclipsed and erased, even in certain decolonial-oriented spaces. Such viewpoints reiterate the need to reflect on hierarchical undercurrents that can cut across decolonial endeavours, as well as structural ones related to issues of equality, diversity, and inclusion in higher education and beyond (Knowles and Lander, 2011).

Feelings of existing "(in)side of the academy", rather than prospering as part of it, were evidenced by the words of other interview participants, including Sasha Barrow. Sasha Barrow is a 36-year old artist and postgraduate research student in Scotland who spoke about how she goes online as a means of harvesting academic content that is otherwise often missing from the spaces she has physical access to. It is here that Sasha Barrow hears about the work and experiences of other people of colour in academia:

> I definitely consider myself an elder, just even through my years [laughs] I'm in a much better position since I did my undergrad [...] and I think that I'm stronger and I've got a better perspective of things, so I must be here to support other women of colour and other people of colour as best I can [...] because I can shoulder it at the moment, and I think that we must all do what we can, to really support each other [...] it's a kind of critical and necessary community [...] it's definitely happening. I watch loads of YouTube things [...] I watch a lot of lectures by bell hooks or different Black academics. I watch *a lot* of lectures. I am also part of a couple of secret groups on Facebook and also some more open pages, such as Why is My Curriculum White?, which I find really helpful.
>
> *(Sasha Barrow)*

As the remarks of Sasha Barrow convey, Black women continue to seek out alternative spaces and autonomous means to think through and communicate their experiences of academia outside of institutional confines.

As there is a strong international community of Black feminists who post content through digital platforms such as Twitter, Tumblr and YouTube (Cottom, 2016; Hull *et al.*, 2015), Black women are engaging in online activity that serves a somewhat compensatory and self-empowering function, in comparison to the infrequency of opportunities to learn about and by Black people in formal educational settings (Ahmed, 2012). The use of Twitter hashtags is one example of this, and has "emerged as an effective way to share information and spur action" (Williams, 2015: 342) concerning the specific obstacles that Black women face.

Present day social interactions are punctuated by the significant role of digital media and associated "strategic hypervisibility" (Cottom, 2016: 211) in the lives of many, including Black feminists. Online content-sharing platforms are increasingly used as part of the recording and sharing of Black feminist discussions, such as those that occurred as part of the Feminist Digital Pedagogies Conference at Rutgers University, in January 2014. This video content enabled individuals from outside of the US, including myself, to learn from and about Black feminists involved in this event.

Black feminist scholars have long been involved in "sustained efforts to establish cross-disciplinary alliances within the academy as well as efforts to develop coalitions outside the academy" (James *et al.*, 2009: xix). How such women partake in these community-building activities has evolved in line with the intensification of digital media platforms. A case in point is the first *Black Feminism, Womanism, and the Politics of Women of Colour in Europe* symposium *(WoC Europe)*, which took place at the University of Edinburgh in September 2016 (Emejulu *et al.*, 2016).

The landmark international *WoC Europe* event brought together academics, activists, creatives, and collectives, from a wide range of countries, with a shared dedication to enhancing the lives of Black women and women of colour. When reflecting on personal experience of attending a different event in 2010, entitled *The Women of Color in the Academy* conference, Clark and Davis (2011: 434) reminiscently write of becoming immediately aware of "the audience composition" when entering the room. Such an observation is followed by the statement: "they surprisingly resembled me" (ibid.).

When participating in *WoC Europe*, I could not help but have similar thoughts to those of Clark and Davis (2011), about the rare and nourishing nature of finding myself in a space which centred dialogue between and about Black feminists and women of colour. The energy at the event was a clear indication of the vibrant activity of Black feminists, whose work has found a home in a whole host of different spaces, including digital ones. Many of the women involved in the first *WoC Europe* conference engaged in discussion with one another online prior to the event, including as part of the Twitter hashtag #WoCEurope. Such digital activity enabled the continuation of conversations long after the inaugural symposium, as well as in the run up to the second one, in Amsterdam in October 2017.

Further evidence of how social media is shaping Black feminist community-building and the global mobilization of Black feminist narratives is the ascent of the Crunk Feminist Collective. When initially launched online, the Crunk Feminist Collective had a modest presence as a low-key blog that provided a platform for the voices of those who "were committed to doing the work of being visible and vocal Black feminists and cultivating a space where feminism – specifically crunk feminism – wasn't an overly theoretical concept held hostage by the academy" (Cooper *et al.*, 2017: 2). The website tagline for the Crunk Feminist Collective, which is "Where Crunk Meets Conscious and Feminism Meets Cool", captures the ethos of a powerhouse group of individuals who often blend discussion of politics, popular culture and current affairs with astute critical observations via a Black feminist lens.

The wide following that the Crunk Feminist Collective has amassed since its inception has enabled such content to move into other offline areas. This includes the arts venue in Scotland where *The Crunk Feminist Collection* (Cooper *et al.*, 2017) was excitedly purchased by myself. Reading *The Crunk Feminist Collection* reminded me of how digital and social media can amplify Black feminist views, voices, and visions, which translate into opportunities to take up space, in and across different online and offline forums (Cottom, 2016).

Conclusion

Indeed, there are examples of the inclusion of Black feminist perspectives in academia, such as the Pembroke Center "Black Feminist Theory Project" at Brown University, which encourages visiting scholars to contribute to the enhancement of Black feminist discourse (Brown University, 2017). Additionally, along with studies of race and gender, curriculum studies have been recognized as being a domain within which Black feminist thought can flourish, albeit one where Black feminism still has a marginal status (Edwards *et al.*, 2016). However, Black feminist perspectives remain strikingly absent from many curricula (Guillory, 2011).

As Mirza (2015: 1) maintains, "in order to tackle race and gender inequality in higher education, it is imperative to understand the nature of power relations and the ways in which racialized, classed and gendered boundaries are produced and lived through black/postcolonial female subjectivity in our places of learning and teaching". If academic curricula do not provide opportunities to interrogate such matters, where else may students and staff turn when trying to disentangle and articulate these structural issues? The internet is one key alternative.

Calls for Black feminist work to be further incorporated into academic curricula are far from being new. Yet, what has changed in recent decades is some of the ways that such scholarship is disseminated, including as part of video blogs (vlogs) and social media content. The expansion of online content-sharing platforms, which have been used for "storytelling, mobilization and citizen journalism" (Emejulu and McGregor, 2016: 7), have fostered increased opportunities for the expression of marginalized voices, as well as facilitating their global reach. Therefore, as efforts to decolonize minds and academia persist, so too will dissemination of Black feminist thought via creative and digital channels, devoid of the constraints of institutional educational outlets.

Whilst academic equality, diversity and inclusion measures often focus on issues related to the representation of people from different backgrounds, the content of curricula is not commonly subject to intense scrutiny, despite the ways that this can tie into issues of access, inclusion, attainment, and retention. Curricula which nurture "the development of integrated analysis and practice based upon the fact that the major systems of oppression are interlocking" (The Combahee River Collective, 2009: 3) can aid decolonial efforts, by highlighting the specificity of people's experiences of marginalization.

Although Black feminist academic work is incredibly varied, at the heart of it often lies a commitment to "initiating curriculum transformation projects in diverse academic settings" (James *et al.*, 2009: xii), as well as an intention to destabilize racist and sexist colonial legacies, amongst others. The wider inclusion of intersectional Black feminist perspectives in academic curricula can encourage students to critically explore self-reflexive approaches to analysing and producing academic work in ways which help them to connect scholarly research to their lived realities and the unique lens through which they view the world.

References

Ahmed, S. (2007) A phenomenology of whiteness. *Feminist Theory* 8(2): 149–168.

Ahmed, S. (2012) *On Being Included: Racism and Diversity in Institutional Life*. Durham: Duke University Press.

Ahmed, S. (2017) *Living a Feminist Life*. Durham: Duke University Press.

Andrews, K. (2016) The black studies movement in Britain. *The Black Scholar* 6 October 2016. Available at: www.theblackscholar.org/?s=The+black+studies+movement+in+Britain (Accessed 3 January 2017).

Arya, R. (2012) Black feminism in the academy. *Equality, Diversity and Inclusion: An International Journal* 31(5/6): 556–572.

Barongo-Muweke, N. (2016) *Decolonizing Education: Towards Reconstructing a Theory of Citizenship Education for Postcolonial Africa*. Wiesbaden: Springer VS.

Bassel, L. and Emejulu, A. (2017) *Minority Women and Austerity: Survival and Resistance in France and Britain*. Bristol: Policy Press.

Baszile, D.T., Edwards, K.T. and Guillory, N.A. (2016) *Race, Gender, and Curriculum Theorizing: Working in Womanish Ways*. London: Lexington.

Benjamin, L. (1997) *Black Women in the Academy: Promises and Perils*. Florida: University Press Florida.

Bhambra, G.K. (2014) Postcolonial and decolonial dialogues. *Postcolonial Studies* 17(2): 115–121.

Bosch, T. (2016) Twitter activism and youth in South Africa: The case of #RhodesMustFall. *Information, Communication & Society* 20(2): 221–232.

Boylorn, R.M. (2006) E pluribus unum (out of many, one). *Qualitative Inquiry* 12(4): 651–680.

Brown University (2017) Black feminist theory project. Available at: www.brown.edu/research/pembroke-center/news/2017-06/black-feminist-theory project (Accessed 12 August 2017).

Busia, A.P.A. (1993) Poetry, transcription and the languages of the self: Interrogating identity as a 'post-colonial' poem. In S.M. James and A.P.A. Busia (eds) *Theorizing Black Feminisms: The Visionary Pragmatism of Black Women*. London and New York: Routledge, pp. 205–216

Carby, H. (1997) White women listen! Black feminism and the boundaries of sisterhood. In H.S. Mirza (ed) *Black British Feminism*. London: Routledge, pp. 45–53.

Clark, S.L. and Davis, D.J. (2011) Women of color in the academy. *Equality, Diversity and Inclusion: An International Journal* 30(5): 431–436.

Coleman, N. (2014) Decolonizing my discipline by teaching research on 'race'. Available at: www.academia.edu/6651199/Decolonising_my_discipline_by_teaching_rese arch_on_race (Accessed 8 January 2017).

The Combahee River Collective (2009) A black feminist statement. In S.M. James, F.S. Foster and B. Guy-Sheftall (eds) *Still Brave: The Evolution of Black Women's Studies*. New York: Feminist Press, pp. 3–11.

Cooper, B.C., Morris, S.M. and Boylorn, R.M. (2017) *The Crunk Feminist Collection*. New York: Feminist Press.

Cottom, T. (2016) Black cyberfeminism: Intersectionality, institutions and digital sociology. In J. Daniels, K. Gregory and T. McMillan Cottom (eds) *Digital Sociology*. Bristol: Policy Press, pp. 211–231.

Crenshaw, K. (1989) Demarginalizing the intersection of race and sex: A black feminist critique of antidiscrimination doctrine, feminist theory and antiracist politics. *The University of Chicago Legal Forum* 140: 139–167.

Crenshaw, K. (2015) *On Intersectionality: The Essential Writings of Kimberlé Crenshaw*. New York: New Press.

Cutts, Q.M., Love, B.L. and Davis, C.L. (2012) Being uprooted: Autobiographical reflections of learning in the [New] South. *Journal of Curriculum Theorizing* 28(3): 57–72.

Edwards, K.T., Baszile, D.T. and Guillory, N.A. (2016) When, where, and *why* we enter: Black women's curriculum theorizing. *Gender and Education* 28(6): 707–709.

Emejulu, A. (2016) Beyond feminism's white gaze. *Discover Society* 1 March 2016. Available at: http://discoversociety.org/2016/03/01/beyond-feminisms-white-gaze-black-feminism-womanism-and-the-politics-of-women-of-colour-in-europe/ (Accessed 5 January 2017).

Emejulu, A., Marcus, G., Sang, K., Sobande, F. and Ward, O. (2016) 1st Black feminism, womanism, and the politics of women of color in Europe. Available at: woceuropeconference.wordpress.com/ (Accessed 2 February 2017).

Emejulu, A. and McGregor, C. (2016) Towards a radical digital citizenship in education. *Critical Studies in Education* 19 September: 1–17.

Essed, P. (1996) *Diversity: Gender, Colour, and Culture*. Amherst: University of Massachusetts Press.

Gay, R. (2014) *Bad Feminist*. New York: Harper Perennial.

Gerbner, G. and Gross, L. (1976) Living with television: The violence profile. *Journal of Communication* 26(2): 172–199.

Grosfoguel, R. (2012) The dilemmas of ethnic studies in the United States: Between liberal multiculturalism, identity politics, disciplinary colonization, and decolonial epistemologies. *Human Architecture: Journal of the Sociology of Self-Knowledge* 10(1): 88–104.

Guillory, N.A. (2011) What's a Hip Hop feminist doing in teacher education? A journey back to curriculum theory in three acts. *Journal of Curriculum Theorizing* 27(3): 20–32.

Hill Collins, P. (2000) *Black Feminist Thought: Knowledge, Consciousness, and the Politics of Empowerment*. New York: Routledge.

Hill Collins, P. and Bilge, S. (2016) *Intersectionality*. Cambridge: Polity Press.

hooks, b. (2014) *Yearning: Race, Gender, and Cultural Politics*. New York: Routledge.

Hull, G., Bell-Scott, P. and Smith, B. (2015) *But Some of Us Are Brave: Black Women's Studies*. New York: Feminist Press.

Jackson, S.J. (2016) (Re)Imagining intersectional democracy from black feminism to hashtag activism. *Women's Studies in Communication* 39(4): 375–379.

James, S.M., Foster, F.S. and Guy-Sheftall, B. (2009) *Still Brave: The Evolution of Black Women's Studies*. New York: Feminist Press.

Johnson, A. (2017) Getting comfortable to feel at home: Clothing practices of black Muslim women in Britain. *Gender, Place & Culture* 24(2): 274–287.

Kilomba, G. (2010) *Plantation Memories: Episodes of Everyday Racism*. Unrast-Verlag: Münster.

Knowles, G. and Lander, V. (2011) *Diversity, Equality and Achievement in Education*. London: SAGE.

Liu, H., Cutcher, L. and Pechenkina, E. (2016) Staying quiet or rocking the boat? An autoethnography of organisational visual white supremacy. *Equality, Diversity and Inclusion: An International Journal* 35(3): 186–204.

Lorde, A. (2002) *The Collected Poems of Audre Lorde*. London: W.W. Norton & Company.

Marcus, G. (2015) Marginalization and the voices of Gypsy/Traveller girls. *Cambridge Open-Review Educational Research e-Journal* 1(2): 55–77.

Means Coleman, R.R. (2011) 'Roll up your sleeves!': Black women, black feminism in feminist media studies. *Feminist Media Studies* 11(1): 35–41.

Means Coleman, R.R. and Yochim, E.C. (2008) The symbolic annihilation of race: A review of the 'blackness' literature. *African American Research Perspectives* 12: 1–10.

Mirchandani, K. (2003) Challenging racial silences in studies of emotion work: Contributions from anti-racist feminist theory. *Organization Studies* 24(5): 721–742.

Mirza, H. (1997) *Black British Feminism: A Reader.* London: Routledge.

Mirza, H. (2015) Decolonizing higher education: Black feminism and the intersectionality of race and gender. *Journal of Feminist Scholarship* (7/8): 1–12.

Mohanty, C.T. (2007) *Feminism Without Borders: Decolonizing Theory, Practicing Solidarity.* Durham: Duke University Press.

Nayak, S. (2014) *Race, Gender and the Activism of Black Feminist Theory: Working with Audre Lorde.* New York: Routledge.

Noble, D. (2017) *Decolonizing and Feminizing Freedom: A Caribbean Genealogy.* London: Palgrave Macmillan.

Noxolo, P. (2009) My paper, my paper: Reflections on the embodied production of postcolonial geographical responsibility in academic writing. *Geoforum* 40(1): 55–65.

Sabaratnam, M. (2017) Decolonizing the curriculum: What's all the fuss about? The School of Oriental and African Studies (SOAS), University of London, 18 January 2017. Available at: www.soas.ac.uk/blogs/study/decolonising-curriculum-whats-the-fuss/ (Accessed 10 July 2017).

Said, E. (1978) *Orientalism.* New York: Pantheon Books.

Simmonds, F.N. (1997) My body myself: How does a black woman do sociology? In H. Mirza (ed) *Black British Feminism: A Reader.* London and New York: Routledge, pp. 226–239.

Spivak, G. (1988) *In Other Worlds: Essays in Cultural Politics.* New York: Routledge.

Williams, S. (2015) Digital defense: Black feminists resist violence with hashtag activism. *Feminist Media Studies* 15(2): 341–344.

7

DENATURALIZING SETTLER-COLONIAL LOGICS IN INTERNATIONAL DEVELOPMENT EDUCATION IN CANADA

Trycia Bazinet

Introduction

Canada, along with other settler-colonial nations such as the United States, New Zealand and Australia, was not an original signatory of the Universal Declaration of the Rights of Indigenous Peoples (UNDRIP). Even when Canada became a so-called "full supporter of the declaration" in 2016,[1] Justice Minister Wilson-Raybould said that in retrospect adopting UNDRIP as Canadian law is "simplistic" and "unworkable" (Canada, 2016). This is only one out of the many statements and policies that demonstrate that settler governments, regardless of the effectiveness of their reconciliatory public relations, are ultimately concerned with reconciling Indigenous' rights as long as they can be reconciled with the Crown's sovereignty (Borrows, 2002; Mackey, 2016). Indeed, one of the recurring reason cited by Canadian authorities in their attempts to justify their refusal to endorse UNDRIP had to do with concerns over sovereignty and the management of natural resources.

Beyond the expectable let-downs of a settler government that is invested in retaining power, do Canadian citizenships, identities and affective structures matter in this vast, ongoing problem of broken promises and relationships? For instance, what does it mean for Canadians to be able to participate in development projects abroad when the original People and Nations of the land on which they stand have their sovereignty and inherent and historical rights trampled upon, as determined by their own legal traditions, by Treaties or by UNDRIP itself? What does it mean for Canadian development workers to become experts on food security and water management in so-called developing countries (whom are struggling against their own colonial and imperial legacies) while many First Nation communities do not have access to clean water (White *et al.*, 2012) and Inuit communities do not have food security "here" because of ongoing colonialism (Burnett *et al.*, 2016)? Furthermore, what is it that has made Canada prosperous enough, and some of its students privileged enough, to assume the role and *burden* of global citizenry? This chapter tackles these questions from a critical settler-colonial perspective by observing the basic premises found upon which the education of the interdisciplinary subject of international development is based. The examples are drawn from a systematic analysis of the syllabi of the mandatory courses of the Master of Arts in International Development at the University of Ottawa from September 2007 to December 2015. My content analysis serves as a precursor for more extensive studies aiming to "uncover settler grammar" in curricula (Calderon, 2014). The discussion included in my chapter can hopefully serve to facilitate future research that examines the syllabi of other disciplines and subjects such as International Relations, Environmental Engineering and Studies, Economics, Community Development and Human Rights. Still, the "subject" of International Development is a popular area of study in Canada; with at least 15 similar Master's Degrees and at least 17 Bachelor's degrees in Canada.[2] It is significant to also point out that this particular Master's degree takes place within "Canada's largest academic unit specializing in international development".[3]

The land beneath Canadian universities

Settler-colonialism as a reality is nothing new. It has been and continues to be resisted and critiqued by Indigenous peoples and Indigenous scholars alike (Macoun and Strakosch, 2013). Indigenous scholars have "created" the field of settler-colonial studies – which is not exactly the same as Indigenous studies (Kauanui, 2016). Still, one of the scholars who is often associated with the scholarly field of settler-colonialism, Patrick Wolfe (1999: 163; 2006), has made many non-Indigenous scholars re-think their understanding of colonialism when saying "settler-colonialism is a structure, and not an event". Settler-colonialism is increasingly theorized as a naturalized and normalized structure of constant occupation that is replicated and supported at all levels of a society in the present, the whole structure taking place within the realm of modernity (Arvin *et al.*, 2013; Bonds and Inwood, 2015). It prioritizes, works for and assumes the future of white settlers at the cost of other futures (white settler futurity) (Tuck and Gaztambide-Fernández, 2013).

Settler-colonial analysis and critical Indigenous studies invite us to push beyond anti-imperial and post-colonial critiques of development and globalization, so as to reveal more proximate and material understandings of actors involved in the international and global networks of relations, as they are by default themselves invested in white settler-colonial relationships. For the particular case of universities located in the capital of Canada, Ottawa, one might be surprised to learn that the very land upon which they sit is unceded and unsurrendered. This means that there are no Treaties in that area designating the nature of the international relations between incoming settlers and local Anishinaabeg (Algonquin) Nations and that Anishinaabe immigration protocols were and are not respected. If we consider the fact that Canada, as an imagined concept and as an ever-forming entity, relies on a "fantasy of entitlement" (Mackey, 2016) to Indigenous territories, cultures and resources that are made to matter and materialize, then technically, we have to grapple with the question of illegitimate occupation or of squatting.

This reality is not taken seriously by Canadian institutions, including universities. Instead, it is generally agreed that both decolonization and colonization are finished business. At best, the multiple assaults of colonialism are spoken of as events of the past that *might* have had repercussions in the present. Is that the case? If one listens to the various Indigenous peoples of what is now known as North America, it becomes impossible to answer this question with a confident affirmative. Moreover, it is crucial to remember that decolonization is not just about decolonizing our minds or our textbooks – in fact, the metaphorization and watering down of what decolonization (Tuck and Yang, 2012) tangibly implies, such as the rematriation of Indigenous lands, is an example of the need for settler-colonialism to absorb the contentious potentialities of what it considers more threatening processes of decolonization. What is Indigenous futurity? To paraphrase Tuck (2011) and Nixon (2016), it is living, restoring and producing life unapologetically, threading traditions and technologies, as if the present and future worlds are Indigenous. It "forecloses settler-colonialism" in its entirety (Tuck and Gaztambide-Fernández, 2013: 80). On the other hand, (white) settler futurity is living life with the assumption that white settler-colonialism will (or should) continue. White settler futurity is the "permanent virtuality of the settler on stolen land" (Baldwin, 2012, as cited in Tuck and Gaztambide-Fernández, 2013: 80). However, contrary to its own self-serving impression, the structure of settler-colonialism is not fatally inevitable. Rather, it is fragile and remains incomplete in the face of Indigenous resistance (Macoun and Strakosch, 2013), refusals and resurgence (Betasamosake Simpson, 2014b; Flowers, 2015). This is best exemplified by the ongoing criminalization of Indigenous land and water protectors in Canada[4] (Palmater, 2015).

Continued colonial occupation does not play out as a simple difference in political opinions; rather, it requires and practises a nonchalant yet violent erasure and dispossession of Indigenous peoples and Nations. Those who theorize living under its condition have likened it to living under a (post)-apocalyptic reality (Baldy, 2015; Johnson, 2015; Violet-Lee, 2016). The settler interest and material practices emerging from the need to secure constant access to the land have apocalyptic repercussions for Indigenous peoples. In fact, settler-colonialism "licenses the disappearance of Indigenous peoples" (Bonds and Inwood, 2015). This is all to say that settler-colonialism is not just a theory but rather that it has urgent repercussions. Thus, the role of universities, whether they uphold or unsettle these structures, indirectly becomes a matter of life or death.

A settler-colonial lens for curricula

Higher education can normalize settler-colonial rule. Academic fields that specifically rely on evidence-based education engage in a colonial discourse in an inadvertent fashion or not, which in turn perpetuates current imperial interests and unequal material relations of power (Shahjahan, 2011). As a stand-alone, evidence-based education relies on categories, measurements and worldviews that were developed by marginalizing Indigenous and racialized knowledges and by using it against these same populations. Producing statistics and quantifying peoples and nonhuman entities, such as forests and lands, assisted in creating categories of difference to be managed by colonial rule (Agrawal, 2005). Evidence-based education is thus often taught and shared within epistemologies that still regard other methodologies or ways of knowing as inferior and invalid, because of a colonial baggage that is rarely even acknowledged. Evidence-based education, when applied in economic development, for instance, is invested in maintaining current and deeply unequal material relations that have their very onset in colonialism and slavery; specifically, some of the foundational concepts in higher education, like modernity and liberalism (and their correlates of seamless progress, industrialization, democracy, secularization, humanism, human rights, linear time, scientific reasoning and nation-state) (De Oliveira Andreotti *et al.*, 2015; Mignolo, 2002), continue to shun Indigenous epistemologies (Kuokkanen, 2007). Thus, higher education curricula becomes entangled in colonization most obviously through what is called eurocentrism (Battiste, 2013; Kerr, 2014), which directly impacts the well-being of Indigenous students in the classroom (Cote-Meek, 2014). Even when curriculums mention or include Indigenous peoples' ideas or issues, they do so in a "partial" way (Littlebear, 2009). Moreover, Indigenous knowledges and stories can problematically be taught by non-Indigenous educators (Violet-Lee, 2015). Higher education and its productions, such as their curricula, continue to play a powerful role in shaping common-sense worldviews that enable and naturalize past and present settler-colonialisms. This is why a settler-colonial lens can and should be applied to them.

Calderon (2014) applies the framework of settler-colonial studies to the study of social sciences curricula through the analysis of a series of presences and absences. Her approach enables us to analyse patterns of absences and presences of keywords and themes that are highlighted by settler-colonial studies. How is the nation talked about? Is colonialism mentioned, and if so, only in the past tense? Do Indigenous peoples only have culture and ethnicity – or do they also have sovereignty? The patterns of what is absent and of what is present in curricula become non-coincidental, but marked as participating in the normative project of late modern settler-colonial occupation. I personally focused on the mandatory part of the curricula because my emphasis is not on what is offered by the programme, but what is institutionalized as required and foundational for one to graduate from this programme. I was less interested in the optional possibilities offered by optional courses, joint programmes, conference and internships but more in the *mandated* lessons to be learned. In other words, what kind of future (Indigenous or white settler) is assumed within the content that a student *must* know in order to graduate from this programme?

Through search strings of relevant keywords in 81 syllabi of four mandatory courses taught over an eight-year period, 2007–2015,[5] I found that colonialism is barely mentioned, and when it is, it is usually either referred to as a *finished* event cast in the past or as something geographically distant – never as here and now. For instance, decolonization would refer to the historical period after the second world war, which naturalizes colonization on Turtle Island. "Post-colonial" would be used to describe the situation in Canada, which begs the question of when did the so-called post-colonial period begin in Canada? In Canada, the settlers never left, in comparison to other Nations who faced European occupation. Of course, the educational experience cannot be limited to the mandatory content of a programme and will also not always mean an automatic career in international development for its students. Still, these core syllabi offer a tangible insight into institutional values and required readings presented in them are usually meant to direct, start off or influence class discussions and activities.

The collapses of white settler management

The degree to which settler-colonialism is naturalized in the content of the core syllabi is not an easily measurable question. Is any reference to colonization or Indigenous peoples to be considered a step towards the "right" direction? No – the context of certain mentions can actually further re-entrench the functioning of white settler-colonialism. The presence and absence of certain topics in the curricula are ordered by a series of collapses that work through the lines of space, time and race. Without offering the breakdown results for each word here,[6] I searched whether words meaning "Indigenous" were present, and when they are, whether they are referred to as if they are Nations and/or international matters or if that mention is collapsed under special/minority/cultural/racial. Second, I also searched whether mentions of "Canada" and/or "Nation-State" are referred to in a way that could possibly unsettle them or not. Third, I searched whether references to "colonization" are relegated to the past (post-colonial and "decolonization" of Africa/Asia). I assessed what I gather to be non-random orderings of the presences and absences of relevant terms and of their context according to the settler-logics of space, time and race.

time. Under naturalized settler-colonialism, the presence of colonialism is solely acknowledged if it is simultaneously associated with past and finished events. At most, colonialism can be acknowledged to have ongoing repercussions in the present, but it is not considered the present. However, normative and public relegation of colonial "wrongs" to the past serves to affirm innocence in the present and to keep open the possibility that there was a "prior sense of good will" (Simpson, 2016). It is a way of saying that "we did not know that it was wrong; now that we know it, we recognize it, sorry". It fails to engage with the shape-shifting nature of a colonial occupation concerned with its own continuation. When references to colonialism did not have to do with being a step in history, they were stored in the suggested readings sections, meaning that possible contestations were not central to the subject matter. If any reference to the present is made, it is a present whose conditions are post-colonial. It is generally implicit that the passage of time has rendered occupation irrelevant.[7] However, settler-colonialism occurs in the present and caters to its (white settler) future, covering its own track while doing so (Tuck and

Gaztambide-Fernández, 2013). For instance, references to "Canada" treat the Nation state entity and its apparatuses as if they are given, and any critical appraisal that could challenge its legitimacy in the present, perhaps in an anti-colonial direction, is scarce.

space. Most forms of institutions indirectly produce, recreate and perform settler governance in everyday ordinary occurrence through the subtle role of non-state, non-institutional and non-Indigenous actors (Rifkin, 2014). Settler geographies, for instance, are naturalized by "repetitive practices of everyday life that give settler space meaning and structure" (Goeman, 2013: 236–237). The repeated use of Canada as a legitimate and uncontested entity in curricula performs that function. The great majority of mandatory courses in international development departments were highly unlikely to question Canada's cohesion and legitimacy. This means that students can graduate from the programme without even coming across texts and ideas that represent and honour Indigenous nations as Nations and/or as international actors.

For the settler-state to not have to recognize Indigenous nations within its borders it must elaborate systems that collapse them within its parameters. To continue the earlier discussion, Byrd (2011: xvii) explains how colonialism, both domestically and abroad, "often coerces struggles for social justice for queers, racial minorities, and immigrants into complicity with settler colonialism". When settler-colonialism is revealed to enact white supremacy, then one can see how white supremacy "is wrapped up in everyday geographies" and that it is the settler-colonial condition that sustains the material privileges of whiteness (Bonds and Inwood, 2015). Settlers' imposition of their geographies depends on another logic – race.

race. Colonization and racialization[8] are two processes of domination that have been conflated in white settler-colonial contexts. Racializations serve a specific role in maintaining not only white supremacy, but in invisibilizing settler-colonial occupation. The "racialization of the 'Indian' is in fact 'the erasure of the sovereign'" (Barker, 2005, as cited in Byrd, 2011, p. xxvi). Settler-colonialism is hidden by a logic that collapses Indigenous peoples as a racial group, so that space and land is erased, which furthers naturalizes settler-colonial geographies mentioned earlier. The argument that ethnicity assimilates the Indigenous, in which Indigenous peoples are categorized as "racial minorities rather than as sovereign peoples seeking decolonization", is not a new one (Simpson and Smith, 2014; Stevenson, 1998). Indigenous peoples, for the sake of settler need for readability and management, are essentially "rendered an identity category [...] rather than members of separate nations", despite legally binding but ignored Treaties (Goldstein, 2014). As policy analyst Hayden King states, the subject of International Relations and the resulting Canadian foreign policies work to excuse, marginalize and categorize Indigenous diplomacies as domestic, even though their struggles for justices, even if taking place within the borders of Canada, are part of international politics. He illustrates that blockades and direct actions of land protection are acts of diplomacy, and that they should be appraised as such (King, 2017).

Again, the primary motive for the constant attempt at elimination, erasure and assimilation of Indigenous peoples in settler-colonies is the land, not their race

(Wolfe, 2006: 338). However, racialization is centrally useful to these attempts. The collapsing of colonialism as racism blurs competing and various understandings of citizenship, identities and their intersections to sovereignty, race and land (Byrd, 2011: 126). These relations have implications beyond settler borders, which is why Byrd (2011) theorizes settler states as "transits of empires". The transit of empire is the basis upon which the new white settler societies, the "motherland" and the displaced settlers of colour, stolen people, refugees and migrants battle and "struggle for justice and equality" (Byrd, 2011: xxv); that same basis serves to enable and facilitate contemporary imperial and colonial relations. In accordance with the logic of erasure, Indigenous people are erased but Indigeneity remains as a sign. Byrd (2011: 192) asserts that, on the settler battleground, "every attempt to found political liberties in the rights of the citizens is, therefore, in vain". Even though non-white coded Indigenous people experience racism based on skin colour, seeking justice and resisting different manifestations of racisms through the rights-based institutions of the settler-state further legitimizes the structures of settler-colonialism as "finished". Justice through this state-sanctioned avenue "is heavily indebted to the settler-state and its legal apparatuses" (Cacho, 2012, as cited in Tuck and Gorlewski, 2015). Through racialization, the struggles of Indigenous peoples are forced to exist and to be intelligible only within the categories established by the settler battleground, through a distinctly liberal and modern racism. These categories ensure that they will seek justice through federal recognition through minority/ cultural claims, in which the settler-colonial relationship is not threatened (Coulthard, 2014). Other struggles for social, environmental and racial justice occurring on the settler battleground will often not only take place within the parameters established by the settler-state, but also within taken-for-granted "conceptions of place, politics and personhood" (Rifkin, 2014: xviii).

Racialization of Indigenous people also serves to make Indigeneity, the quality of being Indigenous, as if it is detached from certain bodies, histories and kinships and thus available for settler "recuperation", or appropriation. White settler societies attempt to instrumentalize Indigeneity, the quality of being Indigenous, for their own sake. Settlers are known to "make themselves Natives" for the purpose of legitimating access to territory (Deloria, 1998; Gaudry and Leroux, 2017). Through this process, Indigenous peoples are made the other on their own land, which is consistent with and required for the simultaneous (parodied) indigenization of the white settler. The white settler becomes the normal, de facto and deserving citizen. Within this theatre, non-whites are also invited and constrained to strive to achieve this status, even though it is never fully granted.[9] The national membership of settlers and the making official of their institutions directly "depends on the extension of a geopolitical claim to Indigenous lands" in which ongoing processes of settlement creates the (settler) "we" (Rifkin, 2014: 2). Control of citizenship and violent imposition of borders – referring back to settlers' geographies and logic of space – are thus also part of settler-nation making.

Thus, tending to how and when the topic of race (or its liberal/multicultural referent) comes up in curricula enables us to trace where Indigenous nations may be reduced to ethnic groups. This racialization performs the role of undermining competing sovereignties while diminishing the possibility for emergent solidarities between differently racialized groups.

Futurity-making in curricula

Here I return to the concept introduced at the beginning of this chapter: futurities. If Indigenous sovereignty was honoured, then Indigenous peoples would not be treated as a racial minority (without implying that anyone should be, whether questions of sovereignty are at hand or not), which means that their land-based rights would be evident rather than constantly under attack. In decolonization, the future is an unapologetically Indigenous one, where Indigenous nations could deal with settlers, "stolen people" (Thomas, 2016), refugees, migrants and more (see Walia, 2013) according to their own protocols. If Indigenous sovereignty is constantly attacked under an assumed white settler future, then Indigenous peoples are seen as a special-case group to be integrated, and land-based rights are watered down to selfies taken by politicians whose actions continue to prove that land dispossession and devastation will continue when required. The future that is currently assumed is a white settler-colonial one, in which the settler-colonial project (Canada) paternalistically deals with "its" Indigenous peoples and other racialized peoples.

Together, these results point not only to reluctance or dismissal of race-based issues or discussions, but that any limited engagement with the topic further takes place in the terms of the multicultural liberal language. First, the avoidance of race altogether is typical of fields related to humanitarianism, which has a strong empathetic colour-blind appeal (Jefferess, 2015b). However, it is again not just a matter of whether issues of race and racism are present in the syllabi. Rather, the absence and presence of these concepts seem to occur within specific liberal parameters. Thus, limited engagement with issues of race immediately evacuate race by using words like "culture", "ethnicity" and "religion". White settler-colonialism delimits which language and terms are intelligible, excluding those that have the capacity to threaten sovereignty and whiteness and including those that have the propensity to be co-opted by Settler logics. At most, even when critical race language and scholarships are present in curricula, with a reference to Indigenous peoples or not, the point is that the function of anti-racist approaches alone is different in settler-colonial context (Nichols, 2014). The collapsing of Indigenous peoples (and by consequence their inherent self-determination and legitimate claim to the land) serves to absorb them and make them only intelligible within the established settler battleground as governance is exercised against a series of differently racialized and othered populations.

For students, academics and others to address settler-colonialism as an obstacle to decolonization in universities which are built on unceded territories, possible steps could be acknowledging the current role in naturalizing settler-colonialism, working to denaturalize the logics of settler-colonialism, and engaging in tangible and deliberate praxis that do not centre white settler futures, while other futures (Indigenous futures) flourish.

Rescuing white settler futurity in curricula

The review of settler-colonial ordering of presences and absences along the lines of time, space and race can forecast what kind of future is assumed in the mandatory curricula. Conceiving white settler-colonialism as strappingly invested in its continuity in the present and future also means that challenges to its logics are met with swift and clever adaptations. When Indigeneity poses challenges, settler futurity is ensured by "the absorption of any and all critiques", but also through the "replacement of anyone who dares to speak against ongoing colonization" (Tuck and Gaztambide-Fernández, 2013). This is so because settler-colonialism is constantly "producing the conditions of its own supersession", even when it becomes self-aware or "announces its passing" (Veracini, 2011, as cited in De Leeuw *et al.*, 2013). Settler moves to innocence are the said adaptations that occur when settler sovereignty or settler capacities to possess land and Indigeneity are perceived to be challenged (Tuck and Yang, 2012). Settler moves to innocence are among the vast range of techniques and reflexes of settler people and settler institutions to reinscribe their non-complicity in the present structure settler-colonialism from which they benefit. With varying outward appearances and manifestations, they all serve to restore settlers' feelings of security, but do nothing to contribute to or even impede processes of decolonization (Battell Lowman and Barker, 2015). White settler curricula are skilled and equipped to "re-occupy the spaces" that anticolonial, anti-racist, queer and feminist critiques struggle to create (Tuck and Gaztambide-Fernández, 2013). They can happen voluntarily but also under the pressure "to collapse decolonization into coherent, normative formulas with seemingly unambiguous agendas" (De Oliveira Andreotti *et al.*, 2015: 22). The good intentions behind the impulses to overcome or diminish the systems of oppression in which we bathe often serve to "most potently re-entrench" these same systems (De Leeuw *et al.*, 2013). White settler curricula, whether well intentioned or not, can absorb, tame and distort progressive language and ideas so that it leaves its own core unaffected. Settler attachments to their futures are turned into material quotidian representations, such as our very school curricula, in which presences and absences of Indigeneity/Indigenous might first appear like a simple result of ignorance or dismissal when in reality it is not. The first three spaces presented below often end up rescuing white settler futurity.

In the table, rematriation is completely invested in Indigenous futurity and takes no time to cater to settler futurity (Tuck and Gaztambide-Fernández, 2013) nor does it offer an answer to settlers while they are left disconcerted for not being the centre of everything for once. Refusal matters to the interruption of settler-colonialism in academia, as it sets standards for what can be researched and leaves settler institutions to introspect about their roles in naturalizing settler-colonialism. The comfort of settlers just cannot be an additional burden that Indigenous peoples have to carry (Battell Lowman and Barker, 2015; Yerxa, 2015). Decolonization is not about the future of settlers, and to imagine decolonization as addressing settlers' anxieties means that it has already failed, including in education and curricula

TABLE 7.1 Settler moves to innocence in curricula and beyond. Inspired by De Oliveira Andreotti *et al.* (2015: 31). Italics are my modifications and additions. Land-based pedagogy refers to a specific article (Betasamosake Simpson, 2014a), whose (non)-relevance for settlers is still in articulation.

Space	Meaning of decolonization	Practices and pedagogy
Everything is awesome	No recognition of decolonization as a desirable project	No practice related to decolonization, *oblivious or conscious maintenance of white settler-colonialism*
Soft reform and multicultural inclusion	No recognition of decolonization as a desirable project, but conditional inclusion of "diversity" in liberal terms	Providing additional resources to Indigenous, racialized, low-income and first generation students so as to equip them with the knowledge, skills and cultural capital to excel according to institutional standards
Radical reform, Browning and Critical Race Theory	Call for: recognition, representation, redistribution, voice, reconciliation	Center and empower marginalized groups, and redistribute and re-appropriate material resources
Beyond reform, rematriation and living decolonization	Dismantling of modernity's systematic violence and understanding of their converging intersections on the settler battleground (capitalism, ableism, racism, cis heteropatriarchy, transmisogyny, freedom of mobilities, citizenship)	Subversive educational use of spaces and resources, hacking, hospicing, teaching of refusal politics, refusal, engage complicities and roles in structures of domination, land based pedagogy, land return (rematriation), directly supports and refers to grassroots voices and struggles, questions futurities

(Gaztambide-Fernández, 2012). Settlers must be willing to be refused and direct their work to imagine relational ways of being that are not dependent on dispossession (Flowers, 2015). Even when settlers do engage in unsettling settler-colonialism, it cannot be for their own instrumental ends (De Leeuw *et al.*, 2013). Settlers' engagement with the process of creating spaces for decolonization cannot happen if it has a pre-determined goal to attain an assumed reconciliation to be reached in their own terms.

Dragging the joy of the global citizen

Scholars are carving conversations about the problematics of Canadian global citizens in the context of international solidarity and development work (Jefferess, 2011; Mahrouse, 2014; Tiessen and Huish, 2014). Mahrouse (2014: 160) advances that "the idea of global citizenship and acting or intervening across, without, or beyond borders has become an axiom so imbued with righteousness that it tends to remain largely outside of the purview of critique". In other words, any intentions or practices geared towards making the global world better is often uncritically celebrated and assumed as "moral". The desire to make the world a better place becomes harmful when the building blocks enabling the possibility for such a desire are not engaged with. When power relationships, as dictated by whiteness and/or citizenship, are brushed aside, complicities in more proximate systems of oppressions can be ignored. It is easier to "have" the expertise to fix someone else's problems rather than the ones we are inevitably imbricated in. The (white) saviour (industrial) complex (Cole, 2012) is activated through feelings of superiority and desires for benevolence, which are states of mind granted by power relations that were established through colonialism and bioracism. These uncritical desires afforded by privileged positions can mask the intimacy of settler-colonial violence. By default, this means that practices of Canadian international development necessarily rest on and are made possible by our own settler-colonial status quo. By not addressing it, or assuming it is outside of its disciplinary realm, the programme of international development often remains in denial when it comes to considering international matters in "its" own backyard.

As such, a great deal of the great Canadian humanitarian or global citizenry spirit becomes unintelligible when one realizes not only what it stands on, but that what it stands on connects it in a complicit and intimate way to the problems that first appeared overseas, foreign and identity-defining (Jefferess, 2013). Not addressing our own settler-colonial status quo contributes to the erasure of the continued destruction of other futures and other worlds. Canada has been elevated as a rich country that holds the supposed expertise to help others exactly through the continued dispossession of the Indigenous peoples and their lands (Jefferess, 2015a; Vimalassery, 2013). In other illiberal words, it is the western control and stealing of Indigenous resources (and knowledges) as well as the ongoing genocide (Henderson, 2017) that allowed Canada to join the ranks of so-called developed countries (Watson, 2014), so that they can start to "do it" somewhere else.[10] Canadian development workers can work everywhere in the world, whether in the field or on paper, while knowing that they can of course be in or comeback to Canada to start or join their safe and comfortable families here, now and in the future. The materialities and localities of development theorizers and workers must not be eclipsed in the critical study of the field.

The modern nation state and contemporary globalization are seen as recent phenomena and as somehow separate from settler-colonialism. Today, the actions of the West are analysed or discussed in isolation of the fact that settler-colonialism has been and is the West's "leading edge by establishing grounds for globalization and universalization of its governance" (Morgensen, 2011). While it is often assumed that matters of Indigeneity or colonization are non-global questions by students and professors of international development, settler-colonialism is not restrained to the borders it created; it was made possible through transnational links in the first place, and today it also remains inherently transnational (Morgensen, 2014). Settler-colonialism is thus a mode of domination that went global (Veracini, 2015). In this global context, the hegemony and dominance of settler nations is directly "dependent on the resources at home" (Goeman, 2013). Byrd's (2011: xix, 149) argument of settler-colonial nations as "transits of empires" is a preliminary inquiry of the roles of the simultaneous control of Indigenous nations at home and the use of settler indigenization in solidifying the boundaries of settler-colonialism so that it can "enact itself as settler imperialism" right now. International matters are not solely imperial ones, but settler-colonial ones. International development practitioners from Canada who engage in projects abroad inadvertently find themselves in this transit of empires, in which constructed presences of Indigenous/Indigeneity "have created the conditions of possibility" for the enactment of imperialism (Byrd, 2011: xv).

Conclusion

The methodology and research questions used for the present content analysis of higher education curricula can be adopted to prompt similar inquiries into other fields of study's own operative role in the transnational and material implications of settler-colonialism. The choices and ordering of mandatory educational material by settler institutions offer a glimpse of their worldviews and of the delineations of what they consider relevant to the topic and to on-the-ground problems. Thus, it is not only the blatant absences of certain topics that contributes to the further naturalization of settler-colonialism in curricula, but also the strategic and sporadic presences of certain themes in relation to others (e.g. liberalization and racialization). As such, when Indigenous issues or perspectives are included, they are most likely to be collapsed under themes that do not challenge whiteness and settler futurity. In the vast majority, the choices and orderings of themes that relate to colonialism, history, racism and nation states do not appear to have the potential to explicitly challenge white settler futurity. Again – assuming white settler futurity is not neutral or impactless – it denies the possibility of alternative pasts, presents or futures (Steinman, 2015). It remains complicit in present and future land dispossession, of violence against the land and waters, and of violence against feminine and queer and/or Two-Spirit Indigenous people (Arvin et al., 2013), locally and globally. If higher education curricula have the tremendous capacity to further normalize settler-colonial occupation, this capacity can certainly be hacked and hospiced (De Oliveira Andreotti et al., 2015) towards other futurities. Tenured professors who have decision-making power in their own department will be crucial to those endeavours.

Notes

1 www.cbc.ca/news/indigenous/canada-adopting-implementing-un-rights-declaration-1.3575272.
2 The number only includes Masters in International Development and excludes Masters in International Relations, International Management, International Business, International Communication, International Politics, International Trade, Comparative International Development Education, International Security. Repertoire available at: www.canadian-universities.net/. For the 17 Bachelor's degree, search bachelorsportal.eu with search string "international development". I only included Bachelors that contain "International/Global Development".
3 http://socialsciences.uottawa.ca/international-development-global-studies/.
4 A current example of this is Elders Beatrice Hunter, Eldred Davis, Marjorie Flowers and James Learning, who all have been jailed in 2016–2017 because of their opposition to a dam project that would contaminate their lands and food sources with methyl mercury, in Muskrat Falls, Labrador, Canada.
5 Keywords include "Nation, Colonialism, Canada, Indigenous, Aboriginal, Racism" along with their different suffixes and synonyms. The context(s) in which these keywords appear were systematically reviewed. For full details on methodology, see Bazinet (2016).
6 Care was taken in verifying the context of every result. Word searches include the synonyms and related words of the one of interest. For instance, Indigenous included Aboriginal, Autochtones (French), First Nations, names of most prominent groups such as Anishinaabe(g), all words falling under "Indig".
7 For "Indigenous forms of time that threaten settler sovereignty", see Rifkin (2017: viii).
8 Racialization is "the social processes by which a population group is categorized as a race" (Scott and Marshall, 2015: 624). It refers to the instances "where social relations between people have been structured by the signification of human biological characteristics in such a way as to define and construct differentiated social collectivities" (Miles, 1989: 75).
9 For more context on the discussion of Blacks and People of Colour in settler-colonialism, see Lawrence and Dua (2005); Day (2015) and Richardson (2017).
10 See, for instance, Canadian mining (Butler, 2015).

References

Agrawal, A. (2005) Forest of statistics: Colonial environmental knowledges. *Environmentality: Technologies of Government and the Making of Subjects.* Durham and London: Duke University Press.

Arvin, M., Tuck, E. and Morrill, A. (2013) Decolonizing feminism – Challenging connections between settler colonialism and heteropatriarchy. *Feminist Formations* 25(1): 8–34.

Baldy, C.R. (2015) On telling native people to just "get over it" or why I teach about the walking dead in my native studies classes. *Sometimes Blogger-Write Cutcha Rising Baldy* 12 November. Available at: www.cutcharislingbaldy.com/blog/on-telling-native-people-to-just-get-over-it-or-why-i-teach-about-the-walking-dead-in-my-native-studies-classes-spoiler-alert (Accessed 5 August 2017).

Battell Lowman, E. and Barker, A. (2015) *Settler: Identity and Colonialism in 21st Century Canada.* Halifax: Fernwood Publishing.

Battiste, M. (2013) *Decolonizing Education: Nourishing the Learning Spirit.* Saskatoon: Purich Publishing.

Bazinet, T. (2016) White settler colonialism, international, development education, and the question of futurity: A content analysis of the University of Ottawa master's program mandatory syllabi in globalization and international development. Master of Arts dissertation, University of Ottawa.

Betasamosake Simpson, L. (2014a) Land as pedagogy: Nishnaabeg intelligence and rebellious transformation. *Decolonization: Indigeneity, Education & Society* 3(3): 1–25.

Betasamosake Simpson, L. (2014b) Not murdered, not missing: Rebelling against colonial gender violence. Available at: www.leannesimpson.ca/writings/not-murdered-not-missing-rebelling-against-colonial-gender-violence (Accessed 5 August 2017).

Bonds, A. and Inwood, J. (2015) Beyond white privilege: Geographies of white supremacy and settler colonialism. *Progress in Human Geography* 40(6): 715–733.

Borrows, J. (2002) *Recovering Canada: The Resurgence of Indigenous Law*. Toronto: University of Toronto Press.

Burnett, K., Hay, T. and Chambers, L. (2016) Settler colonialism, indigenous peoples and food: Federal Indian policies and nutrition programs in the Canadian North since 1945. *Journal of Colonialism and Colonial History* 17(2).

Butler, P. (2015) *Colonial Extractions: Race and Canadian Mining in Contemporary Africa*. Toronto: University of Toronto Press.

Byrd, J. (2011) *The Transit of Empire: Indigenous Critiques of Colonialism*. Minneapolis: University of Minnesota Press.

Calderon, D. (2014) Uncovering settler grammars in curriculum. *Educational Studies* 50(4): 313–338.

Canada, D.O.J. Assembly of First Nations – Annual General Assembly: Notes for an address by The Honourable Jody Wilson-Raybould, PC, QC, MP, Minister of Justice and Attorney General of Canada. 37th Annual General Assembly of the Assembly of First Nations 2016 Scotiabank Convention Centre, Niagara Falls. Government of Canada.

Cole, T. (2012) The white-savior industrial complex. *The Atlantic* 21 March. Available at: www.theatlantic.com/international/archive/2012/03/the-white-savior-industrial-complex/254843/ (Accessed 5 August 2017).

Cote-Meek, S. (2014) *Colonized Classrooms: Racism, Trauma and Resistance in Post-Secondary Education*. Halifax: Fernwood Publishing.

Coulthard, G.S. (2014) *Red Skin White Masks: Rejecting the Colonial Politics of Recognition*. Minneapolis and London: University of Minnesota Press.

Day, I. (2015) Being or nothingness: Indigeneity, antiblackness, and settler colonial critique. *Critical Ethnic Studies* 1(2): 102–121.

De Leeuw, S., Greenwood, M. and Lindsay, N. (2013) Troubling good intentions. *Settler Colonial Studies* 3(3–04): 381–394.

De Oliveira Andreotti, V., Stein, S., Ahenakew, C. and Hunt, D. (2015) Mapping interpretations of decolonization in the context of higher education. *Decolonization: Indigeneity, Education & Society* 4(1): 21–40.

Deloria, P.J. (1998) *Playing Indian*. New Haven and London: Yale University Press.

Flowers, R. (2015) Refusal to forgive: Indigenous women's love and rage. *Decolonization: Indigeneity, Education & Society* 4(2): 32–49.

Gaudry, A. and Leroux, D. (2017) White settler revisionism and making Métis everywhere: The evocation of métissage in Quebec and Nova Scotia. *Critical Ethnic Studies* 3(1): 116–142.

Gaztambide-Fernández, R.N.A. (2012) Decolonization and the pedagogy of solidarity. *Decolonization: Indigeneity, Education & Society* 1(1): 41–67.

Goeman, M.R. (2013) *Mark My Words: Native Women Mapping Our Nations*. Minneapolis: University Of Minnesota Press.

Goldstein, A. (2014) Possessive investment: Indian removals and the affective entitlements of whiteness. *American Quarterly* 66(4): 1077–1084.

Henderson, P. (2017) Imagoed communities: The psychosocial space of settler colonialism. *Settler Colonial Studies* 7(1): 40–56.

Jefferess, D. (2011) Benevolence, global citizenship and post-racial politics. *TOPIA* 25: 77–95.

Jefferess, D. (2013) Humanitarian relations: Emotion and the limits of critique. *Critical Literacy: Theories and Practices* 7(1): 73–83.

Jefferess, D. (2015a) Humanitarian relations and post-racial ideals: An evolving settler logic? *The Canadian Association for the Study of International Development Annual Conference.* University of Ottawa.

Jefferess, D. (2015b) Introduction: 'The White Man's Burden' and post-racial humanitarianism. *Australian Critical Race and Whiteness Studies Association* 11(1): 1–13.

Johnson, J. (2015) *Colonialism and Narratives of the Apocalypse.* Florida: Nothing Arts Centre.

Kauanui, J.K. (2016) "A structure, not an event": Settler colonialism and enduring indigeneity. *Lateral* 5(1).

Kerr, J. (2014) Western epistemic dominance and colonial structures: Considerations for thought and practice in programs of teacher education. *Decolonization: Indigeneity, Education & Society* 3(2): 83–104.

King, H. (2017) The erasure of indigenous thought in foreign policy. *Open Canada* 31 July. Available at: www.opencanada.org/features/erasure-indigenous-thought-foreign-policy/ (Accessed 5 August 2017).

Kuokkanen, R. (2007) *Reshaping the University: Responsibility, Indigenous Epistemes, and the Logic of the Gift.* Vancouver: UBC Press.

Lawrence, B. and Dua, E. (2005) Decolonizing antiracism. *Social Justice* 32(4): 120–143.

Littlebear, L. (2009) Naturalizing indigenous knowledge: A synthesis paper. Canadian Council on Learning's Aboriginal Learning Centre. Available at: www.aerc.usask.ca (Accessed 5 August 2017).

Mackey, E. (2016) *Unsettled Expectations: Uncertainty, Land and Settler Decolonization.* Halifax: Fernwood Publishing.

Macoun, A. and Strakosch, E. (2013) The ethical demands of settler colonial theory. *Settler Colonial Studies* 3(3–04): 426–443.

Mahrouse, G. (2014) *Conflicted Commitments: Race, Privilege, and Power in Solidarity Activism.* Montreal: Mcgill-Queen's University Press.

Mignolo, W. (2002) The geopolitics of knowledge and the colonial difference. *The South Atlantic Quarterly* 101(1): 57–96.

Miles, R. (1989) *Racism.* New York: Routledge.

Morgensen, S.L. (2011) The biopolitics of settler colonialism: Right here, right now. *Settler Colonial Studies* 1(1): 52–76.

Morgensen, S.L. (2014) Indigenous transnationalism and the AIDS pandemic: Challenging settler colonialism within global health governance. In A. Simpson and A. Smith (eds) *Theorizing Native Studies.* Durham and London: Duke University Press, pp. 188–206.

Nichols, R. (2014) Contract and usurpation: Enfranchisement and racial governance in settler-colonial contexts. In A. Simpson and A. Smith (eds) *Theorizing Native Studies.* Durham and London: Duke University Press, pp. 91–121.

Nixon, L. (2016) Visual cultures of indigenous futurisms. *GUTS: Canadian Feminist Magazine* 6. Available at: http://gutsmagazine.ca/visual-cultures/ (Accessed 5 August 2017).

Palmater, P.D. (2015) Bill C-51- An Act to Enact the Security of Canada Information Sharing Act and the Secure Air Travel Act, to Amend the Criminal Code, the Canadian Security Intelligence Service Act and the Immigration and Refugee Protection Act and to Make Related and Consequential Amendments to other Acts. *Standing Committee on Public Safety and National Security.* Ottawa, Ontario

Richardson, W.J. (2017) Understanding the city as a settler colonial structure. Available at: https://decolonialblack.com/2017/06/27/understanding-the-city-as-a-settler-colonial-structure/ (Accessed 5 August 2017).

Rifkin, M. (2014) *Settler Common Sense Queerness and Everyday Colonialism in the American Renaissance*. Minneapolis: University of Minnesota Press.

Rifkin, M. (2017) *Beyond Settler Time: Temporal Sovereignty and Indigenous Self-Determination*. Durham: Duke University Press.

Scott, C. and Marshall, G. (2015) *A Dictionary of Sociology*. Oxford: Oxford University Press.

Shahjahan, R.A. (2011) Decolonizing the evidence-based education and policy movement: Revealing the colonial vestiges in educational policy, research, and neoliberal reform. *Journal of Education Policy* 26(2): 181–206.

Simpson, A. (2016) Whither settler colonialism? *Settler Colonial Studies* 6(4): 438–445.

Simpson, A. and A. Smith (eds) (2014) *Theorizing Native Studies*. Durham: Duke University Press.

Steinman, E.W. (2015) Decolonization not inclusion: Indigenous resistance to American settler colonialism. *Sociology of Race and Ethnicity* 2(2): 219–236.

Stevenson, W. (1998) "Ethnic" assimilates "indigenous": A study in intellectual neocolonialism. *Wizazo Sa Review* 13(1): 33–51.

Thomas, A. (2016) Stolen people on stolen land: Decolonizing while black. *RaceBaitr* 24 May. Available at: http://racebaitr.com/2016/05/24/stolen-people-stolen-land/ (Accessed 5 August 2017).

Tiessen, R. and Huish, R. (2014) *Globetrotting or Global Citizenship?: Perils and Potential of International Experiential Learning*. Toronto: University of Toronto Press.

Tuck, E. (2011) Rematriating curriculum studies. *Journal of Curriculum and Pedagogy* 8(1): 34–37.

Tuck, E. and Gaztambide-Fernández, R.A. (2013) Curriculum, replacement, and settler futurity. *Journal of Curriculum Theorizing* 29(1): 72–89.

Tuck, E. and Gorlewski, J. 2015. Racist ordering, settler colonialism, and edTPA: A participatory policy analysis. *Educational Policy* 30: 197–217.

Tuck, E. and Yang, K.W. (2012) Decolonization is not a metaphor. *Decolonization: Indigeneity, Education & Society* 1: 1–40.

Veracini, L. (2015) *The Settler-Colonial Present*. London: Palgrave Macmillan.

Vimalassery, M. (2013) The wealth of the natives: Toward a critique of settler colonial political economy. *Settler Colonial Studies* 3(3–04): 295–310.

Violet-Lee, E. (2015) "Indigenizing the academy" without indigenous people: Who can teach our stories? *Moontimewarrior* 9 November. Available at: http://moontimewarrior.com/2015/11/09/who-can-teach-indigenous-philosophy/ (Accessed 5 August 2017).

Violet-Lee, E. (2016) Reconciling in the apocalypse. *The Monitor*. Canadian Centre for Policy Alternatives.

Walia, H. (2013) *Undoing Border Imperialism*. Oakland: AK Press.

Watson, I. (2014) Recentring first nations: Knowledge and places in a terra nullius space. *AlterNative* 10.

White, J.P., Murphy, L. and Spence, N. (2012) Water and indigenous peoples: Canada's paradox. *The International Indigenous Policy Journal* 3: 1–25.

Wolfe, P. (1999) *Settler Colonialism and the Transformation of Anthropology: The Politics and Poetics of an Ethnographic Event*. London: Cassell.

Wolfe, P. (2006) Settler colonialism and the elimination of the native. *Journal of Genocide Research* 8(4): 387–409.

Yerxa, J.-R. (2015) Refuse to live quietly! *Settler Colonial Studies* 5(1): 1–3.

8

PLANETARY URBANIZATION AND POSTCOLONIAL GEOGRAPHIES

What directions for critical urban theory?

Simone Vegliò

Introduction

The field of urban studies has acquired significant and increasing attention over the last few decades. The dramatic growth of cities and urban areas has spread new questions among scholars across many disciplines, including sociology, geography, economics, politics, and anthropology. The systematic study of the urban environment can be traced back to the first half of the 20th century, when the Chicago School of Urban Sociology attempted a comprehensive theoretical investigation and used the city of Chicago as the main focus of their empirical work. However, it was only from the 1970s, a period in which the world was dramatically urbanizing, that the investigation of the urban environment acquired "critical" tools with which to frame the "urban question" in relation to the global capitalist system.

Critical urban scholarship challenged the idea of the city as a sort of *natural* environment and established a profound dialogue with critical theory as a whole,[1] becoming a fundamental point of view in the study of contemporary society. In this chapter, by specifically adopting a geographical viewpoint, I interrogate some of the most relevant "critical" literature in urban studies in order to reflect upon the contemporary debates and to draw out some hypotheses regarding the future of the scholarship. However, far from pretending to draw a complete presentation of the variety of critical standpoints, I converse with this literature by exploring some of the crucial concepts currently at stake in urban studies, constructing the discussion over the purpose of calling into question Eurocentric approaches. I contend that the postcolonial perspective offers a sharp critique of Western/colonial epistemologies by revealing the power relations characterizing the geographical hierarchies in urban studies; these classifications habitually position Euro-American urbanization as a privileged case, which indicates universal urban tendencies. Furthermore, postcolonial scholarship has interrogated the ways in which and the

places from which knowledge about the urban is produced, challenging the dominance of Northern urban theorizing. By contrast, neo-Lefebvrian thinkers offer
suggestive insights by conceiving the whole of the global space as "urban", regardless of the number of inhabitants living in the settlements, putting in tension the
very idea of city as a coherent and analytically independent object of study.

An effective urban scholarship, I would argue, needs to take advantage of a constant dialogue between critical perspectives by productively exchanging methods
and practices. In addition, other approaches not directly focused on urbanization,
such as that of decolonial studies, are able to provide original and multidisciplinary
insights that offer a more efficient, flexible, and non-Eurocentric landscape for
urban investigation. For example, these studies identify a variety of practices (*coloniality*) through which colonialism organized the modern world. Although much
research has been concerned with the entrenchment of these practices both in
postcolonial and post-imperial spaces, decolonial critique has given little attention
to how these practices acted in and through the urban environment, substantially
omitting questions regarding the urban as such.

By proposing some directions for theoretical reflection, this chapter aims to add
new and hopefully helpful elements to enrich the contemporary debate in critical
urban studies. I strongly agree with Neil Brenner (2009: 204) when he states that
"because the process of capitalist urbanization continues its forward movement of
creative destruction on a world scale, the meanings and modalities of critique can
never be held constant; they must, on the contrary, be continually reinvented".
However, I would add, in addition to being reinvented, these critiques should also
make value of a constant and productive dialogue between them. In particular, I
stress the necessity for a more systematic approach to postcolonial/decolonial critical perspectives in the study of urbanization as a global phenomenon.

Urban studies and critical theory

The 1970s represented an important decade for the renovation of urban studies.
Inspired by Karl Marx, this work attempted to offer a new and more comprehensive understanding of the urban environment by stressing the spatial dimensions of
capitalism. This change was part of a broader innovation undergone by geography
as a discipline that was crossed by a wave of critical considerations that challenged
its supposed "objectivity" and, as a result, uncovered the power relations underlying
the production of geographical knowledge. With regard to urban studies, the work
of David Harvey represented a watershed in the field. Starting with *Social Justice
and the City* in 1973, his work deeply investigated the spatial nature of capital and
how capital was powerfully related to urbanization processes (Harvey, 1973, 1985,
2001). This critical understanding of the city strongly influenced the research of
several scholars in that period and in the following decades (for example, Castells,
1977; Saunders, 1981; Merrifield, 2002). These contributions gave great impetus
to urban geography, which became a prominent subject within the field, attracting

scholars, theorists, and activists who were engaged in the development of critical understanding accompanied by the desire to transform the contemporary world.

At the same time, the scholarship of the 1970s was heavily influenced by the work of French philosopher Henri Lefebvre. Contributing to an innovative and non-orthodox idea of Marxism, Lefebvre (1991 [1974]) investigated the role of the city, as well as urban life as a whole, and their strategic importance in producing and re-producing the social space in contemporary society. His work was written in French and was not widely translated into English until the early 1990s, when it finally came to circulate widely within Anglo-American debates.[2] Lefebvre's contribution is extremely important, as it continues to inspire both critical urban thinkers and activists. Theorizing the practice of the *right to the city* (Lefebvre, 1996), differently from any nostalgic and pessimistic vision of the urban environment, Lefebvre put the city as a space at stake in the contemporary world. In addition, he conceived of the urban as a wide set of social and material relations which came to exceed the city itself up until, from the 1970s, they invaded the entire surface of human activities, achieving the stage of *planetary urbanization* (Lefebvre, 2003 [1970]).

As a result of these critical approaches that emerged in the 1970s, the field of urban studies was radically renovated and these authors, on the one hand, inspired theories and practices of (urban) struggle, and, on the other hand, they exceeded the field of geography and contributed to the epistemological transformation in social theory known as the *spatial turn* (Warf and Arias, 2009). However, without denying the great importance of this scholarship within urban studies and beyond, this theoretical and empirical investigation was strongly grounded in the West and little attention was given to cities and urban landscapes belonging to the non-Western world. At the same time, these studies became progressively important in universities worldwide, and started to be used as tools for investigating urban areas around the world. At the end of the 1970s these issues would start to be raised by scholars coming from different disciplines and working on different areas of the world, who highlighted the strong partiality underpinning European/Western knowledge that proved to be largely incapable of offering non-colonial conceptual frameworks for the investigation of the non-Western world.

The postcolonial critique

Starting with the groundbreaking work of Edward Said (1978, 1993), postcolonial studies demarcated an innovative field of research in the Anglo-American world, highlighting how the European construction of the *Orient* as *Other* imbued the latter with a sense of inferiority and wildness that contributed to the justification and naturalization of the colonial enterprise. Just a few years later, maintaining its marked Anglophone production, a reappraisal of Indian history was undertaken by the South Asian Subaltern Studies Group, which underlined the *subaltern*'s central role in the making of Indian historical process. The Indian experience is one of the most important to the postcolonial critique (Guha, 1983, 1997; Guha and Spivak, 1988).

The core of the postcolonial approach relies on the conviction that colonialism has been crucial for the shaping and global diffusion of a form of knowledge posited as universal and neutral but which is in fact fully drenched in power relations and is therefore local, provincial, and partial. Insights from postcolonialism progressively expanded into several bodies of knowledge and soon came to be part of the analysis of cities and urban environments. This represented an important move in the scholarship, as postcolonial works were criticized for being disconnected from reality due to their predominant preoccupation with literary works. The focus on urbanization and attention to the physical expressions of postcolonial relationships such as cities, buildings, and architecture gave postcolonial scholarship a welcome "materiality" (King, 2003).

Postcolonial urban studies began to explore a variety of places belonging both to former colonies and empires. For example, the study of India has continued to make an important contribution to critical scholarship, considering both historical and contemporary questions concerning urban areas (Roy, 2009b; Chatterjee, 2012; Datta, 2012, 2015). Similar works have been carried out, for example, in Africa (Simone, 2010), Australia (Jacobs, 1996), the Middle East (AlSayyad, 1991), and South-East Asia (Bishop, Phillips and Yeo, 2003). Postcolonial urban thinkers innovated the field of urban studies by stressing the constitutive importance of colonialism in the shaping of cities both in the former colonies and former empire, and they pointed out that this was one of the most important things that many Marxist scholars had failed to recognize.[3]

Thus, in order to show how urban studies has attempted to *decolonize* itself, I further discuss what I believe are two fundamental aspects of the theme of colonialism. First, I consider colonialism in relation to the aforementioned materiality of the urban environment, namely how colonial cities have been of great significance to the transformations of *the urban*, globally. Second, I take into account the geography of knowledge in two senses: on the one hand, how urban studies participated in the division of global areas of the world, and on the other, how the area of the world where knowledge itself is produced might be relevant for alternative approaches to theory.

Colonial cities and global geographies

Currently there are lively and interesting discussions about what is a city, where are its limits, and how it can be understood within a world which has undergone a dramatic process of urbanization (see, for example, Brenner and Schmid, 2015; Scott and Storper, 2015). However, moving back to colonial times, colonial cities constituted specific urban forms whose design depended exclusively on the colonizer's desires. In addition, it is worth noting that, in relation to urban planning, there was at times a *temporal discrepancy* between metropolis and colony. For example, in the case of Spanish colonial America, the model used in the colonies, the gridiron plan, had not been deployed in European cities since the Middle Ages (Smith, 1955), a fact that highlights the controversies defining the transfer of Western/

colonial knowledge to the colonies (for Latin America's post-independence period, see Hardoy, 1992). Instead of adopting the contemporary urban models used in the metropolis, the empires prioritized the specific needs of the colonial administration and organized its urban spaces accordingly. After all, the establishment of permanent settlements has arguably been the most important action in stabilizing the colonial enterprise (for a set of examples about colonial urbanism, see AlSayyad, 1992).

Within this scene, colonial cities "were used, both consciously and unconsciously, as social technologies, as strategies of power to incorporate, categorize, discipline, control, and reform, in terms of symbolic code and new system of colonization, both the colonial and the indigenous populations" (King, 1990: 9). There are two important elements at stake here. On the one hand, cities had been fundamental for the control, defence, and the very production of the colony. Also, this urban process influenced not only the social, cultural, and material organization of the post-independence period but, as a sort of "boomerang effect" (in the sense of Foucault, 2003: 103), it also had an effect on post-imperial urban spaces, making the violent legacy of colonialism still actively operative on the global scale. For example, Graham (2011) argues that techniques of surveillance and control typical of colonial spaces have been used equally in Western contemporary metropolises, giving birth to what he calls *the new military urbanism* (other examples can be drawn from Davis, 1990; Jacobs, 1996; Merrifield, 2014).

On the other hand, it is possible and helpful to think of the character of cities as dependent on the specific configuration of the international economy. According to Antony King (1993: 262), there are clear distinctions between the roles of "the city (the built environment) of mercantile capitalism, the colonial port city, the city of industrial or monopoly capitalism, the postcolonial city, and so on". There are at least two reasons why this consideration is relevant. First, it intrinsically avoids the methodological mistake of considering the city as a sort of object somehow independent from the international context by reflecting upon the multiplicity of geographies in which the city is articulated. Second, as a result, this challenges the idea of conceiving urbanization in the non-Western areas as something *dysfunctional* and not related to the global dimension of power – thinking, for example, of phenomena such as the diffusion of mega-cities and large informal sectors. I discuss more these questions later in the chapter. Now I would like to consider the problem of how to conceive of urbanization today, as well as to support the idea that the city is no longer a clear and identifiable object.

City/urban

Generally speaking, when talking about cities there is usually a fairly clear concept in mind that relates to certain kinds of buildings, specific social activities, or ways of life. Until recently, it was also relatively easy to think of where their borders were – that is, for example, imagining the end of the urban area with the beginning of the rural area, which was not only a different landscape but also consisting of different social relations, economic connections, and so forth (the most extreme case of bordered city is probably the medieval walled city). Using Lefebvre's (2003 [1970]) reconstruction of the progressive metamorphosis of cities over time, we can see that with the industrial transformations that occurred in the nineteenth century the city not only became fully dominant over the countryside, but it also began to develop social, material, and ideological relationships that spread the urban in the form of "disjunct fragments" (2003 [1970]: 14) beyond the traditional space of the city itself.

In the second half of the 20th century, due to the dramatic changes in the global economy, as well as its relationships and technologies, these fragments finally invaded the global space, clearly exceeding the mere form of the city. In urban studies, these transformations created the need for new concepts in order to understand this ongoing process. In other words, as stressed by the neo-Lefebvrian theorists Brenner and Schmid, at the end of the 1960s urban studies underwent an "epistemic crisis" (2015: 154): the city lost its centrality in favour of a more generally and spatially unstable concept such as "the urban".

Lefebvre (2003 [1970]: 191) famously described this theoretical break in these words: "we can say that the urban [...] rises above the horizon, slowly occupies an epistemological field, and becomes the episteme of an epoch". Yet, at the same time, he stressed that space was not to be understood as "a thing" but actually as "a relation between things" (1991 [1974]: 83), underlining the theoretical shift of the city (and the urban) from object to *concept*. As indicated by the neo-Lefebvrian theorists – who constitute, along with postcolonial thinkers, the most influential contributors to critical urban theory – the urban in the first place is a "theoretical category" (Brenner and Schmid, 2015: 163) which is not easily identifiable through empirical and quantitative elements such as population, agglomeration, infrastructure, and so forth (Brenner and Schmid, 2014).

Breaking any distinction between inside/outside, as well as refusing any set definition of agrarian or non-urban space, these scholars describe contemporary (urban) geographies by stating that "the erstwhile boundaries of the city – along with those of larger, metropolitan units of agglomeration – are being *exploded* and reconstituted as new forms of urbanization reshape inherited patterns of territorial organization, and increasingly crosscut the urban/non-urban divide itself" (Brenner and Schmid, 2015: 154, emphasis added). However, this epistemological move has broader implications for the conception of space itself and goes beyond mainstream divisions of the area of the planet such as, for example, that of Global North and Global South.

Thus, traditional divisions are overcome through the classification of urban areas into three "mutually constitutive" categories – concentrated urbanization, extended urbanization, and differential urbanization – depending not only on quantitative elements (population, infrastructures, etc.) but also on the role (production and circulation of goods) they play within the planetary urbanization scene (Brenner and Schmid, 2015: 166–169). This approach seems very helpful in order to avoid Eurocentric epistemologies that take the Euro-American example as a model and the rest as something somehow *on the way of*, such as in the case of modernization theories which consider the spread of urbanization (thus in opposition to the rural environment) as a necessary step towards development.

Global North/Global South

One of the important results of the planetary urbanization approach is to challenge successfully a very common geographical division such as that of the Global North and Global South. Framing the urban as "a process" and "not a universal form, settlement type or bounded unit" (Brenner and Schmid, 2015: 165) allows these scholars to avoid the mere comparison of urban shapes across the world. This does not mean, however, that the urban (and the city) are trans-historical elements that we can find across the world (such as in Scott and Storper, 2015). As with neo-Lefebvrian theorists, postcolonial scholars also challenged the idea of Global North and Global South but from a different perspective.

First of all, it is worth remembering the postcolonial critique of geography and how it has denaturalized spatial definitions that are used as "proof" of different stages of modernization or civilization. These include North and South, Occident and Orient, West and non-West, and also First World and Third World (Young, 2001). Being framed as spatial dichotomies, the diversity of the world was erased in favour of homogenizing notions of progress and development. As highlighted by Said (1978), these geographies represented stereotyped ideas about "faraway" and "wild" lands waiting to be civilized. Against the pretension of universal thought, Walter Mignolo (2014: 11) noted that "the enunciation cannot be global, it is always local: the 'West' is a local enunciation that creates a global enunciated". Once again, geographical knowledge is far from being the mere description of some ontological reality.

With regard to urban studies, there have been many attempts to reject these spatial differentiations as a framework for the investigation of urban settlements. Especially over the last few years, postcolonial urban scholarship strongly tried to find new theoretical and methodological tools in order to defy these controversial divisions. As these thinkers stressed, non-Western contexts, usually referred to as Third Word or Global South urbanization, are normally seen as distortions, a sort of *immature* configuration of urban development (for example, Rao, 2006). Slums, shantytowns, and urban informality in mega-cities constitute a fashion in the study of non-Western areas, which are often thought about in (negative) comparison with the standard model of city experienced in the Western world. Hence, regarding the global space of the "South", it emerges as a specific image whose immanent dystopia is materialized in the urban environment. As a result,

> this passage from slum as population and terrain to slum as theory is made from within a particular historicist narrative that centralizes empire as the subject and object of history within which the 'global' emerges as an analytic lens and category.
>
> *(Rao, 2006: 226)*

Once again, the non-Western world seems to be permanently locked in that "waiting room" described by Dipesh Chakrabarty (2000: 8), in which history is trapped in this sort of disfigured slowness and incapable of coping with the "Northern" pace of modernization.

On the other hand, instead of focusing on these peculiar aspects of mega-cities in the South, Jennifer Robinson called for an ordinary city approach (drawing the concept from Amin and Graham, 1997), stressing the strong bonds between any city and the global economy and, at the same time, pointing out the unique way in which any city responds to global challenges. Robinson's (2002, 2006) approach stresses the impossibility of any sort of theoretical generalization and attempts to challenge, on the one hand, the developmentalist urban theorists who associated the Global South with the aforementioned urban distortion (for example, see Allen and Thomas, 2000) and, on the other hand, the global-city perspective (Sassen, 1991; Taylor, 2000) – which, although not being framed according to a North/South approach, considers non-global cities as unimportant for the structure of the global economy. Furthermore, taking account of the city's constitutive particularism does not mean one should overlook the relationships and similarities between them; on the contrary, the strategy of exploring cities through comparison gave birth to a methodological school known as *comparative urbanism* (Nijman, 2007; Robinson, 2011, 2015).

A comparative method can be fruitful not only to identify cities' resemblances and similarities, but also, and perhaps in a more interesting way, to concentrate on elements of difference (McFarlane and Robinson, 2012; Jacobs, 2012). After all, postcolonialism as a *condition* (Mezzadra and Rahola, 2006) is necessarily characterized by global elements, and it seems it is perhaps more productive to investigate how colonial methods and relationships are differently rearticulated throughout (postcolonial) space rather than attempting to develop an all-encompassing theory able to explain urban processes worldwide. Having said that, reflections about space are important not only in order to define the areas of investigation through their historical, social, and cultural contexts, but also because they raise questions about *where* theory itself is produced.

Theory and space

In accordance with postcolonial reasoning, Ananya Roy is particularly concerned with the power relations embodied in the making of knowledge. Roy (2009a: 820) stresses how urban scholarship is characterized by the necessity of "dislocating the Euro-American center of theoretical production". She argues that, in order to avoid the reproduction of the "Orientalist" approach in urban studies, namely the Western perception of Third World cities as "the heart of darkness", it is necessary to "move" the production of theory to the "Global South" (ibid.). Instead of applying in a mechanical way theories forged in order to answer questions raised in the North, Roy's call is for a theory which comes from the place of investigation – that is, Southern tools for Southern questions. Only after that move, it will be possible to test these concepts transnationally and see what they are able to say in relation to global urbanism (for example, Simone, 2001). A transnational method is therefore significantly different from a comparative one; if the former explores global socio-spatial links, the latter is concerned with correspondences or dissimilarities between different areas. For example, using a transnational method, given that Latin America's urbanization has been strongly characterized by urban informality, an investigation of this issue in other areas of the world starts with the consideration of the Latin American experience (see Roy and AlSayyad, 2004).

This *area studies* approach aims to subvert the geography of knowledge by *provincializing* Euro-American scholarship and opening the space to other experience coming from elsewhere. This claim is surely important for any attempt to produce non-Eurocentric research; after all, the "global university" still relies quite often on the universalization of Euro-American models – especially the Anglo-American – when it comes to non-Western areas of investigation, and sometimes even in the case of "critical theory". However, this does not mean, quoting Storper's and Scott's (2016: 1122) criticism of Roy, that "an idea developed at place a must invariably fail when transferred to place b", nor that a theory produced in the North is necessarily wrong. The problem is that some areas of the world are thought of as a model, and therefore research based on these places is given more theoretical validity. The Global South is supposed to follow such models and is deemed incapable of providing examples that can be translated to the Global North. Having said that, an important attempt to produce theory from the margins, as a challenge to Euro-American models, occurred in Latin America between the 1960s and 1970s. Known as Dependency Theory, this contribution represented a fundamental stage in the genealogy of Latin American critical thought.

Theory from the South: The Latin American experience

The end of World War II was rapidly followed by a rapid process of decolonization which resulted in a great number of formally independent national states acting in the international scene. Within the new global order, the former colonies were soon confined as "Third" World, according to a discursive construction that used socio-economical elements – those corresponding to the Euro-American standards – as a classifying method (Escobar, 1995). Thus, Western powers established a winning strategy of *development*, predicated upon linear and progressive stages of economic growth that would have allowed the Third World to reach the level of the most advanced countries (Rostow, 1990 [1960]). It was largely agreed that economic growth – the measure of development – was the consequence of a combined process of political modernization, industrialization, and urbanization.

In the 1960s and 1970s, some Latin American thinkers challenged both development theory and orthodox Marxism (both theorizing, from different perspectives, a linear succession of stages characterizing national development) by claiming that the "backwardness" associated with Latin American countries was not due to the lack of social and political structures but, on the contrary, it was the result of the relationship of *dependency* on the "developed" world. In other words, the "development" of the Western world was the cause of the "underdevelopment" of its peripheries (Furtado, 1964; Frank, 1970; Quijano, 1977; Cardoso and Faletto, 1979). This centre/periphery model of the world innovated Latin American critical thought thanks to the production of a theory that moulded the concepts according to its particular *location*.

Within this context urbanization represented an important factor of analysis as Latin American cities were expanding rapidly. The huge migration from the rural areas to the cities signified a historical change that was defined by Aníbal Quijano (1975) as a process of "urbanization of Latin American society". The "urban question" became thus central to the research carried out by dependency theorists (Schteingart, 1973; Hardoy, 1975; Castells, 1977). Urban growth was far from producing the results expected by developmentalists, and something new appeared to be emerging. The dramatic growth witnessed in the urban environment and the emergence of areas populated by poor and marginal populations became a key focus that was framed within the distortions produced by the asymmetrical relationships of the international system (see for example, Quijano, 1977).

Overall, this theoretical move represented an important *epistemological change*, as it demonstrated how good models produced in the centre were not necessarily valid in the periphery. Dependency scholars embodied an attempt to renovate critical theory carried out from the South. However, dependency theorists' experience ended in the late 1970s, due probably to the relative simplicity of its centre/periphery model and to the arrival of neoliberalism, which dismantled the role of the state (a critical actor for dependency theory analysis and action);[4] in any case, their work continued to be an important source for Latin American critical thought in the following years. For instance, it constituted an important antecedent for the *Modernity/Coloniality* group formed in the 1990s – which included Quijano himself – and that continued to investigate the importance of European colonialism in the study of Latin America as well as its very conception (Mignolo, 2005).

Among the theoretical innovations brought by this group of scholars, the concept of "coloniality" is one of the most relevant. Coloniality is not synonymous with colonialism – or more precisely, its meaning goes beyond the formal territorial domination which the latter usually indicates. Coloniality, instead, sets out the *matrix of power* (Mignolo, 2011) which marks the social organization and both the colonial and postcolonial world. The concept was initially conceived by Quijano and is more accurately described by the expression *coloniality of power* (Quijano, 1992, 2000, 2007). It consists of classifying the world's population according to racial categories and of creating a hierarchical division of all human activity, from everyday life to the division of labour. Such a matrix of power started with the colonization and the constitution of the Americas, and through the consequent internationalization of capitalist power, spread all over the world (Quijano, 2007). In keeping with their intention to break with Western Eurocentric epistemologies, decolonial thinkers consider the conquest of the Americas as the beginning of the modern/capitalist world (Dussel, 2000, 1995; Lander, 2000; Mignolo, 2000).

Decolonial studies investigates a significant range of subjects such as Latin America's cartography, literature, and philosophy, yet this scholarship still displays a lack of systematic attention to urbanization, both in terms of theory and practice. This is somehow contradictory given the importance of urbanization in the 1960s and 1970s within a critical school of thought that can be considered fundamental for the development of the decolonial experience. Can decolonial studies provide an original set of tools for investigating the urban environment in Latin America and elsewhere? Are their concepts able to contribute to a further renovation of contemporary critical urban theory? These are the questions I would like to open and put forward for discussion; the Latin American experience surely represents a strong attempt to break with Eurocentrism and the universalization of Western knowledge, with the urban question being one of the main elements at stake. Particularly, future discussion should engage also with the concrete possibilities for a postcolonial/ decolonial urban agenda within the multifaceted world of planetary urbanization. In these terms, far from thinking of infallible recipes, the dialogue between different experiences can enhance new and effective urban practices which aim to produce ruptures within the global space. However, considering important but substantially concluded experiences such as, to mention only two prominent examples, the City Statute[5] in Brazil in 2001 and the Occupy movement in New York in 2011 – in which something as important as the city's material relationships was at stake – it is clear that an alternative urban agenda is still a complex work in progress. Yet this is a path that has started, and I would argue that it needs to be progressively refined through a theoretical investigation which informs and is informed by practices of social and material transformation of the urban environment.

Conclusion

The concepts discussed in this chapter have attempted to offer an overview of the variety of theoretical problems that characterize current analysis of the urban environment. On the one hand, critical scholarship has explored the urban by looking at the role of global capitalism and its geographical dynamics; on the other, colonialism constituted the core from which to develop the investigation. Regardless of the relevant differences between these two approaches, they are able to make important contributions to new urban scholarship, which can successfully investigate the contradictions of the urban environment and at the same time free itself from the Eurocentric trap.

The planetary urbanization approach is extremely helpful in conceptualizing the urban as something that exceeds far beyond the city. The omnipresence of the urban certainly helps better understand its contemporary nature as well as the ways in which it is transformed by the recent changes in capitalism. However, this method seems to overlook the importance of colonial history and the reconfiguration of its systems, methods, and cultures in many parts of the world's urban environments. By contrast, postcolonial scholarship largely explored these elements, calling into question the very production of knowledge and its universalization. Nonetheless, there are still blind spots in countermapping efficiently and effectively urban studies.

Perhaps, as Scott and Storper pointed out (2015), there is the risk of a "new particularism" in the postcolonial approach, especially in relation to the ordinary cities discourse. Despite being sceptical about the possibility of a rigid theory which is able to work equally anywhere in the world, it can also be problematic to theorize the absence of any good theory but the local. Rather, it is more productive to search for a theory that, borrowing Fernando Coronil's (2004: 240) words used to describe postcolonialism, is "fluid and polysemic" and "whose power derives in part from its ability to condense multiple meanings and refer to different locations". In addition, some postcolonial urban theory seems still to be excessively "city-centric" and might fail to understand contemporary urbanizing processes, which are often articulated beyond the traditional form of the city. In any case, it is impossible not to consider the role of colonialism in the making of the contemporary world, and the postcolonial point of view is extremely helpful at pointing this out.

The decolonial contribution seems to be able to add further valuable insights in this direction. If coloniality has been the device marking the shaping of the modern/colonial world, cities have certainly been crucial for the spread of this new form of social organization. In addition, coming to more contemporary times, further research could explore how coloniality is spatially re-organized and spread, to use Brenner's and Schmid's framework, from *concentrated* to *extended* urban areas. It is only by testing and bending our set of concepts, as well as putting them into practice, that we can produce valuable theory that is finally able to move outside the Eurocentric epistemologies.

The final question concerns how to frame these problems in everyday academic work. I suggest that there is not something like an unequivocal answer to this; rather, it is a constant practice in seeking new theoretical challenges to universal knowledge that would be extremely helpful. While teaching and researching subjects related to urban studies, it is necessary to continuously stress the subjective character of knowledge and its inevitable location within multifaceted geographies of power. The specificities of the space under study, as well as of the space in which the academic work is produced, need to be constantly questioned, swaying incessantly between the asymmetric lines that tie the particular with the global.

Notes

1 A valuable definition about the meaning of "critical" in critical theory is offered by Brenner (2009).
2 A detailed overview of Lefebvre's impact in the English-speaking world at the turn of the millennium is offered by Elden (2001).
3 An interesting reflection upon Lefebvre's brief but insightful work on colonialism is offered by Kipfer and Goonewarden (2013). For a recent discussion about Harvey and (post)colonialism, see Gudynas (2015).

4 Grosfoguel (2000) discusses some dependency theorists' failure to overcome Eurocentrism.
5 The City Statute is a constitutional law that resulted from a series of formidable urban struggles. It aims to reduce urban inequality by giving the possibility of land access to large cities' populations.

References

Allen, T. and Thomas, A. (eds) (2000) *Poverty and Development into the 21st Century*. Oxford: Oxford University Press.

AlSayyad, N. (1991) *Cities and Caliphs: On the Genesis of Arab Muslim Urbanism*. New York: Greenwood Press.

AlSayyad, N. (1992) *Forms of Dominance: On the Architecture and Urbanism of the Colonial Enterprise*. Aldershot: Avebury.

Amin, A. and Graham, S. (1997) The ordinary city. *Transactions of the Institute of British Geographers* 22(4): 411–429.

Bishop, R., Phillips, J. and Yeo, W. (eds) (2003) *Postcolonial Urbanism: Southeast Asian Cities and Global Processes*. New York and London: Routledge.

Brenner, N. (2009) What is critical urban theory? *City* 13(2–3): 198–207.

Brenner, N. and Schmid, C. (2014) The 'urban age' in question. *International Journal of Urban and Regional Research* 38(3): 731–755.

Brenner, N. and Schmid, C. (2015) Towards a new epistemology of the urban? *City* 19(2–3): 151–182.

Cardoso, F.H. and Faletto, E. (1979) *Dependency and Development in Latin America*. Berkley: University of California Press.

Castells, M. (1977) *The Urban Question*. London: Edward Arnold.

Chakrabarty, D. (2000) *Provincializing Europe: Historical Thought and Postcolonial Difference*. Princeton: Princeton University Press.

Chatterjee, P. (2012) *The Black Hole of Empire: History of a Global Practice of Power*. Princeton: Princeton University Press.

Coronil, F. (2004) Latin American postcolonial studies and global decolonization. In N. Lazarus (ed) *The Cambridge Companion to Postcolonial Literary Studies*. Cambridge: Cambridge University Press, pp. 221–240.

Datta, A. (2012) *The Illegal City: Space, Law and Gender in a Delhi Squatter Settlement*. Farnham: Ashgate.

Datta, A. (2015) New urban utopias of postcolonial India: Entrepreneurial urbanization in Dholera smart city, Gujarat. *Dialogues in Human Geography* 5(1): 3–22.

Davis, M. (1990) *City of Quartz*. London: Verso.

Dussel, E. (1995) *The Invention of the Americas: Eclipse of "the Other" and the Myth of Modernity*. New York: Continuum.

Dussel, E. (2000) Europe, modernity, and Eurocentrism. *Nepantla* 1(3): 465–478.

Elden, S. (2001) Politics, philosophy, geography: Henri Lefebvre in recent Anglo-American scholarship. *Antipode* 33(5): 809–825.

Escobar, A. (1995) *Encountering Development: The Making and Unmaking of The Third World*. Princeton: Princeton University Press.

Foucault, M. (2003) *"Society Must Be Defended": Lectures at the College de France, 1975–76*. New York: Picador.

Frank, A.G. (1970) *Capitalismo y Subdesarrollo en América Latina*. Mexico City, Mexico: Siglo XXI.

Furtado, C. (1964) *Development and Underdevelopment*. Berkeley: University of California Press.

Graham, S. (2011) *Cities under Siege: The New Military Urbanism.* London: Verso.

Grosfoguel, R. (2000) Developmentalism, modernity, and dependency theory in Latin America. *Nepantla* 1(2): 347–374.

Gudynas, E. (2015) Friendly colonialism and the 'Harvey fashion'. *ENTITLE Blog – A Collaborative Writing Project On Political Ecology* 15 October. Available at: https://entitleblog. org/2015/10/15/friendly-colonialism-and-the-harvey-fashion/ (Accessed 1 October 2017).

Guha, R. (1983) *Elementary Aspects of Peasant Insurgency in Colonial India.* Delhi: Oxford University Press.

Guha, R. (1997) *Dominance without Hegemony: History and Power in Colonial India.* Cambridge: Harvard University Press.

Guha, R. and Spivak, G.C. (eds) (1988) *Selected Subaltern Studies.* New York: Oxford University Press.

Hardoy, J.E. (1992) Theory and practice of urban planning in Europe, 1850–1930: Its transfer to Latin America. In R.M. Morse and J.E. Hardoy (eds) *Rethinking the Latin American City.* Washington: Woodrow Wilson Center Press, pp. 20–49.

Hardoy, J.E. (ed) (1975) *Urbanization in Latin America: Approaches and Issues.* New York: Anchor Books.

Harvey, D. (1973) *Social Justice and the City.* Oxford: Blackwell.

Harvey, D. (1985) *The Urbanization of Capital: Studies in the History and Theory of Capitalist Urbanization.* Baltimore: Johns Hopkins University Press.

Harvey, D. (2001) *Spaces of Capital: Towards a Critical Geography.* London: Routledge.

Jacobs, J.M. (1996) *Edge of Empire: Postcolonialism and the City.* London and New York: Routledge.

Jacobs, J.M. (2012) Commentary – Comparing comparative urbanisms. *Urban Geography* 33(6): 904–914.

King, A. (1990) *Urbanism, Colonialism and the World-Economy.* London and New York: Routledge.

King, A. (1993) Cultural hegemony and capital cities. In J. Taylor, J. Lengellé and C. Andrew (eds) *Capital Cities/Les Capitales: International Perspectives/Perspectives Internationales.* Ottawa: Carleton University Press, pp. 251–270.

King, A. (2003) Actually existing postcolonialisms: Colonial urbanism and architecture after the postcolonial turn. In R. Bishop, J. Phillips and W. Yeo (eds) *Postcolonial Urbanism: Southeast Asian Cities and Global Processes.* New York and London: Routledge, pp. 167–183.

Kipfer, S. and Goonewardena, K. (2013) Urban Marxism and the post-colonial question: Henri Lefebvre and 'colonisation'. *Historical Materialism* 21(2): 76–116.

Lander, E. (ed) (2000) *La colonialidad del saber: eurocentrismo y ciencias sociales.* Caracas: IESALC.

Lefebvre, H. (1991 [1974]) *The Production of Space.* Blackwell: Oxford.

Lefebvre, H. (1996) *Writings on Cities.* Oxford: Blackwell.

Lefebvre, H. (2003 [1970]) *The Urban Revolution.* Minneapolis: University of Minnesota Press.

McFarlane, C. and Robinson, J. (2012) Introduction – Experiments in comparative urbanism. *Urban Geography* 33(6): 765–773.

Merrifield, A. (2002) *Metromarxism: A Marxist Tale of the City.* New York: Routledge.

Merrifield, A. (2014) *The New Urban Question.* London: Pluto Press.

Mezzadra, S. and Rahola, F. (2006) The postcolonial condition: A few notes on the quality of historical time in the global present. *Postcolonial Text* 2(1). Available at: postcolonial.org/ index.php/pct/article/view/393/139 (Accessed 1 October 2017).

Mignolo, W. (2000) *Local Histories, Global Designs: Coloniality, Border Thinking and Subaltern Knowledges.* Princeton: Princeton University Press.

Mignolo, W. (2005) *The Idea of Latin America*. Malden: Blackwell.

Mignolo, W. (2011) *The Darker Side of Western Modernity: Global Futures, Decolonial Options*. Durham: Duke University Press.

Mignolo, W. (2014) The north of the south and the west of the east: A provocation to the question. *Ibraaz Contemporary Visual Culture in North Africa and the Middle East* (8). Available at: www.ibraaz.org/essays/108 (Accessed 1 October 2017).

Nijman, J. (2007) Introduction: Comparative urbanism. *Urban Geography* 28(1): 1–6.

Quijano, A. (1975) The urbanization of Latin American society. In J.E. Hardoy (ed) *Urbanization in Latin America, Approaches and Issues*. New York: Anchor Books, pp. 109–153.

Quijano, A. (1977) *Imperialismo y "marginalidad" en América Latina*. Lima: Mosca Azul.

Quijano, A. (1992) Colonialidad y modernidad/racionalidad. *Perú indígena* 13(29): 11–20.

Quijano, A. (2000) Coloniality of power and Eurocentrism in Latin America. *International Sociology* 15(2): 215–232.

Quijano, A. (2007) Coloniality and modernity/rationality. *Cultural Studies* 21(2–3): 168–178.

Rao, V. (2006) Slum as theory: The South/Asian city and globalization. *International Journal of Urban and Regional Research* 30(1): 225–232.

Robinson, J. (2002) Global and world cities: A view from off the map. *International Journal of Urban and Regional Research* 26(3): 531–554.

Robinson, J. (2006) *Ordinary Cities: Between Modernity and Development*. London: Routledge.

Robinson, J. (2011) Cities in a world of cities: The comparative gesture. *International Journal of Urban and Regional Research* 35(1): 1–23.

Robinson, J. (2015) Comparative urbanism: New geographies and cultures of theorizing the urban. *International Journal of Urban and Regional Research* 40(1): 187–199.

Rostow, W.W. (1990) [1960]) *The Stages of Economic Growth: A Non-Communist Manifesto*. Cambridge: Cambridge University Press.

Roy, A. (2009a) The 21st-century metropolis: New geographies of theory. *Regional Studies* 43(6): 819–830.

Roy, A. (2009b) Why India cannot plan its cities: Informality, insurgence and the idiom of urbanization. *Planning theory* 8(1): 76–87.

Roy, A. and AlSayyad, N. (eds) (2004) *Urban Informality: Transnational Perspectives from the Middle East, Latin America, and South Asia*. Oxford: Lexington Books.

Said, E. (1978) *Orientalism: Western Representations of the Orient*. New York: Pantheon.

Said, E. (1993) *Culture and Imperialism*. New York: Vintage.

Sassen, S. (1991) *The Global City*. Princeton: Princeton University Press.

Saunders, P. (1981) *Social Theory and the Urban Question*. London: Unwin Hyman.

Schteingart, M. (ed) (1973) *Urbanización y dependencia en América Latina*. Buenos Aires: Ediciones Siap.

Scott, A.J. and Storper, M. (2015) The nature of cities: The scope and limits of urban theory. *International Journal of Urban and Regional Research* 39(1): 1–15.

Simone, A. (2001) On the worlding of African cities. *African Studies Review* 44(2): 15–41.

Simone, A. (2010) *City Life from Dakar to Jakarta*. London and New York: Routledge.

Smith, R.C. (1955) Colonial towns of Spanish and Portuguese America. *Journal of the Society of Architectural Historians* 14(4): 3–12.

Storper, M. and Scott, A.J. (2016) Current debates in urban theory: A critical assessment. *Urban Studies* 53(6): 1114–1136.

Taylor, P.J. (2000) World cities and territorial states under conditions of contemporary globalization. *Political Geography* 19(1): 5–32.

Warf, B. and S. Arias (eds) (2009) *The Spatial Turn: Interdisciplinary Perspectives*. London: Routledge.

Young, R.J.C. (2001) *Postcolonialism: An Historical Introduction*. Malden: Blackwell Publishers.

9

DECOLONIZING LEGAL STUDIES

A Latin Americanist perspective

Aitor Jimenez González

The problem of legal studies and the rise of racialized societies

Frontex, the European Agency tasked with the promotion, coordination and development of the borders of the European Union, is more than a policing structure. Its functions by far exceed supporting physical control and vigilance over European borders.[1] Accordingly, in its 2017 risk analysis report (Frontex, 2017a), the agency appears more like an intelligence institution than a policing structure. This entity coordinates existing efforts in migration analysis and surveillance of immigration flows. It is actively collaborating with European Bodies such as Europol or the European Asylum Support Office, with EU Countries, and with third countries such as Georgia, Armenia, Turkey or Azerbaijan. It is effectively the lead node of a vast and influential network within the European defence community.[2] The analysis tasks performed by Frontex are not constrained by European borders. Algeria, Senegal, Morocco and Turkey are among the target countries of its intelligence and surveillance operations. Frontex collects collective and personal data for a broad variety of purposes such as providing advice (for example, to police services), controlling, and perhaps most importantly, creating "a picture of the situation at the EU's external borders and the key factors influencing and driving it" (Frontex, 2017b). We might say that Frontex seeks to define migrant subjectivities. Its agents decide when a subject may represent a risk and determine which subjectivities belong to the "common law" (and thus can be policed) and which, such as refugees and asylum seekers, fall under the scope of humanitarian international law. Therefore we can consider Frontex to be the author of the interpretative framework through which other European agencies such as Europol, European Asylum Support Office (EASO) or the European Maritime Safe Agency (EMSA) will understand the complex reality of immigrants, asylum seekers and refugees. Frontex embodies what Sandro Mezzadra and Brett Neilson (2013) have defined as a border methodology, or a technology of power designed to control, manage, distribute and *multiply* labour. Thus, Frontex, along with the above mentioned European agencies have turned from policing bodies to being the governmental structures that establish the classification of future underprivileged subjectivities. This apparatus announces the rise of a racialized legal structure in Europe – a legal narrative that is articulating privileged and underprivileged categories of population along the colour line. Of course, this does not bring to mind a renewal of the Jim Crow laws in the United States but a newly white supremacist ideology structured through a non-binary colour line.

Those who are placed on the privileged side of the colour line have ensured access to political and economic rights. Moreover, they do not have to struggle with systemic hostility. Those who are on the underprivileged side of the line may be met with violence that ranges from everyday racist attitudes – such as discrimination at work, police racial profiling, racism in the media – and stereotypes to more radical actions such as torture, imprisonment, deportation or even death. Drawing on the work of Charles Mills (1997), I call this hierarchized structure the *European racial contract*. This emerging racial apparatus is *not* a deviation from the supposed humanistic values of the European Union. It does not find representation in the racist and sometimes extremely violent attitudes of individuals committed to the achievement of a sick white supremacist agenda, as manifest, for example, in the case of Anders Breivik. It is the deliberate consequence of a political strategy designed in the highest instances of Europe in order to manage the undesirable but essential influx of migrants that is its workforce. The racial contract in Europe started as migrant governance, but has increasingly moved towards the complete governmentality of entire populations (the German management of the Syrian population during the last refugee crisis represents a good example of the *European racial contract*). Without the collaboration of myriad legal actors from researchers to lawyers, judges, and policymakers, it would not have been possible to put this structure in place. Behind every working paper, document, book, policy or decision that supports the European racial contract there is a large number of legal professionals who by their silence or active positions have been accomplices to the present situation.

The importance of the law for the articulation of racialized social relations has been extensively documented: from the well-known Jim Crow laws to the Nazi legislation, and passing through to the management and repression of Roma populations by almost all European countries. If the law has been a useful tool to construct and maintain white supremacy – that is, to reproduce the power of the privileged over the underprivileged – the same can be said of the constitutive power of law to create reality by hiding racial oppression. The latest is the path chosen by European politics and legal operators, a colour-blind denial of reality where under the generic abstract concepts of "citizen", "migrant" or "refugee" lies a growing reality of racialized social stratification (the increasing racialized structure of the Spanish labour market shows increasing levels of income inequality between Spanish citizens and immigrants) (Muñoz Comet, 2016).

Law schools are clearly representative of what in this volume we have defined as the "westernized university". The generalized use of the term "law" or "Law" masks the fact that the concept we are using is not a universal category but a highly provincial one premised on the westernized legal cosmovision. The generalized use of the western concept does not result from the "peaceful spread" of a superior science, as some still believe, but instead is the result of centuries of colonialism, violent repression against other legal cosmovisions during the colonial periods and the persistence of the process referred to as coloniality.

The colonial powers of the Global North have dedicated extreme efforts to ensure that the only law allowed within their colonies would be their own, although there is substantial evidence around the world that shows that the colonial powers did not always succeed in this endeavour. Any other expressions of law, such as indigenous legal technologies, were often considered traditional uses and customs, forms of cultural backwardness that should be coopted or eradicated. Furthermore, legal scholars such as Antony Anghie (2007) have shown how our common legal understanding of politically relevant subdisciplines such as International Law or Human Rights law have their foundations in the colonial and racializing processes of legitimation surrounding war against indigenous populations, slavery, land dispossession and cultural appropriation. Future politicians and policymakers are then educated in colorblindness. In university law schools there is little willingness to address the challenges posed by the rise of diversity and the crisis of westernized epistemes. We are in need of a sweeping reform of legal studies in order to understand the configurations that define privileged and underprivileged subjectivities in Europe. Legal education cannot avoid the racial question with the worn-out concept of the abstract and universal subject. We need an analytical framework that enables us to understand the rise of racializing technologies of power among the nations of the Global North and, consequently, will offer us appropriate tools to dismantle it. The question is, where to find it?

Racism has been an extremely useful tool for the Latin American colonial and postcolonial regimes. Racist laws and papal bulls enabled the conquest of America. Colonial legislation allowed slavery, exploitation and expropriation. With the arrival of the postcolonial order, the racist legal regime was officially overturned, but did not lead to a concomitant structural transformation. In some countries such as Cuba, blacks and mulattoes remained relegated to the lower classes, while in others such as Argentina indigenous peoples were not only marginalized but also brutally assassinated. Even those Latin American governments considered to be progressive have done little to dismantle hierarchical structures.

In many ways, the arrival of the 20th century forced a number of ruptures with the past. Structural racism was largely denounced in many countries, including Mexico, Chile, Colombia, Venezuela and Bolivia. But it was not until the overthrow of the neoliberal regimes of Bolivia, Venezuela and Ecuador at the beginning of the 21st century that racism was articulated as a problem to be addressed by the state. In the words of Alvaro García Linera et al. (2010: 6) the structure of the state became a "field of dispute". Structural racism is starting to be dismantled as a consequence of a legal and cultural war against the remnants of the colonial order. These transformations have exposed the racism hidden behind the claims of humanistic and universalistic values.

Freedom, equality and fraternity remain the inspirational motto for European liberal regimes. As seen in Bolivia during the 19th century, the apparent legal equality espoused by these words in fact hides the repression and massive imprisonment of racialized populations, high levels of inequality, and structural and institutional racism. Regimes of the Global North need to go through a similar unveiling process as those which occurred in Bolivia and other parts of the Global South, whereby legal structures undergo structural transformation and are brought into a necessary dialogue with the problem of racialization. I believe that recent Latin American discussions and policies on legal pluralism, decolonization of the state and of academia could serve to guide this unveiling process in the Global North.

In other words, Latin America provides a number of epistemic tools that legal scholars and policymakers in the Global North could make use of in their decolonizing project. Such a move is fraught with complexities, however, but it requires accepting the epistemologies of the Global South as equals. In the next section, I provide a number of concrete examples of how such an endeavour could be embraced in contemporary universities, from research groups to the design of a whole curriculum in Law.

In the first part of the chapter, I outline some recent decolonial epistemological contributions from Latin America. Here, I demonstrate the connection between these ideas and the recent Bolivian and Ecuadorian decolonial constitutional transformation. The second part will be devoted to the legal pluralist turn, especially with regard to its potentially influential role in the westernized university; first, as a theoretical frame for critical research and second as the key element of transformative pedagogical legal initiatives. To achieve this aim, I consider the theoretical framework of an academic legal decolonial process, as well as specific examples of its manifestation.

Decolonizing epistemes, decolonizing constitutions

The neo-constitutionalist process that has emerged in Latin America in the past two decades has led to a reinterpretation of many political and legal concepts, including citizenship, nationality, cultural identity, race and diversity. The plurality of the populations has been recognized in several constitutions including those of Ecuador and Bolivia, not only as a matter of cultural difference but as the very essence of new plurinational states such as Bolivia.

This new constitutional framework has literally opened the door to formerly marginalized critical legal scholars. Their voices are now listened to by legal operators who are attempting to implement plurinational legal orders, something for which they were not prepared. Their influence has increased in legally oriented institutions such as tribunals and parliaments, and law schools such as the Plurinational Public Management School in Bolivia or the Intercultural Veracruz University. Critical legal theory is playing a relevant role in Latin American societies such as Bolivia, Ecuador and Mexico as a consequence of the constitutional transformations that have taken place in the last twenty years. A brief analysis of the

Bolivian constitution (2009) helps us to better understand the connection between the decolonization of constitutional principles and the transformation of the legal epistemic order.

The Bolivian constitution emerged from a broad political process, characterized by the mass mobilization of workers, students, indigenous peoples and intellectuals. It was part of a revolutionary political process which had at its core a critique of the colonial order that has excluded these people for centuries. It is important to emphasize at this point that during the colonial and postcolonial periods Liberalism was the hegemonic, racializing ideology among the elites, who impoverished and marginalized the indigenous demographic majority. It also worked as fuel for inner colonialist thought. Liberal institutional manifestations (state, law, culture) were considered the only legitimate ones, and they worked to displace, forbid and prosecute any other form of diversity in education, legality, economic organization or social behaviour. Hence, colonialism, economic exploitation and racialization worked hand in hand with the centralist monolithic state form and its individualistic society. In contrast, communitarianism was widely spread in different forms among indigenous communities, as indigenous cosmovision, communal economic and political structures and legal communal solutions, among workers unions and cooperatives and among persecuted intellectuals. These pluralistic and communal energies that survived colonialism were evident in the mass mobilizations of 2003 and 2005. The neoliberal regime fell at the hands of the masses and as a result the first indigenous, unionized leader, Evo Morales, was elected president. Morales represented both communalism and pluralism and led the creation of a constituent assembly that was by far the most diverse in Bolivian history, composed of activists, indigenous leaders and persecuted intellectuals.

But it is one thing is to question and struggle against the colonial liberal state and quite another to reorganize a colonial structure, decolonize it and create a new paradigm of collective organization. The debates were inspired by indigenous demands. These ideas about legal and political autonomy were soon at the centre of the discussion on how to transform the state, or how to enable new ways to dismantle the colonial apparatus.

The new political situation enabled a widespread recognition that Bolivia had inherited a colonial social structure that was not dismantled after independence and that was accompanied by a set of resilient racist practices and attitudes (García Linera, 2013), but that different communities functioned according to pluralistic paradigms that were in conflict with the colonial legal order. This ongoing radical transformation of the colonial legal culture into decolonizing systems has come after a deep reflection not only on the institutional structures of colonialism, but on the legal ontologies that legitimize them. The Bolivian constitution embodies the very essence of what legal decolonization could mean. The preamble of the Bolivian constitution notes that:

> In ancient times mountains arose, rivers moved, and lakes were formed. Our Amazonia, our swamps, our highlands, and our plains and valleys were covered with greenery and flowers. We populated this sacred Mother Earth with

different faces, and since that time we have understood the plurality that exists in all things and in our diversity as human beings and cultures. Thus, our peoples were formed, and we never knew racism until we were subjected to it during the terrible times of colonialism.[3]

The Bolivian constitutional preamble is more than a vague principle declaration. It represents a counter-narrative against the postcolonial monolithic tradition of the 19th century republic. The Plurinational Constitution of Bolivia seeks to construct political legitimacy by drawing on indigenous origins. Pluralism thus emerges as a powerful concept for organizing a new decolonized economic, political and societal order. The pluralist declaration is enshrined in the first article of the Bolivian Constitution:

> Bolivia is constituted as a Unitary Social State of Plurinational Communitarian Law that is free, independent, sovereign, democratic, intercultural, decentralized and with autonomy for internal territories. Bolivia is founded on plurality and on political, economic, juridical, cultural and linguistic pluralism in the integration process of the country.

Article 8 emphasizes indigenous legal and political ontologies:

> The State adopts and promotes the following as ethical, moral principles of the plural society: ama qhilla, ama llulla, ama suwa (do not be lazy, do not be a liar or a thief), suma qamaña (live well), ñandereko (live harmoniously), teko kavi (good life), ivi maraei (land without evil) and qhapaj ñan (noble path or life).

These plurinational values have been well articulated throughout the constitutional text. The constitution recognizes in its title II chapter IV "Rights of the Nations and Rural Native Indigenous Peoples" a set of important rights: autonomy, effective legal pluralism, the right to adopt its own rules and to elect its own authorities, and the right to be consulted.

Diversity is then recognized as a fundamental aspect of the nation's legal life. It is an organizing and inspiring principle that enables the possibility of an intercultural dialogue between oppressed and racialized populations and the state. A truly plurinational state does not conceive of marginalized groups as peoples to be integrated and assimilated, but as its political strength. Contrary to what happens in multicultural states such as Canada, the plurinational state of Bolivia recognizes the nationhood of Aymaras or Quechuas.

Alvaro García Linera, Bolivian intellectual and the current vice president of Bolivia, notes that the Bolivian state has not only reached a pluralist status, but is also seeking to become a communitarian state. As García Linera (2013) has noted, individualism is the keystone to understanding westernized societies, especially their legal systems. Liberalism has a long, strong and influential tradition in multiculturalism as the political and legal recognition of diversity.

Given as noted that the communitarian idea was widely spread amongst the indigenous peoples, workers and intellectuals, the constituent assembly began to propagate communitarian principles throughout the Bolivian constitution and they appear in a number of articles, including Article 26 communitarian democracy, Article 47 (communitarian forms of production) and Article 54 (communitarian social enterprises). These communitarian principles have their roots in the Aymara and Quechua tradition. From this tradition, the Bolivian constitution is performing a deliberate estrangement from western individualist law. The latest constitutional transformation has enabled a reassessment of the Quechua and Aymara legal communitarian cosmologies and a vindication of a newly decolonized Latin American Marxism.[4]

The new legal framework brings many possibilities as well as great challenges. First of all, it faces the problem of how to learn, and how to build from diversity after having been so accustomed to the mythical order of the state (Fitzpatrick, 2008). Native peoples find that after centuries of resistance and marginalization they are regaining legal and political autonomy. The context is not absent from danger, but, indeed, communities are experiencing a new blossoming. In the current political situation, indigenous peoples are developing new legal strategies. Not only are they resizing their law, and remaking their relations with the colonial nation state, but they are also transforming and decolonizing the very concept of law itself. An example would include the Communitarian Police of Guerrero in Mexico.

I do not deny the problematic contradictions of some of the political ideas and processes emerging from the Global South. Contradictions have reached the point of antagonism precisely in the places where the decolonial option has reached more spaces of influence, such as Bolivia. These contradictions reveal how alive political ideas are in this part of the world. To mention one example, the role of the state in the conflict between indigenous communities and national economic interest in the TIPNIS affair[5] has shown the breakpoints and limits of the state, and even the reproduction of the old premises of colonial order. On the other hand, it also reveals how politically conscious of their own potency the Latin American population is, capable of establishing a negotiation on equal terms with the state and economic powers. It is precisely within these contradictions where the act of *thinking politically* became a material reality. The colonial state remains in Bolivia, but this state must act according to the acquiescence of a highly organized society.

For Bolivia or Ecuador, the two only countries in the world officially defining themselves as plurinational, theories of decolonization theory have come to underpin state policy, and as a result have called into question the very basis of the monolithic nation-state legal order. The changes that have taken place have encouraged a number of Latin American academics to review the role played by law and the state in maintaining structures of discrimination and racial and cultural hierarchy (Gargarella and Christian, 2009; Carpizo, 2009; Clavero, 2008).

Decolonizing legal academia: The legal pluralist turn

Undoubtedly the new Latin American constitutionalism has opened a window of possibility for indigenous and communitarian movements, who have found a narrow but evident gap in westernized legal rationality. The legal, political and economic autonomy of indigenous communities is now a political fact, not exempt from threats, but constituent of a legally validated pluralist reality. It is in this pluralist environment that a legal epistemological revolution is taking place, brought on by a synthesis of social movements, scholars, intellectuals and legal professionals, such as the case of the ONIC National Indigenous Organization of Colombia and its indigenous law schools. The tremendous impact of this revolution can be easily perceived in the multiplicity of research, studies, papers, essays and monographs that have sprung up in the region. From legal anthropology to sociology of law and theoretical debates concerning so called "New Latin American Constitutionalism", to traditional legal studies on recent codification, the contributions of critical scholars to these fields are building a solid and comprehensive counter-hegemonic legal theory. The collective "Crítica Jurídica" (Legal Criticism) based in Mexico, Brazil and Argentina is a great example of this new wave of legal studies. Among some of its young members we can find Alma Melgarito or Daniel Sandoval (Cervantes, 2016; Melgarito, 2012). Issues eluded in the past, such as structural racism, patriarchy, genocide, epistemicide or/and feminicide are being acknowledged by a new, fresh and determined generation of scholars. These include *inter alia* Nilda Garay Montañez (2012), Raquel Yrigoyen (2015), Yacotzin Bravo (2016).

The pluralist legal environment is diverse by its own nature, and does not have a univocal definition. There are, however, some shared defining features. First, the decolonial approach to law questions the hegemonic colonial position of the westernized legal model and proposes instead a dialogue between different understandings and epistemic positions on law (De Sousa Santos, 2014). Second, it challenges the centrality of the state as the main organizer of legal and juridical life. Third, it highlights the necessity of building a new historical narrative grounded in diversity rather than homogeneity. This counter-narrative not only draws attention to the colonial and racist origins of many present-day institutions, it also brings to light legal and intellectual contributions developed by oppressed and marginalized populations. Finally, legal pluralism is not the result of isolated academic contributions, but the product of the collaborations between academy, community and social forces. On the basis of the above, it can be stated that legal pluralism is neither theory nor method but a "liberating frame" (Wolkmer, 2003: 2). Legal pluralism has acquired great importance among critical legal academics as a tool for decolonizing legal education (Baldi, 2014). This is not only the result of recent constitutional events. The foundation of this decolonial proposal rests upon the shoulders of the Latin American tradition of critical pedagogy, a pedagogical model inspired by socialist and communist ideas as well as by indigenous resistances. This long tradition contributed to the building of autonomous educational spaces in which state and individual-centric positions were questioned. This move was not only a gesture concerning academic content but was also part of a broader cultural war between social movements and the state. Thus the decolonial turn has solid foundations, not only because it has strong and revolutionary legal constitutional instruments, but because it is contributing to the construction of a strong and growing educational infrastructure that allows highly critical and community-based legal thinking. This effort is underway within and outside westernized universities. There are many examples outside the academy that are key for the development of decolonial legal structures. But given the specific aim of this book I focus on some of the projects conducted within westernized academic frameworks, specifically in Mexico.

The example of RADAR is particularly remarkable. RADAR is a research group in critical law whose members are academics and researchers from the UNAM (Universidad Nacional Autónoma de México) and UACM (Universidad Autónoma de la Ciudad de México), both in Mexico City. Its members, experts in law but also in disciplines such as political science, anthropology and sociology, have been developing participatory activist research with indigenous and peasant communities threatened by transnational companies, the narcos or even the state. The hydroelectric dams of La Parota or El Zapotillo (Gutiérrez Rivas, 2008) and the fishing issue with the Yaqui people are among the case laws studied and defended by these professionals (Bravo, 2016). Collectively they have organized conferences and meetings between social movements and academics, as well as provided legal education through workshops in indigenous communities, as happened in the Montaña de Guerrero, where indigenous communities are threatened by mining companies. They have collaborated in the design of the *Manual para jueces y juezas sobre la protección de los derechos de las campesinas y los campesinos (Judges Handbook for the Protection of Peasant Rights)* (Emanuelli and Gutiérrez, 2013). This book, intended for Latin American judges, was widely distributed in the Latin American region and is the product of the alliance between Via Campesina (an international association of peasants), the NGO Habitat and the Red Iberoamericana de Jueces (an international association of judges). The impact of this book is twofold. On the one hand, it has proven to be a very useful tool for the promotion of human rights, specifically of campesinos. On the other, it serves as an exemplary academic text that details the range of comparative legal policies in Latin America. The alliance has also completed a research project called "The Indigenous right to be consulted against megaprojects in a context of social and ecological destruction" in 2012, within the most prestigious legal education research centre in México, the Legal Research Institute (Instituto de Investigaciones Jurídicas) of the Universidad Nacional Autónoma de Mexico (Mexican National Autonomous University).

This research project intended to highlight the chronic violations of Indigenous human rights, and especially the provisions related with the right to prior consultation reflected in ILO Convention 169 concerning *Indigenous and Tribal Peoples in Independent Countries*, among other legal instruments. The case studied within this project documented the degrees of complicity between state interests and transnational companies. The investigations carried out had two key dimensions: the scientific analysis of problematics, and the legal accompaniment of the communities involved in it. The second aspect was central to the research project, and the one that most aptly characterizes this kind of collaborative active research methodology. In both the research and the accompaniment, indigenous and critical epistemologies guided the articulation of knowledge as the core of its legal reasoning. Therefore, legal research and social problematics were linked by a counter-hegemonic use of legal instruments and methodologies.

One line of research of the project was devoted to the analysis of the environmental conflicts in the Huasteca Veracruzana. It was developed in alliance with the "Huasteca's defenders", a community-based Nahuatl collective engaged in resistance against the Canal de Chicontepec, a mega-project guided by international investors intended to extract petroleum from the Huasteca Veracruzana. The aim of this research was both to expose and analyse the human rights violations in the Huasteca as well as to collaborate with the Nahuatl decolonial translation for some of the legal instruments, including Convention 169 of the International Labour Convention. One of the main tasks of the Huasteca's defenders is to teach human rights from a Nahuatl perspective to Nahuatl communities. They want them to know and understand that they are entitled to their land, in a context where big, anxious energy companies try by any means to explore and prospect their rich lands. The Huasteca's defenders have also produced a documentary film regarding the Nahuatl conception of law, and provided training on human rights to rural Nahuatl-speaking communities in the Huasteca. The members of the "Defensores de la Huasteca" are all graduates in Intercultural Law from the Universidad Intercultural Veracruzana or UVI (Veracruzan Intercultural University – VIU). The UVI is part of the Universidad Veracruzana (Veracruzan University – UV), one of the most important educational entities in Mexico. Nonetheless, the UVI enjoys a considerable amount of autonomy with respect to the UV. Its mission is devoted to the promotion of intercultural values and the "knowledge-dialogue". It is not centrally located in cities but in small rural areas mostly inhabited by peoples who speak indigenous languages.

Their curriculum has been developed in accordance with the interests and needs of the indigenous peoples through a process of dialogue and negotiation. The UVI not only provides education, it also supports projects such as the "Defenders". The aforementioned training, workshops and the documentary film were funded by the UVI, allowing this courageous and pioneering research on decolonial legal learning, teaching and translation to develop, along with the community workshops. For many isolated communities, with no access to tertiary education, this legal training enables them to gain an understanding of the rights to which they are entitled under international law.

The UVI has recently approved a fascinating curriculum in legal studies in the form of a law degree with an emphasis on legal pluralism. This came after two decades of indigenous struggles such as those waged by the Zapatistas and the Communitarian Indigenous Police of Guerrero that have demanded the recognition of their law as Law rather than as a costume. This development continues the historical struggle of indigenous peoples against colonizers. The importance of this degree lies in its community-oriented ready-to-use knowledge design. Indigenous communities are in need of legal professionals who share their background, who speak the same language and who can understand their cosmovision. Indigenous communities are also well aware of their need not only for legal professionals capable of understanding the law of the state (and who are eventually able to contest it), but also professionals who understand indigenous law and the pluralist legal reality where these communities live.

The Law degree programme itself is a declaration of plurality, interculturality, decoloniality and socially conscious critical legal studies. It affirms the existence of a violent colonial past, responsible for the legal colonial and postcolonial order. It also asserts indigenous peoples' role as protagonists in the constitution of the legal structure through negotiation with and resistance to the colonial order. Therefore, legal pluralism is placed at the core of the legal model, and not as a mere adjacent aspect. The programme is also deeply embedded in the international human rights framework on indigenous rights and its legal and political applications in the Mexican context, establishing a connection between the critique of colonialism and current legal violations. In this regard, the degree's educational programme (Universidad Veracruzana, 2016) justifies its pertinence of the legal pluralist approach by pointing to crucial legal problematics in the region. These include the need for legal training in indigenous languages and for interpreters for indigenous language speakers during legal processes.

As the educational programme emphasizes, these problems have their roots in the absence of professionals educated within a pluralist legal perspective. To educate conscious, critical and community-based researchers and operators is therefore imperative to put an end to the current discriminatory, racist and culturally unconscious legal reality. This pluralist legal approach, applied to the most traditional law education, could be an asset for legal education, as it could provide future legal professionals with useful critical tools. By putting diversity at the core of the curriculum, this programme is undertaking crucial work on one of the increasingly evident problematics of our societies.

Conclusions

Our legal culture has a problem of racism. It is not a question of the misapplication of the law, and neither is it the consequence of a few ignorant people. The relation between racism and law is ideological and it has to do with the colonial origins of our legal culture. Without acknowledging that reality, challenging structural racism will be extremely difficult. The transformation of legal curricula is urgent if we are not going to continue to train professionals who will be responsible for designing, deciding, judging and applying racist laws.

For this transformation to happen, it will be necessary to confront the colonial foundations of westernized legal culture, and consequently reevaluate our educational models. To this end, I propose to follow the decolonial path already undertaken by some initiatives in Latin America. That is to put into practice the "epistemic decolonial turn" (Grosfoguel, 2007) as the first step to start the above mentioned dialogue of knowledge. Taking thinkers from the Global South seriously is the first step to start a process of decolonization in the westernized university. This is more than a professional critical exercise; it also demands a different epistemic and political attitude. This "taking seriously" represents a move that forces the researcher to confront the geopolitics of knowledge. It is a process that involves displacing researchers from their comfortable and well- respected liberal position, making them share political spaces side by side with unexpected and marginalized allies such as critical Muslims, peasant movements or immigrant workers' unions.

I conclude this chapter by urging legal scholars in the Global North to embark on the decolonization of legal education in westernized universities. Such a move starts with fully acknowledging the colonial past of western societies, the ongoing nature of the racializing processes put in place by colonialism and the structuring role played by the Eurocentric legal apparatus. It also involves understanding that Eurocentric law does not represent all the legal conceptions and beliefs that exist in the world and therefore the westernized university should engage in a genuine knowledge dialogue with decolonial projects in the Global South. Schools of law must grapple with the problem of structural racism and include pluralist legal understandings in curricula. Legal academia should also develop channels of communication with social movements, marginalized peoples and racialized subjectivities. In other words, they need to establish contact with social reality, real problematics and everyday-life legal concerns. Finally, affirmative action policies are necessary at any level of legal education. Without these kinds of policies diversity will remain excluded from the constituent spaces of legal culture.

If authors from the Global North stand on the shoulders of giants, decolonial thought from the Global South stands with the multitude. Mobilizing such a critique is to talk from a conviction that other worlds certainly exist and that a vast universe of epistemic possibilities and political potentialities is out there ready to be experienced. The strength of the decolonial option lies in its popular origins, as it is not an academic product but one of the many expressions of Global South peoples. If we, Global North academics, are to achieve the decolonization of education we should proceed with modesty and embrace the power of plurality.

Notes

1 The new European regulation (2016/1624) has substantially expanded the work of Frontex. As a result, its risk analysis work "should cover all aspects of Integrated Border Management and develop a pre-warning mechanism" (Frontex, 2017a: 11).
2 "In order to deliver strategic risk analysis products, Frontex brings together information from a wide range of sources including: border authorities of Member States and non-EU countries, EU partners (such as the European Commission, EASO, Europol, EEAS, EU SATCEN and Eurostat), international organisations (such as UNHCR, IOM, Interpol) and open sources (watchdogs, think tanks, academia and the media)" (Frontex, 2017b).
3 Bolivian constitutional preamble. Retrieved from: www.constituteproject.org/constitution/Bolivia_2009.pdf.

4 While it is true that many Latin American Marxist contributions fall on a classic, Euro-centric perspective, there is also a critical wave that is shaking the tradition. The critical perspective on Marxism is not new. In 1928 Haya de la Torre published his "Seven essays on the Peruvian reality", an adaptation of the Marxist approach to Latin American reality. More recently Enrique Dussel (2007) has dedicated strenuous efforts to review Marx from a Latin American and critical perspective. Above all these intellectual contributions, social movements such as the Ejercito Zapatista de Liberación Nacional (Zapatist National Liberation Army) have embodied the transformation and reappropriation of Marxism.

5 TIPNIS is the acronym for Isiboro-Sécure Indigenous Territory and National Park, also known as the Bolivian tropical lowlands. This remote region is mostly populated by Amazonic indigenous peoples, different to the majoritarian Quechua or Aymara peoples of the highlands. The Bolivian government planned to build a road crossing the Amazonas region connecting Bolivia with Brazil as part of a set of economic and strategic agreements. But indigenous Amazonic peoples said the highway would destroy their way of life, the Amazonic habitat and the fragile equilibrium between the peoples and their territories. Protests resulted, in what was the first serious conflict of the Evo Morales socialist indigenous government with indigenous peoples.

References

Anghie, A. (2007) *Imperialism, Sovereignty and the Making of International Law*. Cambridge: Cambridge University Press.

Baldi, C.A. (2014) Descolonizando o ensino de diretos humanos. *Hendu–Revista Latino-Americana de Direitos Humanos* 5(1): 8–18.

Bravo, Y. (2016) Elementos para comprender los límites y las posibilidades del derecho y los derechos frente al despojo de los territorios indígenas. *Amicus Curiae* 12(2): 204–233.

Carpizo, J. (2009) Tendencias actuales del constitucionalismo latinoamericano. *Revista Derecho de Estado* 23: 7–36.

Cervantes, D.S. (2016) Poder, violencia y derecho: movimientos sociales e historia social del derecho en México. *Panoptica-Direito, Sociedade e Cultura* 11(2): 332–350.

Clavero, B. (2008) *Geografía jurídica de América Latina: pueblos indígenas entre constituciones mestizas*. Madrid: Siglo XXI.

De Sousa Santos, B. (2014) *Epistemologías del Sur*. Madrid: Akal.

Dussel, E.D. (2007) *Política de la liberación: historia mundial y crítica*. Trotta: Madrid.

Emanuelli, M.S. and Gutiérrez, R. (2013) *Manual para juezas y jueces sobre la protección de los derechos de las campesinas y campesinos*. Ciudad de México: HIC-AL, FIAN.

Fitzpatrick, P. (2008) *Law as Resistance: Modernism, Imperialism, Legalism*. Farnham: Ashgate.

Frontex (2017a) *Risk Analysis 2017*. Warsaw: Frontex.

Frontex (2017b) *Risk Analysis*. Available at: http://frontex.europa.eu/intelligence/strategic-analysis/ (Accessed 7 May 2018).

Garay Montañez, N. (2012) La idea de igualdad en el constitucionalismo liberal español: lo racial, las castas y lo indígena en la Constitución de 1812. *Cuadernos Constitucionales de la Cátedra Furió Ceriol* 69(70): 129–158.

García Linera, A. (2013) *Autonomías indígenas y Estado multicultural*. Buenos Aires: Clacso.

García Linera, A., Prada, R., Tapia, L. and Vega Camacho, O. (2010) Prólogo. In A. García Linera, R. Prada, L. Tapia and O. Vega Camacho (eds) *El Estado: Campo de lucha*. La Paz: CLACSO, pp. 5–6.

Gargarella, R. and Christian C. (2009) *El nuevo constitucionalismo latinoamericano: promesas e interrogantes*. Santiago: CEPAL.

Grosfoguel, R. (2007) The epistemic decolonial turn: Beyond political-economy paradigms. *Cultural Studies* 21 (2–3): 211–223.

Gutiérrez Rivas, G. (2008) El derecho al agua y su relación con el medio ambiente. Instituto de Investigaciones Jurídicas de la UNAM 123–143. Available at: www20.iadb.org/intal/catalogo/PE/2010/04747a08.pdf (Accessed 7 May 2018).

Melgarito, A. (2012) *Pluralismo jurídico: la realidad oculta. Análisis crítico-semiológico de la relación Estado-pueblos indígenas*. México: UNAM.

Mezzadra, S. and Neilson, B. (2013) *Border as Method, or, the Multiplication of Labor*. Durham: Duke University Press.

Mills, C. (1997) *The Racial Contract*. Ithaca: Cornell University Press.

Muñoz Comet, J. (2016) *Inmigración y empleo en España: de la expansión a la crisis económica*. Madrid: CIS-Centro de Investigaciones Sociológicas.

Universidad Veracruzana (2016) Licenciatura en Derecho con enfoque de Pluralismo Jurídico. Available at: www.uv.mx/uvi/files/2012/11/160212_LDEPLUJ_v9.pdf (Accessed 7 May 2018).

Wolkmer, A.C. (2003) Pluralismo jurídico: nuevo marco emancipatorio en América Latina. In M. Villegas and C. Rodríguez (eds) *Derecho y sociedad en América Latina: Un debate sobre los estudios jurídico críticos*. Bogotá: ILSA, pp. 247–259.

Yrigoyen, R. (2015) *Pluralismo jurídico y pautas de coordinación*. Available at: http://repositorio.amag.edu.pe/handle/123456789/589 (Accessed 7 May 2018).

10

THE CHALLENGES OF BEING MAPUCHE AT UNIVERSITY

Denisse Sepúlveda Sánchez

Introduction

This chapter[1] explores the experiences of Mapuche people who are the first generation in their families to attend university and how participation in Higher Education shapes their identities. It draws on a sample of 40 life stories recounted by a group of Mapuche students who live in the capital city Santiago or in the Araucanía region in southern Chile and who have gained a level of educational mobility. I argue that participants have to face several difficulties and disadvantages during their attendance at university. They have to negotiate their Mapuche identities, and face the tension of structural racism and ethnic boundaries. However, despite all these difficulties my participants managed to achieve a university education and to experience some degree of upward social mobility in their subsequent occupations. Their stories underscore their experience of agency as well as suffering.

Since the period of the Spanish conquest indigenous groups in Chile have faced economic, social, territorial and cultural inequalities. According to Merino (2007), Chilean society functions in quite a contradictory manner. While a public discourse of tolerance towards indigenous peoples exists, in terms of everyday relations and interactions, the dominant attitude is characterized by distance, suspicion and prejudice, especially when dealing with Mapuche people. As a matter of fact, there has been a long history of conflict between indigenous people and the Chilean State. According to Yopo (2012: 192), this conflict and the denial of the existence of indigenous people is demonstrated through "land deprivation, violence and discrimination, and attempts at co-optation by Chilean society".

It is necessary to make my own positionality clear and to note that I share a similar background with the respondents. I am second generation Mapuche; that means that my father has a Mapuche surname, but I do not. Therefore, people do not usually identify me as Mapuche. I belong to a working-class family from Temuco city and I am the first generation to attend university in my family. I have experience of migration and upward educational mobility. Hence, before I started this research, I often wondered how other Mapuche students had experienced university and social mobility. I wondered if they also struggled with their identities and sense of belonging, how they made sense of class, gender and race, and how they related to Mapuche culture. These were the key questions that motivated my research.

The first part of the chapter explores the broader social context in which indigenous people are situated and identifies the structural educational disadvantages which indigenous people in Chile have to face. Like westernized universities elsewhere, Higher Education in Chile emerges from and is embedded in a hierarchical colonial system from which indigenous peoples and their worldviews are largely excluded and in which they are seriously underrepresented. The second part identifies the different experiences of agency of Mapuche people at university and work, focusing in particular on the interesting ways in which they started to assert their Mapuche identities.

Social context of Mapuche people in Chile

There are nine indigenous groups in Chile: Atacameño, Aymara, Colla, Diaguita, Kawashkar, Mapuche, Quechua, Rapa Nui, and Yagán people. Of these, the Mapuche people are the largest. According to the CASEN survey (2015) 9.0 per cent of the Chilean population identifies as indigenous and 83.3 per cent of this population is Mapuche. In the 16th century when Chile was conquered by the Spanish, the Mapuche occupied a large area, from the Copiapo river to Chiloe Island, and from the Atlantic to the Pacific Ocean. The Mapuche territories (Wall Mapu) made up a large part of what is now the states of Chile and Argentina (Ruiz, 2008). Today, 81.6 per cent of the Mapuche live in urban areas (CASEN, 2015). The language of the Mapuche is Mapudungun, which is an oral language transferred from generation to generation.

Among Mapuche, there is a concern about the loss of Mapuche culture, due to several factors, such as migration, the insertion of Mapuche people into Chilean society and discrimination. According to Antileo (2007), after the occupation of the Araucanía in the 19th century, the Mapuche people were plunged into poverty. The dramatic loss of their land to the Chilean state led to a wave of migration to urban areas. Mapuche have continued to migrate to the capital and other cities up to the present day in search of better opportunities. According to Chernard (2006), when Mapuche people arrive in cities, they frequently hide their indigenous identities, in order to integrate into Chilean society and because Mapuche values and practices are less appreciated in urban areas compared with rural areas. Most of my research participants migrated to cities as children with their parents and suffered discrimination in schools and then in employment. For that reason, they decided not to teach their children about Mapuche culture and language to protect them from discrimination. They also wanted their children to gain a good education that included university study. Consequently, Mapuche people started to attend schools and universities that did not embrace intercultural perspectives and did not include indigenous knowledges in the curriculum. One of my participants described this experience as follows:

> I think that fewer [Mapuche] people are living in the countryside, because maybe they have a profession or choose other things. They see that there are many possibilities [...] even myself, I do not go, I do not participate in We Tripantu [Mapuche new year], so I think, how many people are participating in these things? So, I believe that is decreasing. There are few people who speak Mapudungun [Mapuche language]. For example, before you used to see plenty of women with their Mapuche clothes walking along the road, now I hardly ever see that. It is interesting because when they do wear their traditional clothes, people say "oh! Those are Mapuche clothes!" And yet that should be natural, we are in the Araucanía region [where Mapuche people are originary], it should be normal to see people wearing Mapuche clothes, because there are a big percentage of Mapuche, you know.
>
> *(Angelica, Designer, 25)*

Stavenhagen (2001) explains that there are three kinds of racism present in Latin America: legal, personal and institutional. The first refers to legislation and rules that legitimize unequal treatment and so position some populations at a disadvantage. For example, according to Oyarce *et al.* (2012), the state has responded to indigenous political movements and the re-occupation of Mapuche land with anti-terrorist legislation from the dictatorship period (1973–1989), militarizing Mapuche land and persecuting indigenous leaders. This state of affairs persists despite denunciations from human rights organizations. The second refers to a kind of racism which is related to stereotypes, prejudices and individual or collective preferences that enable more privileged groups to feel a sense of superiority over racialized groups. For example, public discourse of tolerance is accompanied by everyday discourses and practices that understand Mapuche as primitive, ignorant, cognitively retarded, lazy, drunken and as an obstacle to progress (Merino, 2007). The third kind of racism is institutional and involves the differential treatment of ethnic minorities within public and private institutions. According to Matthew and Jeffrey (2015: 860) institutional racism refers to "particular and general instances of racial discrimination, inequality, exploitation, and domination in organizational or institutional contexts, such as the labour market or the nation-state". Universities can therefore be challenging spaces for people who belong to groups considered as "different" or "inferior". Stavenhagen (2001) suggests that in Latin America the manifestation of institutional racism could be subtler than in other countries; nonetheless, racist practices are still sanctioned. The loss of the Mapuche culture is a consequence of the three types of racism that indigenous people have to face in Chile. This chapter focuses on both personal and institutional forms of racism and specifically on those that are produced in the university context.

In the case of Chile, while universities do not deny entrance to indigenous people, there are structural conditions that make university study difficult. For example, the best universities in Chile are situated in Santiago; therefore, people who live in the south of Chile have to migrate to obtain this better level of education. Twenty-five of my 40 research participants were from rural areas in the south of Chile and had to move to the nearest city or the capital to study. Another barrier is cost, as students must either pay university fees or seek a loan or a scholarship to cover tuition and living costs. According to CASEN (2015), 38 per cent of Mapuche people live below the poverty line; therefore, it is more difficult for them to access university education, due to income difference that indigenous people have compared to non-indigenous people. Consequently, Mapuche are three times less likely to complete higher education than the non-indigenous population (INE, 2012).

Furthermore, it is important to recognize that most of the participants have suffered racial and class discrimination since childhood; for their skin colour, their Mapuche last name, their way of speaking or their socioeconomic background. One of the participants explained how his personal characteristics contributed to an experience of segregation:

> I am from a rural school; here there is not a distinction between a Mapuche and another type of person, whereas in the city there is a distinction. As I had good marks I was placed in an elite class [...] so everything was different. I arrived from the countryside, shyer, quieter, and my classmates segregated you sometimes. Or maybe they told me things like "that 'negrito' (black) who is there" or things like that [...] there I felt discrimination from my classmates [...] when the teacher took attendance, I was the only one with a different surname, which sounds different with an "ao" [sound], "ao" that was mine! Everyone has a last name which finishes in "er" or "ir" [...] I don't know [...] like normal. And my surname was the only one that sounded different, so then I felt different.
>
> *(Eduardo, Secondary teacher, 30)*

Following Stavenhagen (2001), indigenous people in Latin America have had to deal with disadvantageous positions for centuries and racist practices have become normalized in many contexts. One of respondents provided his own explanation of racism in Chile:

> Racism is a historic construction that is building while the historical processes are changing. In the 19th century, the "indio" was savage, in the 20th century you were lazy and drunk. Now the "indio" is a terrorist. Those are stereotypes typical of a society that does not have a culture of rights. Chilean society has a type of racism which is not open and is a very conservative society, in political and cultural terms. A society where there are all types of discrimination, including the popular society which is the most racist. People

> discriminate for being fat, for everything [...] Chilean society discriminates
> as a whole.
>
> *(Javier, historian 31)*

Javier explained how racism is a structural issue in Chilean society, but notes that
the forms of discrimination exacted against indigenous people have the ability
to change and manifest in different ways. Racism in Chile is a deeply rooted
issue that can be explicit or implicit, according to the dynamic of power relations. After the end of the dictatorship in 1990, the Chilean government started
to implement new policies and initiatives towards indigenous peoples, including
the creation of indigenous law N° 19.253 (1993) and the formation of National
Indigenous Development Corporation CONADI (1993). CONADI works to
develop policies to improve indigenous peoples' access to quality education. For
example, the government created an indigenous scholarship scheme in 1992, to
enable indigenous people with good marks and a vulnerable economic situation
to stay in the education system (Benavente and Álvarez, 2012). Antileo (2012)
explained that the transformation of the policies towards indigenous people in
the 20th century could be interpreted as a shift in power relations between the
"dominant society" and the "other". Indeed, the context has moved from territorial reduction policies to assimilation policies during the dictatorship to the
current valuation of ethnic diversity. However, Antileo (2012: 11–12, my translation) suggests that

> the new multicultural indigenous policies, developed after dictatorship period
> (1990) would constitute a disguise of openness to the diverse in the present
> times. However, here, we understand that as the representation of a colonial-
> ism historically transformed, that shapes the representation of "other", in the
> terms and limitations established by the colonizer.

In that sense, the government pursues interests of social control and domination
that can be interpreted as the continuity of a colonial perspective.

According to Fajardo and Ramirez (2012), by 2011 around 20,000 students
possessed indigenous higher education scholarships. While these scholarships
were hugely important, statistics for 2009 show that only 18.6 per cent of indigenous people managed to attend university compared with 29.9 per cent of non-
indigenous people in the same year (Blanco and Meneses, 2011). While it is clear
that access to higher education is fundamental for social mobility and employment
opportunities (Navarrete, Candia and Puchi, 2013), a study on intergenerational
social mobility according to ethnic origin in the Araucanía region by Cantero
and Williamson (2009) shows that Mapuche people have less social mobility than
non-Mapuche people. Moreover, they note that the entry to Higher Education by
indigenous people was a recent event and is produced under vulnerable social conditions and discrimination. That is the situation of participants in my research and
most of them acknowledged that they belonged to a very small group of Mapuche

people able to attend university. Armando described what it meant for him to be the only Mapuche person on his course and recognized as Mapuche:

> I mean, there were no people with Mapuche last names, but there were a few people with the physical appearance (of Mapuche). But there was nobody that claimed a Mapuche identity. I did, but I was not wearing a t-shirt saying I am Mapuche […] in that period, in fact I thought that I was the first Mapuche lawyer, then I knew there were more people […] but I thought that I was the only one […] I always felt exotic and that grabbed the attention of my classmates.
>
> *(Armando, Lawyer, 34)*

Here Armando introduces the experience of being part of a social context where Mapuche and indigenous people in general did not belong before, where he was considered the exotic person on the course, without any support of others in the same situation.

Despite the barriers to higher education, some indigenous people have managed to overcome the structural obstacles and go on to experience a degree of educational upward mobility that serves to produce social diversity among Mapuche (see Fajardo and Ramirez, 2012). This group has, however, endured social disadvantage and discrimination that existing social policies have failed to address.

Experiences of agency and resistance of Mapuche people at university and work

I turn now to show how respondents developed their process of re-signification of their Mapuche identities, and how this process contributed to their agency. An emerging group from the sample (18 of 40) was aware of and concerned about the loss of Mapuche heritage, and some of them are taking action to recover the Mapuche culture and the language. Most of the participants from this group are from Santiago, did not have to migrate to attend university, and knew very little about Mapuche culture and language, because their parents did not teach them about them. Respondents from this group identify as Mapuche people due to their Mapuche surname, however they did not have any other links with the Mapuche culture at an early age. That situation changed when they went to university, where they became aware of Mapuche culture and began to learn about it. This is important, because it was during university and later at work that they started to become closer to the Mapuche culture and began to identify as Mapuche. They started to engage in a number of decolonizing practices; starting to learn their native language, coming back to live in the rural area, be part of Mapuche groups and focusing their jobs to help Mapuche people. In the process, indigenous identities are redefined, and the redefinition process calls both capitalism and coloniality into question (see Canales and Rea, 2013). For Loreta, as for most of the respondents, the

Mapuche culture was something unfamiliar. The next quote shows her family resistance to learning about Mapuche culture:

> No, nobody teaches me, nothing, I mean nothing. My grandfather neither. He was very quiet, he never told us, neither their daughters […] now I am talking about that with my family and it is like that I am collecting piece by piece and putting them together […] Also it is complicated for me that my mother now is not accessible to talk about the issue. I am participating in a community [in this case it means a group of Mapuche people who partici-pate together in learning and in practising ceremonies, but in an urban space] […] One day I told her that I was going to dress as Mapuche and she almost died, it was horrible. We had a big fight […] she told me that she never taught me about these practices, "I never taught you that, so why you are doing it" […] I do not care that my grandfather did not teach her about the culture, but I am upset that I cannot talk about the subject, it is like I always have to censor myself with that […] At the beginning was fine, but it is like that the closer I am to Mapuche culture, the more tensions there are.
>
> *(Loreta, linguist, 23)*

The participants faced resistance from their families when they tried to participate in Mapuche culture, during or after their studies in higher education. It is possible to see how difficult it was for Loreta to manage her relationship with her family when she started to be interested in the Mapuche culture. She had to manage her dilemma between her old social context related to working-class and unknown Mapuche culture, and her new social context related to middle-class and her new Mapuche identity. Following Hall (1996) in this case, Loreta was dealing with her multiple identities and multiple contexts. She attempted to avoid showing her new Mapuche identity to her parents, in order to avoid creating more conflict between her family. Despite the fact that respondents encountered resistance from their parents, they often persisted in learning about Mapuche culture. Take the case of Angelica, who is from Temuco and started to learn more about her Mapuche family and the language, in order to fit in with the traditional Mapuche identity:

> I think at university I realise about my Mapuche identity, because I started to […] I always hear that people talked very bad things about Mapuche people, so I started to have interest, why did that happen? Or the language, why is it being lost? Why don't people speak the language? Why is it so discredited? You know. Therefore, that made me become closer, and realise that I am Mapuche, that it is not only a surname, and I can contribute to keeping the culture and learning about it.
>
> *(Angelica, clothes designer, 25)*

The last quote shows how respondents started to cope with Mapuche identities, learning to adapt and engage in Mapuche cultural practices, including speaking the

Mapuche language, spending time living in Mapuche land, participating in diverse Mapuche groups and learning about the culture in their lives, in order to legitimate their Mapuche identities. I argue that the process of identification with Mapuche culture involved other dimensions, such as class and gender, resulting in a complex process. That is the case of Ana, who shows that her process of identification as a Mapuche person can be understood as a process of gaining independence:

> I met a guy who is Mapuche and he was from a Mapuche community, but he lived in Santiago. I was deluding myself with that guy, but he moved to Belgium and I cried a lot. Therefore, I had to do my fieldwork, so I took my bag and I went to a Mapuche community and I said, "to be Mapuche I do not need a man" I am Mapuche, so that was my first statement. After that, I did not care what people say, if someone told me I am not Mapuche, I did not care, but it was a process. However, I felt confident when I knew about the culture, when I knew the cultural codes. When I was living in the countryside, I saw how living in the rural area is. I slept where people slept, I ate what people ate there. Therefore, I do not feel that is a historical situation, maybe if my family would have lived in the south, maybe those things would not have happened, but that is I had to live and that cannot determinate if I am Mapuche or not. Therefore, that was the process that now I can say I am Mapuche.
>
> *(Ana, Anthropologist, 33)*

The last quote shows the process that respondents go through in order to identify themselves as Mapuche people. Ana had to develop a set of strategies to "fit in" with this new social context. Ana practised and lived as a "traditional Mapuche" person, which meant living in the rural area and following all the rituals and practices. Her narrative is contradictory, because she said she did not care about what people said about her Mapuche identity; however, she still wanted to be accepted. This process of identification is not the same for all respondents from this group, but reveals how complex and contradictory the process often is. We can see, following Stuart Hall (1996), that Ana accepted the notion of Mapucheness that others had. In that sense, she acknowledges the practices and custom and started to accommodate and learn the language, the rituals and living in a rural area.

It is necessary to clarify that among respondents there are people who have Mapuche ancestry but have become alienated from Mapuche culture, while others had both ancestry and close ties to Mapuche culture. Mapuche who have not lost these ties with their group have the Mapuche surname (Zapata and Oliva, 2011). They have a social commitment to this group; they openly claim their Mapuche heritage and work for the common benefit of Mapuche society. This distinction is important, because some of participants started university recognizing their Mapuche lineage. But over time, their identities shifted, and they became much more identified with Mapuche culture at university, a shift that results from a growing awareness of the need to reduce the discrimination towards Mapuche people within Chilean society. Take the case of Loreta, who reflected on her social compromise with Mapuche people:

> I feel that the part of cosmovision is important but also the political part is important, and feel that in the lof [many Mapuche communities] that is blurred, and we are far away from the politicized way. It is like everyone live their own process of strengthening our identities, but we are not involved in the political side. The fact that some are going to protest and others not. I would like to have more developed that political part. Like having clear how can I participate actively, but I am still looking how can I do it […] I still do not know the language, and I would like to know and help and I think that the language will revitalize […] I feel that when I will commit with Mapudungun, then I will have a clearer attitude.
>
> *(Loreta, linguist, 24)*

The above quotation points out the need to focus on the more politicized dimensions of Mapuche identity. It seems that respondents have to reach a certain level of knowledge regarding Mapuche culture, in order to have a more politicized point of view. This process of politicitization led to a feeling of responsibility to adequately represent the Mapuche culture in other social spaces as well to a need to have the right kinds of knowledge in order to legitimize their identities to other Mapuche people. They have to form their opinion and adopt a more politicized attitude towards Mapuche culture; specifically, they feel that they have to *know* about Mapuche culture. Moreover, they have the responsibility to demonstrate that they are different, that they are professionals and to show people that the stereotype of the Mapuche terrorist is wrong. Ana points out how the idea of a new generation of professional Mapuche people could help to solve their political problems:

> Here we live a strong claim with the Mapuche topic. I think it's fine. I think it's good that this process comes as revitalization. I like that there are spaces, in fact I would like more. I would like it if many Mapuche could know about their family or other people could live the same the process that I did; they could know where their family are, their context. I have my grandfather's land title […] when I talk to someone, I always say that they have to learn to listen, they [Mapuche people] will challenge you first, but then they will teach you.
>
> *(Ana, anthropologist, 34)*

Here, through respondents' narratives, it is possible to see how empowering it can be to embrace a politicized Mapuche identity and how gaining a university education facilitates that process. Despite needing to overcome social barriers of race and class in order to succeed, some Mapuche students are able to represent Mapuche culture in ways that challenge dominant social perceptions in Chile.

Conclusion

The goal of this chapter was to examine the process of agency and resistance of Mapuche people who attended university. It was demonstrated that higher education in Chile is the result of a colonial system that reworks hierarchical power relations and leads to harmful forms of exclusion. The integration of Mapuche students into a Eurocentric education system undermines indigenous values and cultural identities and frequently results in discrimination. Despite these negative outcomes, the process of gaining a university education strengthened and politicized their Mapuche identities. It enabled some Mapuche students to gain an insight into the internalization of colonialism, and the structural factors that placed Mapuche in a situation of social, economic and political disadvantage. So going to university in spite of its challenges does give Mapuche access to a cultural capital that previously they did not have. Rather than embrace the Eurocentric ideas circulating in their universities, they instead began to reinforce their Mapuche identities, in ways that start to tackle racialized structural disadvantage in Chile.

Note

1 This study was funded by CONICYT "Becas Chile" scholarship programme. It also counted on support from the Centre for Social Conflict and Cohesion Studies (COES-Chile) and the Centre of Intercultural and Indigenous studies (CIIR), Pontificia Universidad Católica de Chile.

References

Antileo, E. (2007) Mapuche y Santiaguino: El movimiento Mapuche en torno al dilema de la urbanidad. *Working paper serie 29, Ñuke Mapu Förlaget.*

Antileo, E. (2012) Nuevas formas de colonialismo: diáspora mapuche y el discurso de la multiculturalidad. Available at: www.repositorio.uchile.cl/handle/2250/112920 (Accessed 3 December 2017).

Benavente, J. and Álvarez, P. (2012) Evaluación de impacto de las becas de educación superior de Mineduc. Facultad Economía y Negocios, Universidad de Chile.

Blanco, C. and Meneses, F. (2011) Estudiantes indígenas y educación superior en Chile: acceso y beneficio. In Inclusión Social, Interculturalidad y Equidad en Educación Superior. *Seminario Internacional Inclusión Social y Equidad en la Educación Superior. 2° Encuentro Interuniversitario de Educación Superior*, Temuco, 29 September – 1 October 2010. Fundación Equitas, Chile.

Canales, P. and Rea, C. (eds) (2013) Claro de luz, descolonización e "intelectualidades indígenas" en Abya Yala, siglos XX-XXI. Santiago: Instituto de Estudios Avanzados de La Universidad de Santiago De Chile.

Cantero, V. and Williamson, G. (2009) Intergenerational Social Mobility by ethnicity: Empirical Evidence Araucanía Region, Chile. *Revista Universum* 24(1).

CASEN survey (2015) *Indicadores de pobreza: Encuesta de caracterización socioeconómica Nacional Gobierno de Chile.* Santiago: Ministerio de Desarrollo Social.

Chernard, A. (2006) La identidad mapuche en el medio urbano. *Meli Wixan Mapu Publicaciones.* Available at: http://meli.mapuches.org/spip.php?article177 (Accessed 3 December 2017).

Fajardo,V. and Ramirez, E. (2012) Proceso de formación universitaria de estudiantes mapuche de la universidad católica de Temuco y sus expectativas de inserción en sus territorios de origen. *Documento de Trabajo N° 1, Serie Estudios Territoriales: Proyecto Jóvenes y educación Superior para Territorios Mapuche.* Rimisp, Santiago, Chile.

Hall, S. (1996) Old and new identities, old and new ethnicities. In A.D. King (ed) *Globalisation and the World System.* London: Macmillan Educational, pp. 42–68.

INE (2012) Census of population and housing. Santiago: Statistics National Institute.

Matthew, C. and Jeffrey, D. (2015) Sociology of racism. In J.D.Wright (ed) *The International Encyclopedia of the Social and Behavioral Sciences* 19: 857–863.

Merino, M. (2007) El discurso de la discriminación percibida en Mapuches de Chile. *Discurso y Sociedad* 1(4): 604–622.

Navarrete, S., Candia, R. and Puchi, R. (2013) Factores asociados a la deserción/retención de los estudiantes mapuche de la Universidad de la Frontera e incidencia de los programas de apoyo académico. *Calidad en la educación* (39): 43–80.

Oyarce, A.M., Carvone Quiepul, M., Melin Pehuen, M., Malva-Marina, P. and Coliqueo Collipal, P. (2012) *Desigualdades territoriales y exclusión social del pueblo mapuche en Chile: Situación en la comuna de Ercilla desde un enfoque de derechos.* Santiago: Economic Commission for Latin America and the Caribbean (ECLAC) and Alianza Territorial Mapuche (ATM).

Ruiz, C. (2008) *Síntesis histórica del pueblo Mapuche (Siglo XVI-XX).* Santiago: USACH.

Stavenhagen, R. (2001) *El derecho de sobrevivencia: la lucha de los pueblos indígenas en América Latina contra el racismo y la discriminación.* Santiago: (CEPAL) Comisión Económica para América Latina y el Caribe and (IIDH) Instituto Interamericano de Derechos Humanos.

Yopo Díaz, M. (2012) Políticas sociales y pueblos indígenas en Chile: Aproximación crítica desde la noción de agencia. *Universum* (Talca) 27(2): 187–208.

Zapata, C. and Oliva, M. (2011) Experiencia de inserción e impacto institucional de los becarios del Programa Internacional de Becas de la Fundación Ford con ascendencia y adscripción indígena en la Universidad de Chile. In P. Díaz (ed) *Pueblos indígenas y Educación Superior en América Latina: Experiencias, tensiones y desafíos.* Santiago de Chile: Equitas, pp. 43–71.

11

LEARNING FROM MAYAN FEMINISTS' INTERPRETATIONS OF *BUEN VIVIR*

Johanna Bergström

Introduction

Scholars such as Arturo Escobar (2010) and Walter Mignolo (2016) have emphasized the importance of indigenous perspectives and the idea of *buen vivir* in contemporary Latin American politics. *Buen vivir* can be seen as an initiative to decolonize the knowledge system of the western(ized) capitalist linear development model and embark on a philosophical and theoretical engagement with sociocultural and environmental issues. I believe that non-indigenous feminist activists and academics cannot afford to ignore this initiative around which many organized indigenous women in Latin America are mobilizing. In other words, it is essential that white non-racialized researchers in western(ized) universities learn from and value indigenous feminist perspectives. In this chapter, I discuss the concept of *buen vivir* through a feminist lens and build upon Ackerly and True's (2010: 464) understanding that "… the future of feminist research lies in ever closer cross-disciplinary and global collaborations". From a decolonial perspective, I see the need to include indigenous initiatives in order to enrich feminist research and deepen its understanding of indigenous people's relationship with nature. By nature, however, it is important to stress that I am not referring to Western models of thought rooted in Eurocentric modernity that separate nature from culture, but rather to indigenous epistemological formations for which such separations make no sense. Inclusion of non-Western perspectives, such as the one based on *buen vivir* presented in this chapter, and engaging in intercultural dialogues about the process of knowledge production, can be one way of unsettling Eurocentrism in the westernized university.

My interest in *buen vivir* started in 2011 during my doctoral fieldwork in Guatemala, when I was researching the ways in which international trade and sustainable development are gendered. In Guatemala, I spoke with representatives from

Guatemalan state institutions, European Union (EU) institutions, private sector interest groups, as well as with people from a number of national, regional and local non-governmental organizations, such as women's organizations, peasant organizations, and co-operatives, about their views on sustainable development. The understanding of the term, as well as the interest in the topic, varied greatly among the different respondents. Something that I found especially interesting, however, was that representatives of indigenous organizations were reluctant to use the term development, and articulated scepticism about the usefulness of what they understood to be a Western concept that does not allow us to think outside the logic of economic growth. Instead, these respondents wanted to talk about *buen vivir*. Already before my fieldwork I had felt uncomfortable with sustainable development as a concept, but these encounters in Guatemala made me reflect further upon the transferability of Western understandings of sustainability to non-Western contexts. This experience taught me a lot, and in retrospect I am grateful that the indigenous representatives did not want to talk about sustainable development with me, since this encouraged me to engage with a different perspective.

This chapter reflects upon what Western feminisms could learn from indigenous feminists in Latin America, using *buen vivir* as an example. Additionally, the chapter intends to bridge the gap between western(ized) and Latin American feminist debates and discussions by contextualizing the Mayan feminists' arguments in a regional as well as global context. The chapter presents what I as a Western feminist understand by the concept of *buen vivir* (good living), the way it is expressed in Guatemala and the way Mayan feminists view the concept. Using decolonial theory, the chapter also reflects upon what these arguments can contribute to the broader frameworks of feminist theory. I attempt to achieve this by doing three things. First, I explain the concept of *buen vivir* and examine some of the ways people conceptualize and practise it in the Andean region and in Guatemala. Second, I introduce the Western concept of sustainable development in order to then compare and contrast this concept with that of *buen vivir*. Third, I analyse what Mayan feminist perspectives contribute to the debates on *buen vivir* and what Western feminisms could learn from these contributions. Before examining the concept of *buen vivir* however, the following section outlines some of the relevant key theoretical concepts within the decolonial theoretical framework.

Analytical entry points

Colonialism generally focuses on politics, economics and military, whereas coloniality also includes epistemologies and ontologies (García Pacheco and Lazarte, 2012: 121) in its understanding of the world. Moreover, coloniality helps us understand the continuation of colonial practices of domination after the end of colonial administrations (Grosfoguel, 2009). Coloniality is an order of global power that has classified populations, their knowledge and their cosmologies hierarchically, based on a European standard (Quijano, 2000). One important contribution within decolonial theory is Aníbal Quijano's concept of the "coloniality of power", which developed hand in hand with the "cultural complex known as European moder-nity/rationality" (Quijano, 2007: 170). "Coloniality of power" constitutes a struc-ture that operates through control over labour, authority, sexuality and subjectivity/ inter-subjectivity (Quijano, 2000); in other words, production and exploitation, political administration, gender relations and reproduction, and dehumanization that attempts "… to turn the colonized [people] into less than human beings" (Lugones, 2010: 745). María Lugones (2008) uses the concept of "coloniality of power" as a starting point, but argues for a wider understanding of "gender" than Quijano does. Thus, she believes that labour (and not only control over sex, its resources and products) is also racialized and gendered (Lugones, 2008). Lugones (2008: 12) explains the link between racialization and gendering, arguing that "the colonial, modern, gender system cannot exist without the coloniality of power, since the classification of the population in terms of race is a necessary condition of its possibility". From a "coloniality of gender" perspective, colonization moreo-ver transformed the indigenous sense of gender relations, while the modernity/ coloniality nexus disseminated a Eurocentric view of gender and sex (Lugones, 2008). "Coloniality of gender" has become an important concept within decolonial feminism. There are three key topics within this strand of feminism: first, women's experiences of colonialism and coloniality; second, racism and classism and the reproduction of these oppressions outside as well as within feminist movements; and finally, men's power over women within indigenous and Afrodescendant com-munities (Gómez Correal, 2014).

As mentioned earlier, the concept of coloniality also emphasizes the assertion of Eurocentric epistemologies that filter what knowledge is accepted in a certain historical and cultural context and what knowledge is not. Boaventura de Sousa Santos (2007: 33) usefully describes this process as a "system of visible and invisible distinctions, the invisible ones being the foundation of the visible ones". The so-called invisible distinctions are separated by the "abyssal division" into two realms: the realm of "this side of the line" and the realm of the "other side of the line". Scientific knowledge originally aimed at converting "this side of the line" into the subject of knowledge and the "other side of the line" into an object of knowledge. Knowledge produced on "the other side of the line" exists outside of the realm of what the recognized conception of inclusion is, and is therefore excluded and con-sidered nonexistent. The implication of this invisible line is a substantial epistemi-cide that has led to the denial of cognitive experiences (De Sousa Santos, 2007: 13).

The invisible distinctions and the resulting epistemicide have implications for environmental governance and policy, since different forms of knowledge are linked to different relationships with nature. I agree with Ulrich Brand (2010: 134), who argues that "knowledge about nature is not neutral and is intensively linked to the appropriation of nature". This leads us to the question as to which forms of knowledge are diffused and become dominant or hegemonic as "universal", and which knowledge is seen as peripheral or even marginal and therefore "local", ignored and forgotten (Singer, cited in Brand, 2010: 131). Drawing on Gayatri Spivak's notion of "sanctioned ignorance", Rauna Kuokkanen (2008) frames this problem of marginalization as "epistemic ignorance". For Kuokkanen (2008: 60), epistemic ignorance refers to "academic practices and discourses that enable the continued exclusion of other than dominant Western epistemic and intellectual traditions". The consequence of this ignorance is that western(ized)academia does not hear nor understand indigenous people who speak from the basis of their own "epistemic conventions" (Kuokkanen, 2008: 60).

While Western science has made a valuable contribution to the production of knowledge in a number of arenas, there are many epistemic and intellectual projects in the world today in which modern science plays no role. The preservation of biodiversity made possible due to indigenous and rural knowledge is a good example of such an intervention. The irony here is that these forms of knowledge, so crucial in today's world, "paradoxically, are under threat because of increasing science-ridden interventions" (De Sousa Santos, Nunes and Meneses, 2007, in De Sousa Santos, 2007: 14). The relationships that indigenous peoples have developed with nature are a consequence of living off, in and with the land and depending on its abundance. The underlining understanding in this relationship is that, '... the well-being of the land is also the well-being of human beings" (Kuokkanen, 2007: 42) and this is something that the concept of *buen vivir* encompasses.

Interpreting the concept of *buen vivir*

The common English translation of *buen vivir* is good living, but according to Eduardo Gudynas (2011: 441) the term is difficult to translate into English, since it "includes the classical ideas of quality of life, but with the specific idea that well-being is only possible within a community". Furthermore, in most approaches the community concept is understood in an expanded sense, to include nature. Here nature is interpreted as the constitutive conditions and practices – sociocultural, territorial, spiritual, ancestral, ethical, epistemic and aesthetic – of life itself (Walsh, 2010: 18). The difficulty with the translation of *buen vivir* is related to the inability of Western academics to understand indigenous people when they speak from their own "epistemic conventions", as Kuokkanen (2008) has noted, since Western modernity has taught us to separate culture from nature and to disconnect ourselves from the latter.

Buen convivir and *buen vivir* are Spanish translations of *Suma Qamaña* (Aymara) and *Sumac Kawsay* (Kichwa) from indigenous Andean languages. *Buen convivir* and *buen vivir* play important roles in the Bolivian and Ecuadorian constitutions.[1] Article 71 of the Ecuadorian constitution illustrates this: "Nature or Pachamama, where life is realized and reproduced, has the right to integral respect of its existence and preservation as well as its vital cycles, structures, functions and evolutionary processes" (Ecuador, 2008: 52). Rather than treating nature as property under the law, this article acknowledges that nature has rights. In addition, we – the people – have the legal authority to enforce these rights on behalf of ecosystems. The ecosystems themselves can be named as the defendant and this constitution subsequently becomes less anthropocentric.

Even though this chapter does not examine policy implementation in Bolivia and Ecuador, it is relevant to mention that regardless of the inclusion of the *buen vivir* philosophy in the Bolivian and Ecuadorian constitutions, it has not been easy to enact these ideas in practice. In practice, neither existing mining legislation nor agricultural policies are in line with *buen vivir* (Acosta, 2009; Walsh, 2010). It can be very difficult to articulate economics, culture, the environment and society in these non-Western ways that promote social as well as intergenerational justice and give rights to nature. As Catherine Walsh writes (2010: 17–19), *buen vivir* and development are presented as interchangeable in the Ecuadorian National Plan for development from 2009–2013. The basic problem with the implementation of *buen vivir* is that it is done within the framework of development, even though it has its roots in conceptual categories, indigenous languages and cosmovisions that do not entail the idea of development itself. While policy implementations might be fraught with complexities, the mobilization of *buen vivir* as a concept proves to be politically, culturally and epistemically valuable.

Communalities across different cultural and ecological landscapes

While acknowledging that indigenous peoples are not a homogeneous group and that their cultures, histories as well as socio-economic conditions vary greatly, I agree with Kuokkanen (2008: 65) when she maintains that, "underpinning these differences is a set of shared and common perceptions and conceptions of the world related to ways of life, cultural and social practices and discourses that foreground and necessitate an intimate relationship with the natural environment". This relationship is considered one of the central features of indigeneity which the ideas of *buen vivir* illustrate. Although *Suma Qamaña* and *Sumac Kawsay* have been influential in academic debates about development, ideas of *buen vivir* are present in many indigenous cultures and languages throughout *Abya Yala*.[2] For example, in the Guaraní language, we find concepts such as harmonious living (*Ñandereko*), the land without evil (*Ivi Maraei*), the path to noble life (*qhapaj ñan*) and good life (*Teko Kavi*). In the Mapuche language in Chile there is also a concept of harmonious living; namely, *Küme Mongen* (Gudynas, 2011: 443). These ideas are also expressed in Mayan languages; for example, the K'iche word *Utz K'aslemal* (Macleod, 2011: 249) and the Tseltal word Lekil Kuxlejal (Ávila, 2011: 488) can both be translated as *buen vivir*. The focus on interconnection and balance between humans and earth that we find in *buen vivir* philosophy can also be found in other indigenous cosmovisions outside of Latin America. Other examples are central Australian indigenous populations, such as the Aranda and Walpiri peoples, who have a concept called *jukurrpa* (dreaming) with similar characteristics as *buen vivir*. An all-encompassing principle in *jukurrpa* links flora, fauna, natural phenomena and people into one gigantic interfunctioning world (Bell, 2002: 91). Susan Hawthorne, who is inspired by *jukurrpa*, links feminist, ecological and indigenous thinking and calls it wild economics. In the same way as indigenous cosmovisions and the *buen vivir* philosophy varies in different cultural and ecological landscapes, "the wild is also multiple and can only be understood in relation to the things and processes around it. So, context and relationship are critical to understanding the wild" (Hawthorne, 2009: 99). Alberto Acosta (2016: 138) argues that *buen vivir* as a culture of life has existed and is practised in different parts of the world and suggests that we may include the word *Ubuntu* from Southern Africa and the word *Swaraj* from India.

Returning our focus to the concept of *buen vivir*, Gudynas (2011: 445) argues that it "is more than a simple coexistence or juxtaposition of different cultures, because they interact in dialogue and praxis focused on promoting alternatives to development". Although the different expressions of *buen vivir* are specific to cultural and ecological landscapes, there is a core set of ideas that unify them. First, *buen vivir* is a platform where people share critical outlooks on development, and where the focus is a "replacement of the very idea of development" (Gudynas, 2011: 445). This idea of an alternative to development, as supposed to yet another development model, is a standpoint that several of the organized Mayan civil society actors expressed when I interviewed them (Bocel, 25/09/2012; CONAVI-GUA, 21/09/2012). Second, all the approaches within the philosophy of *buen vivir* react against the way economic values dominate all aspects of life and the resulting commodification of nearly everything. Moreover, the concept of *buen vivir* recognizes that there are various ways to give value to something, including cultural, historical, aesthetic, spiritual and environmental value (Gudynas, 2011: 445). Finally, nature is seen as constitutive of social life (Escobar, 2010: 23) and becomes a subject (Gudynas, 2011: 445).

It is worth mentioning that this view of the economy, and more specifically value, is not completely divorced from some Western critiques of capitalism. Rather, there are linkages between the ideas that spring from indigenous cosmovisions and critical positions within modernity (Gudynas, 2011: 445). Gudynas mentions three examples of Western critique that resonate with indigenous knowledges. These are the concept of post-development developed by Arturo Escobar and others, radical environmentalism that opposes the anthropocentric perspective of modernity and feminist perspectives that question existing social hierarchies and the masculinist domination of nature. These critical perspectives within modernity that can be linked to *buen vivir* are more often critical of the concept of sustainable development (Beckerman, 2002; Redclift, 2005).

Sustainable development and *buen vivir*

One of the most famous and established interpretations of sustainable development is that of the Brundtland Report, which defines sustainable development as "development that meets the needs of the present without compromising the ability of future generations to meet their own needs" (WCED, 1987: 43). The report emphasizes the importance of poverty reduction in the move towards sustainable development, arguing that it is a "precondition for environmentally sound development" (WCED, 1987: 44). Here, the understanding is that economic growth is necessary to fight poverty and that poverty is a cause as well as an effect of environmental damage. In turn, economic growth is presented as a solution to eliminate poverty, with the objective of protecting the environment (Escobar, 1995: 13). The Brundtland Report has its origins in modern Eurocentric and liberal thought and as a result is rooted in a particular set of assumptions, including the belief in objective scientific knowledge, in economic growth as a driver of progress and in the notion that the economy is a "real" autonomous sphere with its own laws, separated from the political, the social and the cultural (see Escobar, 1995: 9).

The concept of sustainable development is, then, rooted in the western(ized) paradigm of progress and in the "common belief that development is not merely one option among many, but is the unavoidable ontological unfolding of universal history: progress, evolution, and development are all members of the same family" (Mignolo, 2016: 8). Within this paradigm the conquering of nature is equal to "development" (Quijano, 2007: 169). In contrast, *buen vivir* promotes the practice of living within nature rather than dominating it, while trying to protect it for future generations. As we can see, both sustainable development and *buen vivir* perceive it as a moral obligation to care for the wellbeing of future generations. Nevertheless, *buen vivir* is critical of a linear model of development and emphasizes complementarities, harmonious balance as well as rights to nature (Hernández, interview, 12/06/2012). Sustainable development, on the other hand, promotes linear development that includes economic, social and environmental sustainability. Juan Tiney at the indigenous peasant organization Coordinadora Nacional Indígena y Campesina (CONIC) noted that

> The sustainability that I am looking at is human sustainability, sustainability of nature, but if they [the EU commission and the US government] see it as economic sustainability, that is cheaper in this sense. Well, I think we have two ways of thinking, two completely different visions, we have two entirely different interests …
>
> *(Tiney, interview, 13/09/12)*

Moreover, social justice plays a more important role within *buen vivir* than it does with sustainable development. The phrase "to live well but not better" than other people, or at the cost of other people nor nature itself in *buen vivir* philosophy illustrates this importance. This phrase might be confusing in the English language though, since "better" and "well" have similar meanings (Thomson, 2011: 451). One consequence of the development paradigm is the idea that "having more is better and being fast and not wasting time is preferable to going slow, having time to think, be creative, and enjoy life" (Mignolo, 2016: 9).

The programme coordinator Petronila Bocel at the National Union of Guatemalan Women (UNAMG) expresses a critique of the Western development paradigm from an indigenous point of view in a clear way when she states that

> They [Western governments and businesses] sell us an understanding of development that we definitely do not share, since we have completely

different visions of the world. Here for example, from Mayan cosmovision
we talk about how to live in balance with everything that we produce with
Mother Earth, and not that humans are superior and have all the power. The
kind of development that "they" sell to us says that if you do not consume
this, if you do not do this, you are underdeveloped. Therefore, I very much
question it, when they call us underdeveloped countries or developing countries based on measuring parameters or indicators that they have established.
We will definitely never manage to advance beyond this logic that they have,
because it is not our logic ...

(Bocel, interview, 25/09/2012)

Here I see the ideas of *buen vivir* as a decolonial tool that can be used to delink
from the coloniality in which sustainable development is entangled. Delinking in
this context means a "de-colonial epistemic shift and brings to the foreground
other epistemologies, other principles of knowledge and understanding and, consequently, other economy, other politics, other ethics" (Mignolo, 2007: 453).

Buen vivir in the Mayan context

Spirituality is central to the Mayan cosmovision, and is therefore present in Mayan
interpretations of *buen vivir*. In this cosmovision it is not possible to disconnect
spirituality from other aspects of life and it forms part of the heritage from the
ancestors that should be passed on from generation to generation. One word that
exemplifies this in the Kaqchikel language is *Raxnaqil*, which can be translated as
happiness and physical, mental and spiritual health. It represents a perfect state of
harmony and balance in the private as well as community sphere, and we need to
keep in mind that these spheres vary from the westernized understanding of them.
Finding *Raxnaqil* requires an awareness of the relationship between people, nature
and the cosmos (Tedlock, 1992).

The Mayan cosmovision emphasizes the importance of (non-Eurocentric)
nature and seeks balance between people, Mother Earth and the universe. A Mayan
person is born with their *nawal* (their character or personality) that combines with
the energies of the other people in the community. Tito Medina (2000: 13, my
translation) writes, "The *nawal* is the main connection between the individual
and her/his own consciousness, between the individual and her/his surrounding,
between the individual and her/his relationship to cosmos". The specific day a person is born on the 260-day Mayan calendar determines a person's *nawal* (Tedlock,
1992: 466). There are 20 *nawals*, one for each day of the month in the calendar, and
these represent different energies and each relate to a different animal.

This association with an animal moreover illustrates the interconnection between
humans and nonhumans. The idea is to keep the energy balanced between the members of the community, with nature, the cosmos and the creator (Bolaños *et al.*, 2008:
303). These energies complement each other and work in pairs, such as masculine
and feminine, negative and positive, day and night, rain and sun as well as old and new.

Duality and complementarity

Macleod argues that the concepts of duality,[3] complementarities and balance can be difficult to understand for someone who is used to Western epistemology. Virginia Ajxup, a Mayan spiritual guide, a school-teacher focusing on intercultural education as well as an active member of various Mayan organizations, undertakes the task of translating and communicating these concepts from her mother tongue K'iche to Spanish. Duality can, according to Ajxup, be explained as

> the integration of two elements for the conservation and continuation of life [...] It is not the opposite, but rather the other me, this completely different me, pertinent, imperative, comprehensive. This principle of understanding is what makes a person complete. The [different] elements are not contrapositions or excluding, but rather necessary, they express themselves at all levels and in all spheres of natural and social life.
>
> *(Ajxup, cited in Macleod, 2011: 124)*

Ajxup understands complementarity as

> the process and understanding of integration to reach totality or fullness. No "being" and no action exists by itself, but always in coexistence with its specific complement, which completes the corresponding element. Complementarity is the total experience of reality. The ideal is not the extreme of one of the opposites, but the harmonic integration of the two, one unit, as a dynamic and reciprocal union. K'ulaj Tz'aqat is the principle of life, of organization. Two energies so different in their essence, nature. But they are necessary for life and existence.
>
> *(Ajxup, cited in Macleod, 2011: 124)*

Petronila Bocel, a UNAMG programme director, argues that by talking about gender equality we do not question an androcentric world in which the man is in the centre. Therefore, UNAMG maintains that men should accept women with the "differences that make us distinct from them" (Bocel, interview, 25/09/12). The CONIC representative Juan Tiney also focuses on the equilibrium and the participation of women, as supposed to gender equality. Women's participation is primarily valued in CONIC, because women represent 50 per cent of the organization's strength. Without women's participation CONIC would lose strength to fight against the enemy and to sustain the communities' own development (Tiney, interview, 13/09/12).

Feminism and *buen vivir*

The concept of feminism is still not well represented in the political discourses of organized indigenous women in Latin America, despite the fact that the struggle for equality between women and men has become central to their activism. Many indigenous women continue to associate the word feminist with urban liberal women. These presumptions are beginning to change however, and some indigenous women's groups in Mexico as well as in Guatemala have appropriated the term feminist (Hernández Castillo, 2010). As Aída Hernández Castillo (2014: 287) explains, women are confronting the indigenous movement, questioning the "dichotomy between tradition and modernity" that disguises inequalities between women and men within the indigenous communities. Likewise, these women also question the generalizations of what it means to be a woman within western(ized) liberal feminism. Much of organized indigenous women's struggle takes place around the theme of dignity and social justice. Social justice in the *buen vivir* context builds on a critique of economic growth as a goal, the importance of the local and the understanding of as well as respect for specificities of local cultural/ecological landscapes. This perspective encourages us to move away from the problems of epistemic ignorance (Kuokkanen, 2008), since it takes other than dominant epistemes into account. Except for dignity and social justice, organized indigenous women also highlight issues of environmental justice and the respect for Mother Earth.

The International Forum of Indigenous Women (FIMI) makes a strong link between women and nature:

> Indigenous traditions and Indigenous women themselves identify women with the Earth, and therefore perceive degradation of the Earth as violence against women. This conviction is more than a metaphorical allusion to Mother Earth. It is rooted in indigenous cultural and economic practices in which women both embody and protect the health and well-being of the ecosystems in which they live.
>
> *(FIMI, 2006: 16)*

The above statement could be seen as portraying women as inferior to men in a Western context, in which material and spiritual realms of life are dichotomized. This is not the case in indigenous cosmovisions however.

Interconnection and balance between humans and earth

As Macleod (2011: 2) points out, FIMI's identification with Mother Earth undermines the "Is Female to Male what Nature is to Culture" (Ortner, 1974) discussion, since indigenous cosmovisions do not consider being close to nature as inferior to being close to culture. Moreover the *buen vivir* philosophy argues for the interconnection and balance between humans and earth, rather than a relation based on domination as in modern western(ized) thinking. This is a view shared by some environmental feminists and ecofeminists.

For example, Mack-Canty (2004: 155) argues that the nature/culture dualism is, just like the private versus public divide, a "distinction central to Western ideas" that can be traced back to the Greeks. In this understanding of the world, man represents culture in the nature/culture dualism and the understanding is that humans need to take control over and dominate natural processes. When it comes to gender, masculinity is associated with disembodied characteristics such as order, freedom, light and reason. These characteristics are moreover seen as better than and opposite to the characteristics associated with femininity and women. In the nature/culture dualism, the features associated with women, such as physical necessity, passion, darkness and disorder, are perceived as "natural" and/or embodied (ibid.: 155). Ortner (1974: 87) stresses that the linking between women and nature "is a construct of culture rather than a fact of nature".

Early modern philosophers, like Hobbes and Locke, continued the association of women with nature and men with culture when developing classical liberalism (Mack-Canty, 2004: 155). These ideas were further expanded on with the development of science, which "broadened the concept of culture to include an even more enlarged notion of the human capacity to dominate nature" (ibid.: 155–156). Additionally, this period was marked by the rise of capitalism and the spreading of colonialism,

> in which the view of the colonised [...] begins to intersect more thoroughly with the perceived "otherness" of women and nature. Man's (i.e., white, Western, and middle- or upper-class man's) freedom and happiness [...] depended on an ongoing process of emancipation from nature, both human embodiment and the natural environment.
>
> *(Mack-Canty, 2004: 156)*

However, I agree with Agarwal (1992), who as part of a wider critique of ecofeminism as essentialism (see Biehl, 1991) and in line with post-colonial/decolonial feminist thinking, criticizes the idea of "woman" as a unitary category. As she explains,

> the processes of environmental degradation and appropriation of natural resources by a few have specific class-gender as well as locational implications [...] "Women" therefore cannot be positioned [...] as a unitary category, even within a country, let alone across the Third World or globally.
>
> *(Agarwal, 1992: 150)*

Moreover, Agarwal (1992: 151) stresses the importance of "women's lived material relationship with nature".

It is important to remember that some Mayan women live in cities but still identify as indigenous. One member in the Mayan feminist group "Kaqla" (Rainbow) in Guatemala states, "Yes, I am Indian, I come from the capital, and what! I am still indigenous" (Mujeres Mayas Kaqla, 2004: 52). Many indigenous women do live in rural areas, however, and therefore have a closer lived material relationship with nature than those who live in urban areas where nature is less predominant. One example of women with a close material relationship to nature are the women of the Maya Mam communities in San Miguel Ixtahuacán in San Marcos in Guatemala, who protest against open-pit gold-mining in their territory. These women see the mining as a violent act, since it destroys the environment, using enormous quantities of water and toxic chemicals. Moreover, hills that used to make up part of their collective identity no longer exist (Macleod, 2012: 9). This has significant consequences for people who highly value community and collective identity, and one board member of the National Coordination of Widows of Guatemala (Coordinadora Nacional de Viudas de Guatemala, CONAVIGUA) shared similar thoughts about her community with me.

The organized Mayan women (Bocel, interview, 25/09/12; CONAVIGUA, interview, 21/09/12), who shared their perspectives on sustainability and equality with me during my doctoral fieldwork in Guatemala, expressed a fight for dignity and respect as women within their communities and indigenous movement while at the same time emphasizing their struggle for territory and the respect for Mother Earth as indigenous people. These different forms of struggle are as Gómez Correal (2014) notes, key issues within decolonial feminism.

Conclusions

The alternative to development that organized indigenous women in Guatemala express and relate to – the concept of *buen vivir* – is something that western(ized) feminisms could engage with and learn from. This alternative includes social sustainability and women's dignity as well as respect for the planet based on the idea that we as humans make up part of nature, rather than being separated from it, as the nature/culture dualism within modernity suggests. It would enable us to rethink our relationship to a non-dualistic nature and develop a broader sense of what it means to live well that is not constrained by Eurocentric thought. We as Western feminists interested in the decolonization of academia can find indigenous women's voices in documents produced in their meetings and workshops, in interviews in feminist journals, in their own academic publications (Hernández Castillo, 2010: 544) and on YouTube and other social media sites. It is crucial that such publications, knowledges and perspectives are engaged with, cited and find their way onto university curricula. As Gómez Correal (2014: 366) writes, one of the main assets of *Abya Yala* is that of "human diversity, options, different visions of the world and alternatives to the hegemonic model". This asset offers a good basis for intercultural dialogue at a time when climate change is a growing problem and all of nature, including ourselves, is affected as a result. This chapter emphasizes the value of the concept of *buen vivir* as a decolonial tool that can be used to delink from the coloniality in which sustainable development is entangled. At the same time, indigenous feminists have made clear that not everybody interprets indigenous cosmovisions as compatible with feminist ideas and there are ongoing epistemological struggles underway in indigenous communities over questions of gender equality and complementarity and the role played by *buen vivir* in these debates over which there is substantial disagreement. Engagement with indigenous cosmovisions must also mean acknowledging their heterogeneity and dynamic character. But doing so can facilitate a non-hierarchical knowledge production that moves beyond the "abyssal divide" (De Sousa Santos, 2007) and through which we can make a small contribution to the unsettling of Eurocentrism in the westernized university.

Notes

1 English language versions of these constitutions can be found at www.constituteproject. org/constitution/Bolivia_2009.pdf and http://pdba.georgetown.edu/Constitutions/ Ecuador/english08.html.
2 AbyaYala means "land in its full maturity" in the Kuna language, spoken by the Panamanian Kuna people, and refers to the American continent since before the arrival of Columbus.
3 Duality is also present in the Andean cosmovisions.

References

Ackerly, B. and True J. (2010) Back to the future: Feminist theory, activism, and doing feminist research in an age of globalization. *Women's Studies International Forum* 33: 464–472.

Acosta, A. (2009) *Interview in Quito, August 18.* In A. Escobar (2010) Latin America at a crossroads. *Cultural Studies* 24(1): 1–65.

Acosta, A. (2016) El Buen Vivir, una propuesta con potencialidad global. *Revista de Investigaciones Altoandinas-Journal of High Andean Research* 18(2): 135–142.

Agarwal, B. (1992) The gender and environment debate: Lessons from India. *Feminist Studies* 18(1): 119–159.

Ávila, M.B. (2011) The Mayan indigenous women of Chiapas: Lekil Kuxlejal and food autonomy. *Development* 54(4): 485–489.

Beckerman, W. (2002) *A Poverty of Reason: Sustainable Development and Economic Growth.* Oakland, CA: The Independent Institute.

Bell, D. (2002) *Daughters of the Dreaming.* Melbourne: Spinifex Press.

Biehl, J. (1991) *Finding our Way: Rethinking Ecofeminist Politics.* New York: Black Rose Books.

Bocel, P. (2012) Coordinator of the Economic Justice Program at UNAMG, interview, Guatemala City, 25/09/2012.

Bolaños, G., Arias, J.E., Julián, O.L. and Yumay, M. (2008) Centro Maya Saqb'e, 'Hacia el Buen Vivir: experiencias de gestión indígena en Centro América', Colombia, Costa Rica, Ecuador y Guatemala. La Paz: Fondo Indígena.

Bolivia (2009) *Constitución Política del Estado Plurinacional de Bolivia.* La Paz: Ministerio de la Presidencia. Available at: www.presidencia.gob.bo/documentos/publicaciones/constitucion.pdf.

Brand, U. (2010) The fragmented hegemony of sustainable development – Gendered policy knowledge in global environmental politics. In C. Scherrer and B. Young (eds) *Gender Knowledge and Knowledge Networks in International Political Economy.* Baden: Nomos.

CONAVIGUA board member, interview, Guatemala City, 21/09/2012.

De Sousa Santos, B. (2007) Beyond abyssal thinking: From global lines to ecologies of knowledges. *Revista Crítica de Ciências Sociais* 30(1): 3–46.

Ecuador (2008) *Constitución de la República del Ecuador.* Ciudad Alfaro, Montecristi: Asamblea Nacional Constituyente. Available at: www.asambleanacional.gob.ec/sites/default/files/ documents/old/constitucion_de_bolsillo.pdf.

Escobar, A. (1995) El desarrollo sostenible: diálogo de discursos. *Ecología política* (9): 7–25.

Escobar, A. (2010) Latin America at a crossroads. *Cultural Studies* 24(1): 1–65.

FIMI (International Indigenous Women's Forum) (2006) Mairin Iwanka Raya: New beginnings for women, Miskito, copyright, FIMI 2006. Available at: www.fimi-iiwf.org/archivos/7ffd8ee2807b42a0df93d25d70c9cfdb.pdf.

García Pacheco, J.C. and Lazarte, P.C. (2012) *Politizando cuerpos, deconstruyendo, descolonización y despatriarcalización: Los espacios de poder en Oruro.* Bolivia: Latinas Editores.

Gómez Correal, D.M. (2014) Feminismo y modernidad/colonialidad: Entre retos de mundos posibles y otras palabras. In Y. Espinosa Miñoso, D. Gómez Correal and K. Ochoa Muñoz (eds) *Tejiendo de Otro Modo: Feminismo, epistemología yapuestas descoloniales en Abya Yala.* Popayán, Colombia: Editorial de la Universidad del Cauca, pp. 353–370.

Grosfoguel, R. (2009) A decolonial approach to political-economy: Transmodernity, border thinking and global coloniality. *Kult* 6: 10–38.

Gudynas, E. (2011) Buen Vivir: Today's tomorrow. *Development* 54(4): 441–447.

Hawthorne, S. (2009) The diversity matrix: Relationship and complexity. In A. Salleh (ed) *Eco-sufficiency and Global Justice: Women Write Political Ecology.* London: Pluto Press, pp. 87–106.

Hernández, G., Asociación Latinoamericana de Organizaciones de Promoción al Desarrollo (ALOP), interview, Brussels, 12/06/2012.

Hernández Castillo, R.A. (2010) The emergence of indigenous feminism in Latin America. *Signs: Journal of Women in Culture and Society* 35(3): 539–545.

Hernández Castillo, R.A. (2014) Entre el etnocentrismo feminista y el esencialismo étnico: Las mujeres indígenas y sus demandas de género. In Y. Espinosa Miñoso, D. Gómez Correal and K. Ochoa Muñoz (eds) *Tejiendo de Otro Modo: Feminismo, epistemología y apuestas descoloniales en Abya Yala.* Popayán, Colombia: Editorial de la Universidad del Cauca, pp. 279–294.

Kuokkanen, R. (2007) *Reshaping the University: Responsibility, Indigenous Epistemes, and the Logic of the Gift.* Vancouver and Toronto: UBC Press.

Kuokkanen, R. (2008) What is hospitality in the academy? Epistemic ignorance and the (im)possible gift. *The Review of Education, Pedagogy, and Cultural Studies* 30: 60–82.

Lugones, M. (2008) The coloniality of gender. *Worlds & Knowledges Otherwise* 2 (Spring): 1–17.

Lugones, M. (2010) Toward a decolonial feminism. *Hypatia* 25(4): 742–759.

Mack-Canty, C. (2004) Third way feminism and the need to reweave the nature/culture duality. *NWSA Journal* 16(3): 154–179.

Macleod, M. (2011) *Nietas del Fuego, Creadoras del Alba: Luchas politico-culturales de mujeres mayas.* Guatemala City: FLACSO.

Medina, T. (2000) *El libro de la cuenta de los nawales.* Guatemala City: Fundación CEDIM.

Mignolo, W.D. (2007) Delinking. *Cultural Studies* 21(2): 449–514.

Mignolo, W.D. (2016) Sustainable development or sustainable economies? Ideas towards living in harmony and plenitude. DOC Research Institute. Available at: http://e-library.doc-research.org/publications/detail/57.

Mujeres Mayas Kaqla (2004) *La Palabra y el Sentir de las Mujeres Mayas de Kaqla.* Guatemala: Cholsamaj.

Ortner, S.B. (1974) Is female to male as nature is to culture? In M.Z Rosaldo and L. Lamphere (eds) *Woman, Culture, and Society.* Stanford, CA: Stanford University Press, pp. 68–87.

Quijano, A. (2000) Coloniality of power and Eurocentrism in Latin America. *International Sociology* 15(2): 215–232.

Quijano, A. (2007) Coloniality and modernity/rationality. *Cultural Studies* 21(2–3): 168–178.

Redclift, M. (2005) Sustainable development (1987–2005): An oxymoron comes of age. *Sustainable Development* 13(4): 212–227.

Tedlock, B. (1992) The role of dreams and visionary narratives in Mayan cultural survival. *Ethos* 20(4): 453–476.

Thomson, B. (2011) Pachakuti: Indigenous perspectives, buen vivir, sumaq kawsay and degrowth. *Development* 54(4): 448–454.

Walsh, C. (2010) Development as buen vivir: Institutional arrangements and (de)colonial entanglements. *Development* 53(1): 15–21.

World Commission on Environment and Development (1987) *Our Common Future*. Oxford: Oxford University Press.

12

OTHER KNOWLEDGES, OTHER INTERCULTURALITIES

The colonial difference in intercultural dialogue

Robert Aman

Introduction

In December 2012, a letter of complaint arrived to the headquarters of the BBC in London. Sent from Paraguay, the author expresses outrage over how the Ayoreo, the indigenous population to which he belongs, had been represented in the documentary television series "History of the World". Set to cover 70,000 years of world history, the series is noted for its elaborate recreations of people and events on which the narrator frames the story. One of these chapters involves the staging of a recent, 1998, encounter of two members of the Ayoreos allegedly making their first contact with westerners. Not only does the voice-over refer to the indigenous population as "primitives", but the representation of them as left untouched by modernity is further emphasized by suggesting that encountering westerners put them "face to face with the 20th century". It is against such a Eurocentric view of history moving along a linear development track that the letter from the Ayoreos set out to explain how the Western ideas underpinning "progress" and "modernity" are not universal. Instead they are as provincial and context-bound as the practices of the Ayoreos. This by way of emphasizing that "the words modern, or not modern, are not useful to us. We the Ayoreos, and the people of the mountains, live as we want to live. Our culture has its own development".

In positioning different paradigms of modernization against each other, the story above offers a compelling point of entry to discuss a severely overlooked feature in endorsements for intercultural dialogue. Policy makers, academics and practitioners around Europe propagate the importance of interculturality as a strategy for coping with ethnic, cultural, and religious diversity, citing the *Sage Handbook of Intercultural Competence* (Deardorff, 1993: xiii) as a means "to better understand others' behaviours to interact effectively and appropriately with others and, ultimately, to become more interculturally competent". In contrast to multiculturalism, which according to its critics functions as a descriptive term for the factual co-existence of people of diverse cultures in a given space with the aim to encourage hospitable attitudes towards new generations of immigrants, interculturality is said to charac- terize actual interaction between people once impediments to relations have been removed (Camilleri, 1992; Gundara, 2000; Meer and Modood, 2012) – that is, to approach *them*, to speak with *them*, and even learn from *them* (Aman, 2015a). Hence, at the heart of interculturality is dialogue.

Nevertheless, the purpose of this chapter is not to uncover what necessarily constitutes such intercultural knowledge that allows us to, in Leeman's words (2003: 31), "learn to live in an ethnically and culturally diverse society". Instead the aim here is to trouble some of these calls for intercultural dialogue. Without rejecting the ideal that a genuine and politically informed dialogue across various forms of differences is a shared goal, my focus in this chapter is on certain limitations of the desire for intercultural dialogue – especially the lack of attention paid to the colo- nial difference. After all, even if intercultural dialogue acts as a code for a fluctuating and unbordered world brought about through a commitment to inclusiveness, it seems unlikely that it would have the same signification and equal appeal to all of us. In his seminal book, *Pedagogy of the Oppressed*, Freire (1972: 61) describes dia- logue as "the encounter between men, mediated by the world, in order to name the world. Hence, dialogue cannot occur between those who want to name the world and those who do not want this naming". What emerges with particular salience in the anecdote on the representation of the Ayoreos on British television is not merely the encounter between two parties that have different ways of nam- ing the world; as Ayoreos themselves point out in their letter, their "culture has its own development". But this can also be seen as an example of the power in play between those parties about whose representation carries the most weight; the privilege to define that other people live in a distant past, a period from which the modern world has advanced (Aman, 2015b).

Another pressing issue in any dialogue is the question of language. Most notably, Wieviorka (2012: 225) has criticized research on interculturality for being Anglo- centric, as he questions the possibility to write sincerely about interculturality "relying exclusively on authors who write in English or by referring to historical experiences that are only accessible through this language". Without disputing the need to move towards an understanding of interculturality that does not restrict itself to the English palette, several decolonial and postcolonial theorists alike have pointed out that there is a tendency to neglect dimensions of power in relation to languages and knowledge systems, where the colonial difference reveals itself in the hierarchies between modern European idioms (languages of science and knowl- edge) and those of the indigenous populations (languages of religion and culture) (Grosfoguel, 2013; Mignolo, 2005; Quijano, 1992).

In taking the hierarchies within epistemologies as a starting point, I seek to address the unequal positions from which participants in an intercultural dialogue may encounter each other. This by pushing for the possibility of other ways of thinking about the concept of interculturality itself depending on where (the geopolitics of knowledge) and by whom (the bodypolitics of knowledge) it is being articulated. In order to make a case for the importance of always considering the geopolitical and bodypolitical dimension of knowledge production in calls for intercultural dialogue, this chapter shifts geographical focus to the Andean region of South America. In that part of the world the notion of *interculturalidad* (often translated as interculturality) is not only a subject on the educational agenda; it has also become a core component among indigenous social movements in their strive for decolonization. Empirically, this chapter relies on data gathered through interviews with teachers and students from a pan-Andean educational initiative on interculturality – or to be more precise: *interculturalidad* – run by indigenous movements with a particular focus on *what* the concept of *interculturalidad* means to the interviewees, *why* they use it, and how *they* see it being accomplished. The argument advanced here is that *interculturalidad*, in contrast to interculturality, actualizes a question of epistemological rights rather than cultural ones, as the difference that straddles the geopolitical contexts from where the concepts are articulated goes beyond cultural differences, as they are above all colonial – that is, they historically encounter one another on asymmetrical, unequal terms, terms of domination or subordination. In conclusion, I argue that a truly genuine intercultural dialogue needs to be inter-epistemic.

Different interculturalities

Where interculturality in a European context emerged as an educational response to a shifting demographic make-up (Meer and Modood, 2012) and, in the case of the EU, also as part of forging common ground between member states (Aman, 2012; Hansen, 2000), the historical backdrop to the emergence of *interculturalidad* is distinctively different. The term evolved in tandem with indigenous people's emergence as an increasingly powerful force on the political arena in the Andean nations during the 1980s and early 1990s; an event in history that Albó (1991: 299) famously dubbed "*el retorno del indio*" ("the return of the Indian"). This was due to these movements' focus on reclaiming their identity as indigenous and revaluing their culture in which *interculturalidad* was adopted as a new watchword (Aman and Ireland, 2015). In the case of Bolivia, the ascension of Evo Morales to power meant that the notion of *interculturalidad* became as significant in state discourse as it historically had been for indigenous movements in their efforts to move towards decolonization. Morales proclaimed a radical shift on gaining office as head of the Movimiento al Socialismo (MAS) party: his presidency was to be seen as a first step for the indigenous populations toward subverting the legacies of colonialism, emphasizing the need to decolonize the educational system. The purpose was, on the one hand, to break down the racial structures imposed by colonialism and, on the other hand, to implement the knowledge systems, histories and languages of the indigenous communities as an integral part of the curricula to put an end to the privileging of European thought as a universal model. According to the first article of Nueva Ley de la Educación Boliviana (The New Bolivian

Education Act), education is now centred on the objectives of decolonization and multilingualism under the name of *interculturalidad* – translation: interculturality. It is intercultural "because it articulates a Multinational Educational System of the state based on the fortification and development of the wisdom, knowledge and belonging to our own languages of the indigenous nations", the article reads; it is intercultural "because it promotes interrelation and living together with equal opportunities with appreciation and mutual respect between the cultures of the Multinational State and the world". For another definition, organizations such as The Confederación de Nacionalidades Indígenas de Ecuador (Confederation of Indigenous Nationalities of Ecuador) and Federación Nacional de Organizaciones Campesinas, Indígenas y Negras (National Federation of Peasants, Indigenous Peoples and Blacks) interpret the principle of *interculturalidad* as respect for the diversity of indigenous peoples, but also as a demand for unity in order to transform the present structures of society, which, they argue, have been preserved from the time when an alien power established itself as the ruler and imposed its own laws and educational system (Walsh, 2009).

What becomes clear from the above is that several definitions of interculturality are simultaneously in play: *interculturalidad* seems intertwined with an act of restorative justice for the way in which the nation-state for centuries has turned the indigenous populations into its blind spot with a particular focus on epistemic change (Aman, 2017), whereas UNESCO, for instance, advocates interculturality as a method of facing the cultural challenges of every multicultural society by uniting around "universally shared values emerging from the interplay of these cultural specificities" (2009: 43). Succinctly put, where *interculturalidad* has its roots in the singular and with strong reverberations of the historical experience of colonialism, interculturality is argued to encapsulate universal principles. Particularity versus universality. The differences between the concepts become even more apparent when focusing on the practical role of language as part of an intercultural dialogue: the EU identifies conditions for interculturality in the cultural and linguistic heritage of the member states, claiming that this serves as a foundation from which "to develop active intercultural dialogue with all countries and all regions, taking advantage of for example Europe's language links with many countries" (European Commission, 2007: 10). Those local languages to which the EU ascribes importance became global through colonialism and, in another part of the world, those very languages echo the imperial order that *interculturalidad* is an attempt to overcome; languages in which the very act of speaking immediately connects the postcolonial subject to a history of violence and subjugation. Reading the rhetoric surrounding interculturality and *interculturalidad*, respectively, in the light of each other, seemingly uncovers the privileged locus from where interculturality makes meaning through its assumed universality. This, in turn, gives flesh to Jones' (1999) observation that all too often discussions on interculturality start from the assumption that all participants sit at an even table, one at which all parties have an equal say. This hierarchy between participants in an intercultural dialogue is what in this chapter is referred to as a "colonial difference".

The colonial difference

A possible avenue to discuss the importance of always keeping the colonial difference in view in relation to interculturality is through the paradoxical argumentation that there is nothing so ethnocentric, so particularist, as the claim of universality. This in turn contributes to uncovering the limitations of intercultural dialogue by drawing attention to the geopolitics of knowledge – that is, to where knowledge is produced in a world that is, according to decolonial scholars (e.g. Dussel, 1993; Grosfoguel, 2002; Mignolo, 2005), simultaneously modern and colonial. According to Quijano (1992), the European arrival in the Americas meant the abolition of existing rationalities on the American continent, which he contends are an alternative epistemology attuned to the experiences of the indigenous peoples of the region. What Quijano pinpoints is the geopolitical and bodypolitical dimension of all knowledges; a dimension that often tends to be overlooked, as it goes against the grain of what Mignolo (1999: 41) has called "the 'normal' procedure in modern epistemology to delocalise concepts and detach them from their local histories". Apart from pointing to the importance of conducting a geopolitical analysis in relation to knowledge – not least, as in the case of interculturality, where the inherent purpose is the respecting of diversity and commitment to equality – Mignolo targets how modern western epistemology carries within itself a privilege to universalize, meanwhile other knowledge systems are considered as particular and context-bound. For Mignolo (2005), the analysis of epistemology must be done in relation to its function in conforming to and sustaining a hierarchy of knowledge and knowers particularly adapted for colonialism, in which the most relevant distinction concerns one's cultural identity. This hierarchy between the various groups depending on their geopolitical and bodypolitical location is the reason that makes decolonial theorists hesitant about the conceptualization of "cultural differences", which is predominant in, for example, intercultural and multicultural discourses (Aman, 2015c); they are suspicious of the ways in which these discourses frame difference merely in cultural terms. In their view, a focus restricted to cultural differences occludes the colonial dimension.

> The "differences" between Latin America and Europe and the US are not just "cultural"; they are, well and truly, "colonial differences". That is, the *links* between industrial, developed, and imperial countries, on the one hand, and could-be-industrial, under-developed, and emerging countries, on the other, *are* the colonial difference in the sphere where knowledge and subjectivity, gender and sexuality, labor exploitation of natural resources, and finance, and authority are established. The notion of cultural differences overlooks the relation of power while the concept of colonial difference is based, precisely, on imperial/colonial power differentials.
>
> *(Mignolo, 2005: 36)*

Fundamentally, the conceptualization of "colonial difference" recognizes the power dynamics at work in how Europeans have represented their Others – that is, a form of hostility to difference embedded in the normative and teleological project of modernity, which is the basis of dominant Western epistemologies. The simultaneous operation of modernity alongside coloniality – they are, according to Mignolo (2005) two sides of the same coin, as you cannot have one without the other – implied the establishment of specific parameters of validity and recognition, not only in regard to conceptualization of humanity, human nature, progress and development but also on what can be known and how this is to be communicated. In a summarizing gesture, Chakrabarty (1995: 757) has pinpointed that the problem with modernity is that it produces opportunities for relationships and dialogue that are "structured, from the very beginning, in favour of certain outcomes"; outcomes that, seen from this perspective, inevitably privilege certain geopolitical spaces, bodies and knowledge systems over others. The representation of the Ayoreos at the beginning of this chapter is an apt illustration of how certain spaces and bodies are perceived as modern in relation to others that are deemed as not.

In comparison to a culturalist language of differences, the analytical advantage that the term "colonial difference" brings is the acknowledgement of knowledge as instrumental to domination, as on the other side of epistemic privilege is epistemic inferiority (Grosfoguel, 2013). In thinking of interculturality in terms of colonial differences rather than cultural ones, I seek to draw attention to the risk that participants in an intercultural dialogue encounter one another on highly asymmetrical, unequal terms, terms of domination or subordination. Without tunnelling into the nuances of the diversity within each concept, the argument I seek to advance here is that in relation to interculturality, *interculturalidad* reveals the necessity to always keep the colonial difference in view, as the two concepts mark two sides of an epistemological divide (Aman, 2013).

Empirically, I draw upon material from a course on *interculturalidad* provided by an indigenous organization spread over the Andean region of Bolivia, Ecuador and Peru. Founded in 1999 as a social movement with the aim of establishing indigenous educational models, the organization provides courses on *interculturalidad* to adult students. With each course spanning over a year, the students, who all self-identify as indigenous, study part-time and are given academic credits on completion of the course. According to the syllabus, the aim of the course is to retrieve and construct knowledge in direct relation to Andean culture and identity in local languages and terminology based upon indigenous methodology. Both the heterogeneity encapsulated by the terms "Andean" and "indigenous" and the common experience of negated identities, ways of thinking and interpretations of the world are acknowledged. Interviews were conducted individually with the three teachers and eight of the students from the course, focusing specifically on *what* the notion of *interculturalidad* means to the interviewees, *why* they use it, and how *they* see it being accomplished, while placing particular attention on articulations that run counter to a framework deemed to be Western.[1]

Different languages

From the outset conceptual terminology emerges as key. If this chapter has started from the assumption that *interculturalidad* is something different from interculturality, what emerges out of the material is that *interculturalidad*, like interculturality, is also a notion that carries different understandings. A student interviewed in Cuzco, Peru, pedagogically explains that although the concept of *interculturalidad* "is nowadays seen everywhere it may nevertheless have few overlapping points with the ways in which the term is being deployed among certain indigenous groupings.

> Currently there are two levels existing [of *interculturalidad*]: the utopian one and the real one. The utopic one would be something that we are still unable to achieve. This would be a superior level where all cultures are able to coexist horizontally, mutually respected, mutually tolerated, accepting each other. It doesn't exist yet which is the reason why I would call it "*utopic interculturalidad*". "Real *interculturalidad*" is what we're practically living nowadays. There's a certain relationship between cultures, but there are still these situations of placing oneself on top of another culture.

This statement offers a compelling understanding of the ways in which *interculturalidad*, at the moment of writing, offers an alternative vision, a horizon to strive for, rather than necessarily already achieved concrete and radical processes of change. What the student emphasizes as an obstacle to fulfilling the ideas of a different vision of society as invested in *interculturalidad* is not limited to clashes of cultural differences. Indeed, the terminology in play in the quotation may allude to such understanding, yet the description of how power structures relationships between different cultures highlights the colonial difference. While echoing these sentiments, other participants are more concrete in their definitions. *Interculturalidad*, explains a middle-aged female student interviewed in Urubamba, a small town in the Peruvian highlands, allows different indigenous cultures to view and interpret the world through the lens of their own beliefs in their own languages. The importance of this manoeuvre of reconstruction appears to stem from the interference of colonial residues in the initiatory pedagogy of school and society. According to the same student:

> On a general basis we have sometimes rejected our culture, we who come from indigenous cultures. This is because of prejudices, of ignorance; we believe that we're inferior, we become ashamed of our culture, we become ashamed of our language, ashamed of our mother tongue. They have taught us (*nos han enseñado eso*) that the European culture is the superior one, that it's the most developed, supposedly. Education here clearly has an occidental format wherein they teach us to value what is European and not what is ours.

By diagnosing core symptoms of the effect of European influence on life in the Andes, the interviewee describes a colonial difference in which being indigenous is equated with lack, synonymous with inferiority in relation to what is ascribed to Europe. Although she is recounting these issues in a predominantly general manner, the student's articulation of negative emotions in relation to being indigenous – an experience of shame leading to gradual rejection – is significant. The process explained is that of identification and disavowal, in which pretensions to be part of the nation's univocal subject require assimilation through the adoption of a perspective on, among other things, life, knowledge and subjectivity derived from modern European models. In locating the dissemination of European texts in an impersonal "they" – that is, the educational system, the student depicts a two-stroke process: the schools bind pupils to a state written in and from the language of the colonizers, which in turn continues to exacerbate the colonial wound (Aman, 2017).

This background emerges as essential to understand what differentiates a certain articulation of *interculturalidad* from another. In contrast to state-sponsored initiatives around the Andean region under the name of *educación intercultural bilingüe* that allow the teaching of indigenous languages alongside Spanish in public elementary schools (Gustafson, 2009), for many indigenous alliances the request for educational rights in indigenous vernaculars in the name of *interculturalidad* extends beyond language learning. This demand is a call for the inscription across subjects and curricula not only of languages but also of knowledge systems, values and beliefs that have been silenced within official discourses ever since the conquest. Since its inception, a stern critique has been directed towards the exclusive focus on languages of *educación intercultural bilingüe* and its disregard for other epistemologies and logics. This is not to dismiss the possibility of important advancements under the name of *educación intercultural bilingüe* that have been reported by academic commentators. Among others, Hornberger (1987) argues that these educational initiatives help indigenous languages to endure. However, as Angel and Bogado (1991) point out, a general tendency was that Spanish continued to be the *lingua franca* of the nation, as indigenous idioms were merely transformed into yet another school subject, similar to the study of a foreign language. Subsequently all school children, regardless of background, remained subjected to the study of Spanish, and only the indigenous populations were expected to become bilingual – not anyone else.

In reaction to state policy initiatives, indigenous alliances across the Andean nations began to develop their own intercultural education referred to as *casa adentro* (in-house). A concrete example of such a course is the one under scrutiny here. According to one of the initiators, Juan García (Walsh and García, 2002), the objective of these courses is to strengthen the ties of belonging and the building of a collective memory among the indigenous populations. This carries a specific purpose: "to unlearn the learned and relearn *lo propio*, 'our own', as a way to understand life, our vision of history, knowledges, and of being in the world" (Walsh, 2011: 51). Against this background of the relational and epistemic violence of coloniality, Spivak has introduced the concept of "strategic essentialism", which has the purpose of allowing subordinated groups to foster their agendas in order to "speak back" to hegemonic powers. Nevertheless, Spivak has also lamented in retrospect that the "strategic" dimension of the notion tends to be neglected among those who are in the process of promoting agendas that could challenge imperial powers (Danius and Jonsson, 1993). Although evaluating what is and is not purely essentialism in accounts such as the one above may seem premature at the present time, learning "our own" can be seen as drawing upon a certain essentialist construction in the idea of something as either "ours" or "theirs". However, it can also be seen as targeting the colonial difference in emphasizing ways of knowing the world from epistemological premises other than the ones sanctioned by modernity. As explained by a student interviewed in El Alto, *interculturalidad* offers:

> tools to re-recognise in my memory what my grandparents had: the language, the forms, the traditions. [...] Thus, to live my reality and accept myself a little bit more for who I am and not try to copy ways of life that are outside of our reality. I think that this is *interculturalidad*, to accept ourselves as we are.

In this account, the participant seems to view the importance of *interculturalidad* as an action that allows the indigenous population to recover traits of identification deemed "extinct" as a result of colonial extirpation since the Spanish arrival. Resistance occurs in the form of claiming particularity, a way of being that, as the argument goes, differs from those who were originally external: "indigenous communities are losing their identity in learning Spanish", another student interpolates. Speaking in a single European language becomes not merely a reinforcement of historical power structures that oblige the addressee to communicate in the idiom of the *metrópoli*: the very act of speaking is a continuous reminder of the imperial legacy the postcolonial subject carries within – *lengua*, Spanish for both language and the physical tongue. Symptoms of coloniality are not limited to the language itself; colonial vestiges are equally ingrained within languages. In the case of Spanish, imperial attitudes have found a home in the realm of the idiomatic negative imperative – *¡No seas indio!* (Don't be Indian!) – in everyday speech that encourages the recipient to stop acting ignorantly and instead be civilized. Thus, to call someone an *indio* to their face is as much of an insult in the Andean region as pretty much anywhere else in Latin America. Behind that term are centuries of discourses, images and unequal power relations rooted in racism.

Recognition of other languages, however, does not necessarily signal the undoing of the linguistic legacy of Spanish that persists in the Andean nations. Rather, in this context, to make use of a collective "we" can be seen as part of the struggle for acknowledgement of the existence and contribution of languages that have been disqualified as tools for thinking. Instead, *interculturalidad*, according to Aymara intellectual (Ticona, 2012, cited in Walsh, 2012), activates the discourse on "*lo propio*" as part of a radical claim for epistemic rights rather than cultural ones – or put differently, for *interculturalidad* rather than *educación intercultural bilingüe* or even interculturality, whose recognition of cultural or linguistic diversity does not necessarily translate into epistemological diversity.

Different knowledges

Besides the demands for the recognition of indigenous languages, Morales' accession to power revealed another longstanding request: right to the land. Confronting the chronicles of the colonial archive, Morales proclaimed already in his inaugural speech as the new president of Bolivia that "we have achieved power to end the injustice, the inequality and oppression that we have lived under. The original indigenous movement, as well as our ancestors, dreamt about recovering the territory". In the final part of this sentence, "recovery" emerges as fundamental to continuing action. A term laden with loss, this word's presence is intimately linked to past experiences of colonial subjugation, of having been stripped of self-determination of the territory over which the various Andean nation-states extend their arms. Morales' words on the importance of recovering the territory are echoed in interviews with the students:

> We've always been fighting for political decisions about the land. The basis of life (*la base de la vida*) is in the territory and it defines everything. Of course, it also has its proper manner of expression; in this case it also signifies a way of life and the conception of life itself and this we express in our own languages (*nuestros propios idiomas*). The major problem has been one culture's negation of all other cultures.

Succinctly, the interviewee begins with historicizing the ways in which the indigenous populations found themselves on the other side of the wall which separates the visible from the invisible part of the social space, officially excluded from the national borders now crossing the land they had cultivated before the arrival of Columbus. Demands to be fully recognized as citizens through various acts of insurgence have had varying degrees of success throughout the history of Bolivia in turning the indigenous populations into equal citizens – from Túpac Katari's siege of La Paz in 1780 to the Katarista indigenous movement in the 1970s; from the Chayanta uprising in 1927 to the "Water War" in the beginning of the millennium which erupted in response to the privatization of Cochabamba's public water service (see Webber, 2012; Angel and Bogado, 1991). In addition, the interviewee conveys how lifeworlds and knowledge systems have been buried under centuries of colonial, Eurocentric and racist dust. In targeting the colonial difference by referring to how this is a consequence of "one culture's negation of all other cultures", the interviewee at the same time produces a counter-narrative in describing a holistic view in which the ground is inseparable from languages, knowledge systems and even life. "It's my territory that gives me my identity," she explains, before underlining the importance of *interculturalidad* as a return to one's identity and to respecting Mother Earth (*la Pacha*) because "she is our mother who provides us with our food. We also respect our water without contaminating it because the water is life, it has life (*el agua es vida, tiene vida*)".

Notable here is the repeated emphasis on points of identification that were equally apparent in the previous section on the struggle over language that stems from the indispensable interrelation of ways of life and the territory. A claim for the existence of life in the waters and protection from los Apus – symbolically, *Apu* is an honorific for a person in Quechua – signals not only interaction with the landscape and dependency on it. The statement also reveals the colonial difference by introducing an indigenous perspective, which, contrary to western epistemology, does not treat nature as an object but rather as a subject (Quijano, 1992). For a concrete case in point, a student interviewed in Cochabamba, Bolivia, describes a logic of resistance to the dominant paradigm of capitalism in relation to the land:

> In the big world (*el mundo mayor*) the land is valued as a piece of merchandise. In the indigenous Andean world it isn't, rather we care for it with respect, as something that gives us life, that is part of ... like another person (*como una persona más*).

In short, what the student accounts for is a counter-narrative against Eurocentric, reductionist notions of development and economic growth. This is done by way of a planetary metaphor that underlines a subjugated position by contrasting "Andean" and "Big" – an inclination that bears traces of the dictum *the West and the Rest* – which draws sharp boundaries between the agents and the silenced in a hierarchy both of ontology (European versus indigenous) and epistemology (science versus beliefs) determined by geopolitical location. In eschewing the binaries alleged to be central to modernity (see Escobar, 2010), the quotation highlights the way in which the common Western opposition between nature and humanity lacks a signifier. Instead the interviewee opposes such duality that splits nature from culture through ascribing agency to the land as knowledge. If humans, living systems, nature and – in Western eyes – lifeless objects are not distinguished, as Mignolo and Schiwy (2003) suggests in reference to Andean cosmology, but are rather all conceived as part of a network of living interactions, it draws attention to the epistemological dimension of *interculturalidad*. Or as succinctly captured by another student in Bolivia: "*[i]nterculturalidad* isn't a concept that solves humanity, rather it permits debating what the human is."

Different state models

Over the past decades, a new paradigm for human progress has been emerging in Latin America referred to as *buen vivir* ("living well"), which is the result of many years of political organization and mobilization of indigenous groups. Before going further, it is important not to confuse "living well" with "living better", as they are set apart by epistemological differences: where "living better" is confined to European modernity with its emphasis on development, consumerism and progress, Morales himself summarizes *buen vivir* as "to live in harmony with everyone and everything, between humans and our Mother Earth; and it consequently implies working for the dignity of all" (cited in Canessa, 2014: 157).

Moreover, the promotion of *buen vivir* is incorporated in the 2008 Constitution of Ecuador and the 2009 Constitution of Bolivia. However, a similar scenario to that of *educación intercultural bilingüe* seems to repeat itself also here, despite the constitutional changes. Differently put, there is no guarantee that *interculturalidad* for the government now means the same thing as for the grassroots movements that supported Morales' campaign. On the contrary, Escobar (2010) claims that the Morales administration have failed to accomplish profound and satisfactory changes in line with the radical programmes proposed by several social movements, which, he continues, highlights how *interculturalidad* as an attempt to transform the existing order is more likely to be struggled for from below than above. In reference to this issue, a student in Bolivia interpolates that "[a]ll the documents of the state nowadays have '*interculturalidad*' all over the place – they breath *interculturalidad*". Yet she identifies a discrepancy between policy discourse and practical implementations when stating that "in concrete practice with racism and coloniality that is crazy, it's [racism and coloniality] super present and they are re-actualised in other forms when the key question is the colonial structure".

Besides the possibility of *interculturalidad* having lost some of its subversive edge in the hands of the state, the student pinpoints an additional obstacle in terms of how the continuously colonial structure of the state prevents the implementation of *interculturalidad*. In viewing the construction of the state with its argued colonial character, the student suggests that the struggles invested in *interculturalidad* move beyond a liberal acceptance of cultural pluralism. While this recognition of cultural differences within the frontiers of the state would possibly allow for cultural rights and educational reforms, it would not necessarily translate in equality of difference within the framework of the nation-state. After all, as Balibar and Wallerstein (1991) remind us, the processes of nation-state building has always been violent, as it was accompanied by the exclusion of national minorities. Applied to the Andean context, as well as in many other parts around Latin America, an important reservation needs to be made: in contrast to Europe, national minorities were not necessarily pushed to the corners in the process of nation-state building. Quite the reverse, when the descendants of the conquistadores founded the respective republics of Bolivia, Peru and Ecuador between 1820 and 1830, it was a national minority that excluded a majority of the population: the indigenous peoples (Prada, 2010).

As one of the course teachers explains in his office in Quito, the indigenous populations have been made aware of the colonial difference ever since its construction: "From the conquest onwards the state has wanted to assimilate the indigenous population and insert them into the state, yet without understanding their processes, without knowing their cultures." The colonial difference reveals itself in the fact that the Creoles were without a demand to adjust, to abolish their cultural identities, in order to acquire full citizenship. Through their fortuitous character as a class, they are already fully considered as citizens. On the direct question of how *interculturalidad* can target the deemed colonial structure of the nation-state, a student in La Paz responds: "we're fighting for a plurinational state (*un estado plurinacional*)". In using the word "nation", the same rhetorical weapon wielded by the state itself, indigenous movements seemingly couch their ideological positions in identical terms – x is our language, x is our territory, x is our religion – for recognition as a political entity. Shifting from the singular to the plural is not only a manoeuvre to emancipate from the authority of the prescribed positions of subjectivity enabled within the state's framework but also a communal manoeuvre to decolonize the Western-modelled state by rewriting it from another viewpoint. Although it was indigenous movements in Ecuador that initially began to use the term "nationalities" to refer to themselves as distinct people within the Ecuadorian state, Bolivia has made the furthest advancement in being the first state in Latin America to recognize itself in the National Constitutional Assembly as plurinational (Gustafson, 2009).

Naturally, such a radical move inevitably produces its own critics. Where some academic commentators have dismissed a plurinational state as a process of "balkanization" in creating territories on ethno-linguistic grounds (see Mayorga, 2007), others are more hopeful in suggesting that it may allow for achieving real democratic pluralism, as it provides a new model of citizenship that challenges former colonial and postcolonial injustices (see Gustafson, 2009). While at the present time of writing it may seem premature to evaluate the practical effects that the refounding of the nation-state as plurinational has had for the invoked populations, Walsh (2009) contends that, albeit underdeveloped on an theoretical level, it has undeniably opened up new avenues of possibilities for decolonization, as it signifies a clear shift from the uni-national framework of the nation-state to the one with more adequate structures to include its people. This is characterized by a new form of nation-building process that sets out to incorporate difference (indigenous languages, knowledges, cosmologies) into sameness (nationhood, modernity, the state apparatus) while also allowing for sameness to be transformed by difference. Succinctly put, the idea of the plurinational – as a central component of *interculturalidad* – finds its primal sustenance in the historically repressed and negated literal plural character of the national. If *interculturalidad* offers, in Escobar's (2010: 25) words, a move away from "the monocultural, monoepistemic, and uninational state", part of the altered link to the state can bring about a new sense of citizenship and entitlement. Granted, a citizenship produced within the framework of *interculturalidad* is not merely a new model of citizenship for indigenous people; it seems fair to suggest that it is a new model of citizenship *per se*.

Final words

In shifting focus from a policy discourse on interculturality produced by supranational bodies orientated towards cultural differences, and instead engaging with the sibling discourse of *interculturalidad*, has allowed an enhanced understanding of the importance to consider epistemology in a project set to bridge cultural difference through intercultural dialogue. Succinctly put, what I have highlighted in this chapter is the importance of the geopolitical dimension of knowledge production and the potential pitfalls of not taking the colonial difference in consideration of interculturality. Quickly returning to the goal of intercultural dialogue as described by the *Sage Handbook of Intercultural Competence*: "to better understand others' behaviours to interact effectively and appropriately with others", UNESCO advocates interaction across cultural differences around "universally shared values", and the

EU underlines the importance to "develop active intercultural dialogue with all countries and all regions, taking advantage of for example Europe's language links with many countries". Juxtaposing them against the backdrop of *interculturalidad* would allow for a profound questioning of not merely the way in which the pragmatic identification of a dialogue held in imperial languages into which subjects in erstwhile colonies continue to be born illustrates a continuing exaltation of the colonial difference, but also a profound questioning of what constitutes such alleged "universal values". What we have seen in this chapter is that there are seemingly few exceptions to the conceptual and terminological premises to *interculturalidad*, in contrast to interculturality, privileged enough to pass as universal. Part of the challenge in achieving an intercultural dialogue with the purpose of, in the words of UNESCO, promoting "respect, understanding and solidarity among individuals, ethnic, social, cultural and religious groups and nations" involves understanding the social-historical power relations that imbue knowledge production. As Quijano (1992: 19–20) puts it: "[e]pistemic decolonisation is necessary to make possible and move toward a truly intercultural communication" – that is, to create a space open to a horizontal dialogue between epistemes from different traditions – from intercultural to inter-epistemic.

Note

1 All interviews were conducted in Spanish. Although I am aware of the limitations of such an approach, the reasons for this undertaking are related both to my own linguistic limitations in Quechua and Aymara and to the use of Spanish as the official language of the course. The explanation for this is that, on the one hand, students may carry different languages with them, meaning that Spanish offers a common ground, and on the other hand, that there are those who identify themselves as, for example, Quechua without having training in the language, because of the dominance of Spanish throughout the educational system. As Morales lamented in a recent interview, when enrolled in school, he gradually lost his earlier fluency in Aymara (Peñaranda, 2011). Although contradictory to the course's aims, support can be found in Mignolo's (2005) writings, which stress the importance of thinking in and from a language historically disqualified as a tool for thinking, such as Quechua or Aymara, while still writing in an imperial language, in order to subvert the geopolitics of knowledge.

References

Albó, X. (1991) El retorno del indio. *Revista Andina* 1(2): 299–345.

Aman, R. (2012) The EU and the recycling of colonialism: Formation of Europeans through intercultural dialogue. *Educational Philosophy and Theory* 44(9): 1010–1023.

Aman, R. (2013) Bridging the gap to those who lack: Intercultural education in the light of modernity and the shadow of coloniality. *Pedagogy, Culture & Society* 21(2): 279–297.

Aman, R. (2015a) In the name of interculturality: On colonial legacies in intercultural education. *British Educational Research Journal* 41(3): 520–534.

Aman, R. (2015b) The double bind of interculturality, and the implications for education. *Journal of Intercultural Studies* 36(2): 149–165.

Aman, R. (2015c) Why interculturalidad is not interculturality: Colonial remains and paradoxes in translation between supranational bodies and indigenous social movements. *Cultural Studies* 29(2): 205–228.

Aman, R. (2017) *Decolonising Intercultural Education: Colonial Differences, the Geopolitics of Knowledge, and Inter-Epistemic Dialogue.* London: Routledge.

Aman, R. and T. Ireland (2015) Education and other modes of thinking in Latin America. *International Journal of Lifelong Education* 34(1): 1–8.

Angel, M. and D. Bogado (1991) *Proyecto: Bosque de Chimanes, apoyo al fortalecimiento y preservación de la territorialidad y el medio ambiente.* Cochabamba: PROBIOMO.

Balibar, É. and I. Wallerstein (1991) *Race, Nation, Class: Ambiguous Identities.* London: Verso.

Camilleri, C. (1992) From multicultural to intercultural: How to move from one to the other. In J. Lynch, C. Modgil and S. Modgil (eds) *Cultural Diversity and the Schools.* London: Falmer Press, pp. 141–151.

Canessa, A. (2014) Conflict, claim and contradiction in the new 'indigenous' state of Bolivia. *Critique of Anthropology* 34(2): 153–173.

Chakrabarty, D. (1995) Radical histories and question of enlightenment rationalism. *Economic and Political Weekly* 30(14): 751–759.

Danius, S. and S. Jonsson (1993) An interview with Gayatri Chakravorty Spivak. *boundary 2* 20(2): 24–50.

Deardorff, D.K. (ed) (1993) *The SAGE Handbook of Intercultural Competence.* London: Sage.

Dussel, E. (1993) Eurocentrism and modernity (Introduction to the Frankfurt lectures). *boundary 2* 20(3): 65–76.

Escobar, A. (2010) Latin America at a crossroads. *Cultural Studies* 24(1): 1–65.

European Commission (2007) *Communication from the Commission: A European Agenda for Culture in a Globalizing World.* Brussels: EC.

Freire, P. (1972) *Pedagogy of the Oppressed.* Harmondsworth: Penguin Books.

Grosfoguel, R. (2002) Colonial difference, geopolitics of knowledge and global coloniality. *Review* 25(3): 203–224.

Grosfoguel, R. (2013) The structure of knowledge in westernized universities. *Human Architecture* 10(1): 73–90.

Gundara, J. (2000) *Interculturalism, Education and Inclusion.* London: Paul Chapman Publishing Ltd.

Gustafson, B. (2009) *New Languages of the State: Indigenous Resurgence and the Politics of Knowledge in Bolivia.* Durham: Duke University Press.

Hansen, P. (2000) Europeans only? Essays on identity politics and the European Union. Diss. Umeå: Umeå University Press.

Hornberger, N. (1987) Bilingual education success, but policy failure. *Language in Society* 16(2): 205–226.

Jones, A. (1999) The limits of cross-cultural dialogue: Pedagogy, desire and absolution in the classroom. *Educational Theory* 49(3): 299–316.

Leeman, Y. (2003) School leadership for intercultural education. *Intercultural Education* 14(1): 31–45.

Mayorga, F. (2007) *Encrucijadas: Ensayos sobre democracia y reforma estatal en Bolivia.* La Paz: Editorial Gente Común.

Meer, N. and T. Modood (2012) How does interculturalism contrast with multiculturalism? *Journal of Intercultural Studies* 33(2): 175–196.

Mignolo, W. (1999) *Local Histories / Global Designs: Coloniality, Subaltern Knowledges, and Border Thinking.* Princeton, NJ: Princeton University Press.

Mignolo, W. (2005) *The Idea of Latin America.* Oxford: Blackwell Publishing.

Mignolo, W. and F. Schiwy (2003) Double translation: Transculturation and the colonial difference. In T. Maranhao and B. Streck (eds) *Translation and Ethnography: The Anthropological Challenge of Intercultural Understanding.* Tucson: University of Arizona Press, pp. 6–28.

Peñaranda, R. (2011) Evo Morales. *ReVista: Harvard Review of Latin America* 11(1): 12–15.

Prada, R. (2010) Al interior de la Asamblea Constituyente. In M. Svama, P. Stefanoni and B. Fornill (eds) *Balance y perspectivas: Intelectuales en el primer Gobierno de Evo Morales.* La Paz: Friedrich Ebert Stiftung, pp. 37–97.

Quijano, A. (1992) Colonialidad y modernidad/racionalidad. *Perú indígena* 13(29): 11–20.

Spivak, G.C. (1988) Subaltern studies: Deconstructing historiography. In R. Guha and G.C. Spivak (eds) *Selected Subaltern Studies*. Oxford: Oxford University Press.

UNESCO (2006) *Guidelines on Intercultural Education*. Paris: UNESCO.

UNESCO (2009) *Investing in Cultural Diversity and Intercultural Dialogue*. Paris: UNESCO.

Walsh, C. (2009) *Interculturalidad, Estado, Sociedad: Luchas (de)coloniales denuestra época*. Quito: Universidad Andina Simón Bolívar/Abya Yala.

Walsh, C. (2011) Afro and indigenous life-visions in/and politics: (De)colonial perspectives in Bolivia and Ecuador. *Bolivian Studies Journal* 18: 49–69.

Walsh, C. (2012) 'Other' knowledges, 'Other' critiques: Reflections on the politics and practices of philosophy and decoloniality in the 'Other' America. *Transmodernity: Journal of Peripheral Cultural Production of the Luso-Hispanic World* 1(3): 11–27.

Walsh, C. and J. García (2002) El pensar del emergente movimiento afroecuatoriano: Reflexiones (des)de un proceso. In D. Mato (ed) *Estudios y Otras Prácticas Intelectuales Latinoamericanas en Cultura y Poder*. Caracas: Consejo Latinoamericano de Ciencias Sociales.

Webber, J. (2012) *Red October: Left-indigenous Struggles in Modern Bolivia*. Chicago: Haymarket.

Wieviorka, M. (2012) Multiculturalism: A concept to be redefined and certainly not replaced by the extremely vague term of interculturalism. *Journal of Intercultural Studies* 33(2): 225–231.

13

POETICAL, ETHICAL AND POLITICAL DIMENSIONS OF INDIGENOUS LANGUAGE PRACTICES IN COLOMBIA

Sandra Camelo

The violent normalization of language and the colonization of America

During the colonization of America, missionaries and Latinists were given the authority to translate and convert Indigenous languages into "proper" languages by creating grammars and alphabets based on the Latin language (Zimmermann, 1997). The definition and construction of Indigenous languages according to foreign alphabets and grammars mark the simultaneous invention and colonization of Indigenous languages (Mignolo, 1992, 1994a).

Generally speaking, the study of Indigenous languages served the colonial purpose of "transposing" the foreign alphabets and grammatical rules of strangers to the Indigenous languages. From this perspective, Indigenous peoples and their languages were defined and colonized according to a presumable "ontological lack" of literacy, or alphabetic writing. Most of the codices and glyphs were burnt and broken into pieces and the few codices that were preserved were redefined as "books" in an attempt to normalize their alien otherness. The term "book" (*libro*) was chosen mainly to bring these productions "closer" to the writing practices of the colonizers (Mignolo, 1992).[1]

As Mignolo (1992) writes, by translating the Nahualt *amoxtli* and the Mayan-Quiché *vuh* as "book", the colonizer-missionaries neglected other Indigenous written practices such as quipus, maps, calendars, pottery, stone engraving, wall painting and body painting, as well as textile weaving. Therefore, making Indigenous *amoxtli* and *vuh* into "books" is not simply an act of translation in the sense of finding an equivalent (which is rather unattainable), but of translation in the sense of dislocation and displacement. *Amoxtli* and *vuh* were displaced, and even "erased" according to Mignolo (1992: 312), by the book, which was believed to be the *normal* carrier of the written forms constituted by the letters of the alphabet. In contrast to the Amerindian texts, the Bible, the colonizers' sacred book, was written in alphabetic characters and printed on paper. If the Bible was believed to have been dictated by God, the colonizers believed the Indigenous books had been dictated by its enemy, "the devil". As a result, the evil Indigenous "books" were condemned to burn in flames.

In 1521, the chronicler Bernal Díaz del Castillo wrote about the natives of today's Guatemala and Mexico. He referred to the "evil" inks and charcoal used to "scrawl" "malicious" words and stories on whitened walls and fabrics (Díaz del Castillo, 1521, cited in Rama, 1998: 50). Similarly, in what is present-day Peru, Alonso Carrió de la Vandera, official of the Spanish Crown, described the Amerindian inscriptions on walls as unsophisticated "graffiti" in contrast to the "cultivated alphabetic writing" of the Spanish (Carrió de la Vandera, cited in Rama, 1998: 51). In this context emerges the imaginary of the savage and illiterate cannibal (Rama, 1998: 191, 205–206), whose "ontological lack" and "deviation" was supposed to justify the colonial enterprise: their persecution, enslavement and later on their conversion to the language and faith of the colonizers.

Grammaticalization and alphabetization operated as colonial technologies that defined Indigenous languages as "illiterate" incomplete languages. Since the early 16th century, alphabetic transcriptions, vocabulary books and grammars have operated as normalizing technologies tasked with making the incomprehensible comprehensible according to the standards of the colonizers, which were the standards of modernity, dissociated from their locality and defended as universal and desirable. This standardization of Indigenous languages involved specific knowledge practices, institutions, discourses and technologies producing and maintaining colonial epistemic violence.

Epistemic violence is understood here as the undermining and exclusion of knowledge practices that are produced outside a specific circuit, privileged in the production and distribution of knowledge. This circuit results from widely accepted and often naturalized colonial power-knowledge relations that confine academic knowledge to a handful of (European) languages and a smattering of centres and institutions, mostly based in the North and yet widely replicated in the Global South (Mignolo, 2005; Grosfoguel, 2013, 2014). This epistemic violence makes it more difficult to hear the voices of other subjects who are producing knowledge outside the circuit. They have become subaltern or subordinated to those whose voice is strengthened by power-knowledge relations rooted in the history of colonialism and imperialism (Spivak, 1988). This epistemic violence can lead as well to *epistemicide* or the constant suppression and eventual disappearance of invalidated knowledge practices, as Grosfoguel (2013, 2016) has indicated.

After the independence of what today corresponds to the Republic of Colombia, in the early 19th century, the scribes and literates of the new ruling elite – known as *criollos* – adopted cleansing policies designed to eliminate the "unwanted", and migratory policies aiming to attract Europeans who would supposedly bring their "civilized" taste and "developing nature" with them (Castro-Gómez, 2005). They also normalized the behaviour of the population, their ways of speaking, religious beliefs and their conditions of citizenship according to handbooks of good manners, grammars, the Catholic Bible, and constitutions (González-Stephan, 1996).

After independence, defenders of the Hispanic legacy in America, such as Andrés Bello and Miguel Antonio Caro, engaged in a passionate crusade to regulate the Spanish language spoken in the emergent nations, hounding the "irregular multitude of dialects" emerging in southern America and the former colonies. They argued that the "unity of language was God's blessing and source of strength" in opposition to the multiplicity of dialects, which was a "curse that led to ruin" and destruction for the nations (Caro, 1871: 132, cited in Zuleta Álvarez, 1997: 94, my translation). These "language defenders" were inspired by the unifying grammar of Antonio Nebrija, published in 1492, and which instituted the Castilian (the royal variety) as the norm to be adopted by all speakers in the Empire (Mignolo, 1992, 1994b).

Grammars were not simply descriptive books or sets of rules; they were colonial technologies that served to normalize languages and produced subjects who were not only under surveillance but also exercised surveillance over others. Grammars operated as disciplinary interventions not only in schools, but also in the public language of the city and the emerging nation (González-Stephan, 1996: 28–29). Grammars regulated the languages of the street and the home, unifying them under a unique project of citizenship (Zuleta Álvarez, 1997: 32). The institutionalization of the written, and the dissemination of a non-face-to-face kind of communication, also led to the institutionalization of a new kind of learning. Children were no longer educated by their communities through daily and spontaneous interactions with their relatives, but were extracted from the community and placed in an institutionalized school system directed by unfamiliar teachers and "special authorities".

In the Americas, Indigenous children were imprisoned and removed from their families in "residential schools", where they were forbidden to speak their Indigenous languages.[2] Oral exchanges in Indigenous languages were for the most part undermined as the illiterate practices of "savages" (Mignolo, 1992: 326). A cultural genocide took place in the Americas as Indigenous peoples were subjected to "decontextualization" and the imposition of the "non-oral" communication (see Goody, 1987: 184). Indigenous children were removed from their families and placed in foreign schools where they were not allowed to nor could communicate in their language, as they were among other children who did not share their language. However, despite this violent persecution that was initiated in the 15th century and continued up to most of the 20th century, in Colombia alone there are 65 identified Indigenous languages still spoken – some by larger populations, others spoken only by a few individuals.

Indigenous languages and colonial epistemic violence

Coloniality is founded on the myth of the "universality" and "neutrality" of knowledge located in a non-place and enunciated by an anonymous "universal" subject of knowledge (Dussel, 1999, 2000). This is to say that the colonial experience transcends colonialism (Maldonado-Torres, 2007). Its experience is not restricted to the physical exploitation and subjugation of the bodies of the colonized and their lands by "a metropolitan administrative and military apparatus" (Restrepo, 2014: 306, my translation). The colonial experience has established regimes of validation of knowledge, language and humanity that produce distinctions between knowledge, language and humans, on one hand, and non-knowledge, non-languages and non-humans, on the other. *Coloniality* remains even after the dismantling of the colonial institutions that sustained colonialism. Not being necessarily concentrated or territorialized in a body or an institution, coloniality is subtle and insidious. It tends to be entangled in ways of living and relating (both locally and globally), in economic exchanges and policy-making processes – and indeed, in the production of knowledge and the construction of subjectivities.

Due to these intimate complexities and the overlapping of global dimensions, coloniality has not been successfully eradicated by rebelling against a king, defeating an army, contesting a set of laws or electing a politician, despite how important these achievements may be. In Colombia, like in many other countries that became independent from the Spanish Crown, the former centres of colonialism and imperialism have adapted to a certain way of living that is preserved by coloniality. While the National Constitution of 1991 recognizes the autonomy and self-determination of the Indigenous peoples in Colombia, still today, coloniality sustains the hierarchical privileges granted to new centres and denied to new peripheries. The former colonies continued to imitate the privileged centres of culture, civilization, knowledge and development they admired. To this date, Western ideals of culture, civilization, knowledge and development continue to mark the agenda in now independent countries that were under colonial regimes, as with the case of Colombia.

Coloniality is particularly difficult to defy because it constitutes subjectivities. This does not mean it cannot be challenged by border thinking resistance and decolonial alternatives, as will be discussed later in this chapter in regard to (po) ethical language practices. Defying coloniality implies rethinking who we are and opening ourselves to the possibilities of being otherwise.

Coloniality operates on the intimate level of subjectivities, symbols, affects and knowledge practices (Castro-Gómez, 2011: 260, 267). From this perspective, colonial knowledge – which is the knowledge that is produced within colonial relations – is not impartial, but inevitably biased. It reproduces these colonial relations and reinforces colonial values. Colonial knowledge is framed by "abyssal lines" that create *ontological* and *epistemic* distinctions between what and who *is* valuable. They separate what is knowledge and what is not, who is "human" or "subhuman", creating and negating "the other side of the line" (De Sousa Santos, 2010: 19).

Around the globe, alphabetical writing and the culture of the printed text made possible the emergence of the romantic concepts of "originality" and "creative spirit", which set apart the individual narratives from anonymous collective narratives (see Ong, 2002: 132). This belief gave rise to a new subjectivity based on authorship and ownership of knowledge including specialists, or "savants" (Goody,

1987: 75). Additionally, as Goody (1987) states, the emergence of writing affected the stratification of the channels of communication and placed itself at the top. It gave a higher status in the social hierarchy to those who wrote.

There are more than 94 Indigenous peoples in Colombia, speaking 65 languages and 300 dialects (Landaburu, 2009), for many of whom the encounter with the Spanish language has been violent and dismissive. This is mainly because the Colombian state and its institutions are "illiterate" or ignorant in regard to the traditions and knowledge practices of the Indigenous peoples (Rocha Vivas, 2012: 63). Like in many other countries, the so-called "minority" groups are forced to adapt to a presupposed national and even "universal" model of learning and producing knowledge. This model is openly and even proudly ignorant of the knowledge practices coming from other traditions, which are only praised in the name of multiculturalism.

The Misak poet and human rights activist Edgardo Velasco recalls his experience in the university as difficult, because there was no dialogue with other knowledge practices: "[...] there were four other Indigenous classmates but the curriculum was completely Western" (Velasco Tumiña, 2015, my translation).[3] Multiculturalism is limiting when considering language planning, as it is not able to effectively engage with Indigenous languages. It is necessary to consider an alternative ecological perspective that engages with the poetical, ethical and political dimensions of Indigenous languages.

The limits of language planning and multiculturalism and alternative (po)ethical language ecologies

The multicultural model not only promotes and celebrates difference, but also defines it. It embraces difference in a restrictive, normalizing and marketable fashion. It is driven by the exploitation of difference as a profitable good that attracts tourism and international investment. Multicultural policies are often a new form of racism, which is a "racism without race" (Balibar, 1991: 23). While they have abandoned openly discriminatory language and embraced more respectful and politically correct expressions, they continue to be complicit with racist structures which are "camouflaged" in the occasional "celebration of difference" (Gunew, 2004: 43). Multicultural states continue to maintain the unequal access to resources and remain silent before the violent expropriation of lands by multinationals and extractive industries, to name a few examples. Multiculturalism remains racist unless it embraces "anti-racism and political equality" (Bannerji, 2000, cited in Gunew, 2004: 41).

In Colombia, the advances made towards the recognition of differential rights do not go beyond the paternalistic and condescending acceptance of a different "minority" by a supposedly tolerant and omnipresent "majority". This creates a situation in which indigeneity must be performed according to a juridical framework devised by the state to govern and limit the claims of minorities (Gros, 2000). Who is Indigenous and who speaks as an Indigenous person or people is one of the most problematic issues in the multicultural model, obliging communities to dialogue with external figures under numerous presuppositions and a legal language that, despite its specificities, remains ambiguous and unsuited to the translation of communities' dynamics and organization.

Certainly, one of the most problematic aspects of the multicultural model is that it not only produces violent ideals of indigeneity, but it also urges Indigenous peoples to adopt those ideals. Some of these ideals include the notion of the noble savage that is obedient and would never affect its natural ecosystem (see Escobar, 1999), or the monolithic authentic replica of a frozen past (see Fabian, 1983). These ideals are particularly violent because they have had the capacity to shape expectations regarding the organization of their communities, as well as influencing the construction of their subjectivities.

Overall, the multicultural model is violent despite its mobilization of an "inclusive" language of recognition and protection. Indigenous peoples' autonomy is not only granted by external institutions and according to external criteria; it is also "controlled by the dominant culture" (Lemaitre Ripoll, 2009: 294, my translation). In the multicultural model, the "state produces, reproduces, establishes and legitimises an ethnic border when it is apparently recognising Indigenous peoples and their autonomy" (Gros, 2000: 105, my translation). Generally speaking, the multicultural discourse and its legal tools do not simply protect or celebrate difference, but produce it according to a framework that promotes the state's interests. It is in this context that Indigenous peoples and ethnic groups are multiculturally re-indigenized and re-ethnicized (Restrepo, 2004, 2007).

Now that Indigenous peoples have been redefined and redefined themselves as subjects of rights, they are expected to defend not only their traditions but perform by example their global ecological mission (Ulloa, 2007). Conceptions of Indigenous peoples as "noble savages" are frequent nowadays; they even serve to judge Indigenous peoples' use or abuse of natural resources and to argue for their responsibility in the extinction of fauna and flora from their territories (see Escobar, 1999; Ulloa, 2007). Similarly, Indigenous peoples are expected to be advocates of peace and harmony and any rupture of these ideals might become a risk for the validity of their claims and demands. Another problematic myth about Indigenous peoples is that they constitute a wide global community, diverse in regards to the nation state, yet undifferentiated from community to community (Gros, 1991, 2002). It is thus often presumed that agreement and consensus reign within Indigenous communities, in contrast to the presumed majoritarian *mestizo* or mixed *society* which has inherited the "Western right" of individuality.

These imaginaries and expectations do not only come from the outside, they are also internalized and recirculated. In fact, Indigenous leaders and representatives often refer to their communities as authentic and naturally ecological. As they do so, they appeal to what Spivak (1987: 205) describes as *strategic essentialism*, a positivist essentialism used strategically as it suits a specific situation, in a scrupulously political situation. This tendency is informed by their need to validate their demands (see Ariza, 2009). Yet these imaginaries and expectations can be violent. When they clash with real practices, for instance, they can paint a less paradisiac picture that can in turn be used to invalidate Indigenous claims. However, we must not forget that these claims were compelled by an invented and imposed legal language that was partially produced by these same imaginaries and expectations. Despite this dangerous route, some of these ideals of community have also served to unite peoples, defend their value and promote genuine ways of living and ethics (see Chikangana, 1995; Kidd, 2000; Overing and Passes, 2000).

The celebration of diversity is mostly deaf to "the voices and political and social proposals of the historically marginalised sectors" (Uzendoski, 2015: 6), which is why "recognition politics" tend so often to "subtly reproduce non-mutual and unfree relations rather than free and mutual ones", as Coulthard (2014: 17) has identified in his analysis of the Canadian context, but which is also true for the Colombian context. Overall, multicultural and recognition politics are deeply problematic (Gutmann, 1994) in at least three ways. First, they do not even question the racism behind the *minorization* of some groups called "minorities" in an undermining language. Second, they maintain economic and political asymmetries and do little against the exploitation, disposition and even assassination of Indigenous communities by states, corporations, multinationals and extractive industries. Third, they permit the commodification of Indigenous cultural and symbolic heritage by tourism, fashion and cultural industries. Additionally, there is very little or no investment at all to support students from Indigenous communities and give them access to higher education. This is also true for supporting their access to schools, universities, medical centres and even sanitary services such aqueduct networks, sewer and drainage systems (see KAS, 2009).

The presumed "majoritarian" society not only portrays itself as a "majoritarian" but also as a "universal" and "tolerant" culture that accepts other "minorities". As Castro-Gómez (2015: 165, my translation) argues, this amounts to a politically correct brand of tolerance which is in fact "intolerant and repressive" because the Other is required "to continue to be tied to its particular identity and not to move from the place it was given […], to happily accept to live in a society that gives it *rights* to be 'what it is'". Generally speaking, recognition is fragmentary because it is given – as if it was some kind of gift or donation – by one who has no need to be recognized in return (see Coulthard, 2014).

Even today, the landscape of language planning and linguistic policies towards Indigenous languages seems stark, since it is circumscribed by forms of interventionism and patronization that basically reproduce colonial power relations, now incorporated also by the Indigenous communities and their leaders. One should not ignore the fact that the bilingualism of Indigenous communities takes place under conditions of linguistic inequality. Nor should one ignore the fact that the policies aimed at intercultural and bilingual education are part of a historical conflict between ethnic societies and their languages, on the one hand, and the official culture and language of the ruling class and the state, on the other (Truscott de Mejía, 2007: 34–35).

While the situation might be different in other countries, in Colombia, Indigenous languages have remained outside the classrooms of secondary schools, professional colleges and universities (Benavides, 2008: 245). Generally speaking, Spanish is the language of instruction in Colombia, and in most programmes of higher education students must use that language to express their acquisition and production of knowledge. This is generally true except for a few classes for learning Indigenous languages and the permission of Indigenous students to submit bilingual versions of their dissertations in a reduced number of institutions.

Given these asymmetries and the constant reduction of speakers of some Indigenous languages in Colombia, linguists as well as Indigenous leaders have defended the need to protect, study and promote these Indigenous languages as "endangered languages". This has been crucial for the destination of public and private funds for linguistic research as well as for initiatives and programmes promoting the production of didactic printed, digital and multimedia materials in Indigenous languages. However, there are relevant critiques of the discourse of endangered languages and the essentialization of Indigenous identities. Linguists have somehow given themselves the task of protecting Indigenous languages. This task is often carried out in a paternalistic way, under the assumption that Indigenous peoples need to be helped and that one of the justifications for the existence of disciplines such as anthropology and linguistics is to protect Indigenous interests. As the Californian linguist and anthropologist Katherine Woolard (1998: 17) and some of her colleagues have observed, minority language activists often "find themselves imposing standards, elevating literate forms and uses, and negatively sanctioning variability in order to demonstrate the reality, validity, and integrity of their languages". Despite how useful grammars and alphabets can be for "elevating the status" of a language and promoting its use among its speakers, language standardization is undoubtedly problematic, as it always privileges one dialect over others (ibid.). New hierarchies are created and again some speakers and their dialectal variations become socially less desirable and undermined.

Revitalization cannot be confused with the attempt to simply return or recover a pristine, authentic past. Indigenous languages and their living, performative, creative and transformative practices exceed linguistic registration and analysis, since they mobilize and reinvent community filiations, affects, memories and ways of living. From this perspective, Indigenous languages are conceived as living ecological bodies produced by complex articulations, and producers in turn of re-articulations of meanings, affects, filiations, territories, cosmologies, poetical forms, ethical engagements, social practices, creative resistance and political transformations. This notion of language ecologies expands understandings of Indigenous languages, writing and orality as practices embedded in Indigenous cosmologies and daily life practices.

Language practices are embodied and related to stories, memories, experiences and affects articulated to a territory, a community, a shared history, and very often a project seeking the transformation of current violence. The exchange of stories and deeply felt narratives can build filiations of affect and facilitate healing. Communities are in this context built from filiations of affect expressed in a desire of conviviality and sharing. This sense of togetherness, which is felt and promoted as what brings the community to life, can be seen as a sense of friendship or as a "co-belonging of nonidentical singularities", as Leela Gandhi suggests (2006: 31).

Here I refer to these practices as (po)ethical. (Po)ethical practices create communities, bringing them together while healing the pains each one carries. Velasco Tumiña (2015) says that in listening to one another "the poetic reveals itself" (my translation). This curative power of listening and storytelling has also been identified by the Vietnamese artist and theorist Trinh T. Minh-ha. She has found that among the Basaa people in Cameroon, the storyteller heals and gathers the community, and the source of power is in "listening and absorbing daily realities" (1989: 140). Minh-ha also explains that "what is transmitted from generation to generation is not only the stories, but the very power of transmission" (ibid.: 134), that is the capacity for "listening to others" and even "reading the other's eyes" (ibid.: 30). Both Minh-ha and Velasco Tumiña seem to agree that poetry is not only about writing, but also about listening.

(Po)ethics go beyond the literary forms of the poetic. They are relational, and this relationality involves ethical practices of care and healing for the narrating voices and the listening bodies involved. (Po)ethics are not abstract; on the contrary, they take place in the body, which is woven with stories, memories, experiences and affects; a body capable of feeling the other's stories and pain. Therefore, it is through the body that (po)ethics can heal. Minh-ha (1989: 121) has noticed the corporality of stories, remarking that they are "transmitted from mouth to ear, body to body, hand to hand", in a materiality that is "seen, heard, smelled, tasted, and touched". (Po)ethics are preserved and guarded in the body, and when someone dies, their body of knowledge leaves a hole.

In Senegal and Nigeria as well as in other communities in West Africa, there is a woman called the *griotte*, who embodies, keeps and shares the knowledge of the community. "It is said that every griotte who dies is a whole library that burns down (a 'library in which the archives are not classified but are completely inventoried')" (Hampate Ba, cited in Minh-ha, 1989: 121). The body is a container and the place where the words of knowledge, memories and stories of these communities reside; it is also the place where they are received, nurtured and felt. The basket (*kirigai*) that represents knowledge for the Uitoto people in Colombia connects language and knowledge to the body. It is in the body of the wise guiders and the apprentices where knowledge and language practices take place (Vivas Hurtado, 2011, 2013). As the Argentinian anthropologist and philosopher Rodolfo Kusch acknowledges, "everything seems to indicate that there is no *ethos* without a *pathos*" (Kusch, 2000: 43, my translation). Following this argument, it is possible to argue that (po)ethics are only possible because there are bodies and lived corporeal experiences; therefore, the body is both the place of the (po)ethical and its condition of possibility.

(Po)ethical dialogues: Agonistic translations

Understanding Indigenous languages as (po)ethical practices implies going beyond the reductive models of alphabetic literacy and grammaticality and acknowledging that Indigenous language practices carry complex systems of ethics in which "speaking beautifully" is related to "thinking and living beautifully", as the Camëntsá poet Hugo Jamioy Juagibioy (2010: 59) recites. (Po)ethical language practices involve the bodies of the practitioners and the experiences, memories and meanings engraved in them. (Po)ethical language practices imply living and feeling language and its historical and political contingencies through the body. This means experiencing the violence and asymmetries of the world and finding ways for creative resistance. It is work that involves healing and rebuilding communities along with their traditions, filiations and affects.

The Uitoto people in the Colombian Amazon believe in a complex articulation of language, knowledge, nature, community and celebration. They refer to *rafue* as the articulation of word and thing which are understood as one entity, or "Word-Thing" (Urbina Rangel, 2010: 55, my translation). *Rafue* is also the "activity that transforms words into things" (Candre and Echeverri, 2008: 28, my translation). *Rafue* is knowledge and power. *Rafue* inhabits the plants of wisdom and facilitates memory and the accurate narration of the myth of origin. *Rafue* transforms in the mouth and becomes Word, Word of knowledge and power. It travels as it is pronounced to those who listen attentively. The "plants of wisdom" – coca, tobacco, yagé, to name but a few – are beings, and when they are chewed, smoked and imbibed, they transform. In the mouth of the wise elders (*abuelos sabedores*), they transform into the Word.[4] Language and knowledge are from this perspective bridges that connect beings, making rhizomes (see Guattari, 2000), and not simply mental abilities or processes, as Noam Chomsky (1995) argued, shaping a still dominant paradigm in the field of linguistics.

An ecological perspective of language and its (po)ethical practices must consider the bridges and articulations that define not only their praxis and *telos*, but also their ontological nature. In his poem "Binÿbe oboyejuayeng mondmën" (We are Wind Dancers), the Camëntsá poet Hugo Jamioy Juagibioy says that the words of the dancing wind are poetry that feeds the people from one generation to another.[5] Poetry is in this sense the "Word of knowledge", learnt from the elders, actualized in the dance and the ways of living of those who learnt from elders' teachings. As Jamioy Juagibioy (2010: 61) states in his poem: "Poetry is the speaking wind traversing the ancient footsteps" (my translation).

Indigenous language practices involve an ethical and political aesthetic; their beauty is not simply in the beauty of the words, but more importantly, in the beauty of thinking and acting according to a set of values. In his poem "Botamán cochjenojuabó" (Beautifully you must think), Jamioy Juagibioy (2010: 59) says: "Beautifully you must think, then beautifully you must speak, now, right now, beautifully you must do" (my translation). Speaking beautifully and acting beautifully implies embodying the beliefs, teachings, traditions and values of one's people. Indigenous language practices carry a profound ethics, which is "an integral vision that does not place itself over other living beings, but it takes place under the cosmos' cape [...] in which life manifests" in different forms and bodies (Mora Púa, 2010: 146).

The ecological (po)ethical understanding of language I am defending in this chapter recognizes the ethical, poetical and political materiality of language. It recognizes the knowledge embodied in leaders such as the griotte in the communities of Senegal and Nigeria, and the *abuelos sabedores* in the Uitoto communities in

Colombia, who are not simply sensitive to nature but capable of understanding its meanings and bringing them in to the rituals and its narratives. (Po)ethical language practices enable the understanding, healing and collective production of knowledge among human beings as well as non-human beings, and the creation of a beautiful way of thinking and living in which the poetical, ethical and political articulate.

When the Yanakuna poet Fredy Chikangana is asked about the relation between his poetry and ethics, he argues that his poetry is not simply a literary genre, but an expression of ethics. In his own words, he says that "doing poetry is a way of living with a certain calm, even during the most difficult times" (Chikangana, 2016, my translation). Similarly, Misak poet Edgar Velasco Tumiña (2015) says that his poetry is informed by the stories of origin of his community, as well as the teachings of his ancestors and family. Velasco Tumiña engages with the Misak communities in Cauca, but he is also an active participant of national organizations such as Autoridades Ancestrales, Gobierno Mayor (Ancestral Authorities, Major Government) and Comisión Étnica para la Paz y la Defensa de los Derechos Territoriales (Ethnic Commission for Peace and the Defence of Territorial Rights). He refers to his role in these two organizations as both integrally poetic and political.

Velasco Tumiña has worked with many Indigenous communities that have been persecuted, displaced and abused by legal and illegal military groups for years. Velasco Tumiña's work is a (po)ethical practice that involves poetics, ethics and politics. This (po)ethical practice consists of the narration and exchange of stories, memories and ways of living and resisting. It also includes listening and communicating through different languages, voices, cultures and experiences in order to enable collective healing and resistance. (Po)ethical practices involve the active practice of listening and enabling the production of narratives to convey experiences of violence in powerful words and make visible the erased and silenced stories of those who have disappeared, been persecuted, displaced, disposed and murdered. They also involve the collective rebuilding of community filiations as part of the process of listening and healing.

Chikangana, *Wiñay Malki*, undertook a quest that he calls *archaeology of the word*. This quest consists, he explains, in a constant dialogue with the elders, not only of the Yanakuna people, but also of other peoples, speakers of Kichwa in Peru, Ecuador and Bolivia. A "dialogue that sets the landscape for a memory journey" (Chikangana, 2016, my translation). Chikangana grew up in the *resguardo* of Río Blanco (Yurak Mayu, in Kichwa) in Cauca, where the Kichwa language was no longer spoken. He was conflicted about not having access to the language of his ancestors and being forced to use Spanish – a "borrowed verb" – to tell the memories and dreams that inhabit him and his people. He lived in Cuzco and Cochabamba where he re-learnt the Kichwa language. Having found his language, he found the "seed of dreams" of his stories and songs, as he says in his poem "Nuqa taki" (I am a singer) (Chikangana, 2010: 106).

As a poet/singer, Chikangana composes new songs, new stories that straddle the border between oral traditions and his writings. These songs are deeply interwoven with the territory, which Chikangana describes as a territory that is hidden within each "I". Each Yanakuna singular individual forms the plurality of the Yanakuna people (Chikangana, 1995: 19, my translation). This territory is not only the background of the community's life, is not simply the geographical borders of a particular settlement; for the most part, Chikangana explains, the territory is a major source of learning for the formation of the "Yanakuna self", as it is the surrounding material symbolic nature and the mythical beings that move around that territory. The Yanakuna people, he adds, are mainly those who walk through their territory and construct a community of care and support: "Yanakuna *Mitmaqkuna* means people who help each other in the darkness" (ibid.). *Mitmaq* means walker (Ceballos Ramírez, 2015: 70, my translation).

(Po)ethical practices do not only convey experiences of violence in powerful words but also enable healing, as well as the rebuilding of community filiations and the creation of new communities. A certain resistance emerges from the articulation of affects (see Zibechi, 2006), and the shared experiences of trauma and healing that unite against the world of death and dehumanization that is at the core of coloniality (Mmembe, 2001, 2003). The experiences of violence that these (po) ethical practices narrate are located, but not isolated from the many experiences of violence taking place around the world. This is why they can dialogue with other calls against violence and the dehumanization and destruction of ways of living. Local (po)ethical practices can form a multi-vocal and plural *(po)ethics.* In this sense, these (po)ethics are both *singular* and *plural* (Nancy, 2000). They convey a multiplicity of voices in a singular poem composed by a poet in dialogue with situated and shared experiences. These situated experiences are not isolated from the experiences of other individuals and collectives around the world who have experienced coloniality. Ultimately, (po)ethics can be define as a lingua franca for connecting and sharing experiences of colonial violence and for producing collective networks of creative resistance.

Since (po)ethics are both personal and communitarian, local and global, they demand to be translated. Translation here means letting their fluxes and intensities gravitate towards new localities so they can articulate with other (po)ethical practices and enable collective healing through the dialogue of shared experiences and traumas delineated by coloniality. This translation allows that voices from different localities meet in order to create a world that is affirmative of life and humanity. Rather than being obstacles to translation, their *singular-plural* embodiment, and their bordering between the plural and universal – their concrete universality (Castro-Gómez, 2015), or *pluriversality* (Mignolo, 2000) – calls for it. However, this translation is not an easy task; it is, on the contrary, agonistic and demanding.

This translation cannot simply be linguistic. It needs to carry their poetical, ethical and political dimensions. It needs to communicate their pain, their healing and rebuilding of community filiations and traditions. It also needs to set the conditions for the creation of larger communities of affect across the globe. This translation cannot have the pretension of merging these communities and erasing their differences, nor of eluding the conflictive relations that abide in them and in their relation to one another. Instead, (po)ethics demand an *agonistic translation.* This is through the plural acknowledgements and feelings of one another's existence and pains, producing a plural and visceral disruption that truly challenges the naturalization of violence and dehumanization of the present world.

An agonistic perspective goes beyond the simplistic opposition between local and global, specific and general, yet does not attempt to dilute locality, specificity and singularity in an over-comprehensive totality. An agonistic perspective acknowledges the struggles, tensions and disagreements between adversaries; it does not promote the elimination of enemies for the resolution of conflict (Mouffe, 2013: 14). (Po)ethics and their agonistic translation can dialogue with many experiences of violence perpetrated in the name of civilization, modernity, progress, profit and so on. They produce a singular and plural space of *ethos* and *pathos*, where epistemic violence and coloniality can be challenged.

Notes

1 Mignolo (1992) argues that the Greek term *biblos* was closer to the Aztec *amoxtli* because it referred to the handwritten papyrus, stored in a roll, and unrolled to be read aloud to an audience.
2 Any language different from Spanish was forbidden in the schools or in any other public areas outside the Indigenous communities, including urban areas such villages, towns and cities, where even the street names had to be in Spanish as proclaimed by Law 17 in 1927 (Pineda Camacho, 1997: 113).
3 The West needs to be understood as positionality. As Grosfoguel (2016) explains, the West is a particular positionality in power relations that promotes a distinctive epistemology around the world, one that is both racist and sexist, as it only values the knowledge produced by United States, England, France, Germany and, to a lesser extent, Italy, and despite the provinciality of the knowledge they produce, this knowledge is disguised as universal while other knowledge practices are accused of being too "particularistic" or local.
4 The *abuelo sabedor* is the grandfather, the wise guardian who takes care of the family and the community. The *abuelo* (grandfather) and *abuela* (grandmother) teach the younger generations the myths of origin of the community, the traditional weaving, pottery, chants, dances and rituals, as well as facial and body painting and adornment. They also teach words of the Indigenous language(s) of the community and correct intonation and rhythm. In general, age is associated with "maturity and knowledge" (Mora Púa, 2010: 127).
5 The Camëntsá people call themselves *Camuentsa Cabëng Camëntsá Biya*, though the short form Camëntsá is commonly used among them, as well as by other Indigenous peoples and non-Indigenous peoples referring to them.

References

Ariza, L.J. (2009) *Derecho, saber e identidad indígena*. Bogota: Siglo del Hombre Editores.
Balibar, É. (1991) Is there a neo-racism? In É. Balibar and I. Wallerstein (eds) *Race, Nation, Class: Ambigous Identities*. C. Turner (trans.). London: Verso, pp. 17–28.
Benavides, C. (2008) Democratizar va más allá de acceder a la educación: Pensamiento propio y alternativa al desarrollo. In G. Dietz, R.G. Mendoza Zuany and S. Téllez Galván (eds) *Multiculturalismo, educación intercultural y derechos indígenas en las Américas*. Quito: Ediciones Abya-Yala, pp. 245–266.

Candre, H. and Echeverri, J.A. (2008) *Tabaco frío, coca dulce: palabras del anciano Ki-nerai- de la Tribu Cananguchal para sanar y alegrar el corazón de sus huérfanos.* 2nd edn. Colombia: Colcultura.

Castro-Gómez, S. (2005) *La hybris del punto cero: ciencia, raza e ilustración en la Nueva Granada (1750–1816).* Bogota: Editorial Pontificia Universidad Javeriana.

Castro-Gómez, S. (2011) *Crítica de la razón latinoamericana.* 2nd edn. Bogota: Editorial Pontificia Universidad Javeriana.

Castro-Gómez, S. (2015) *Revoluciones sin sujeto: Slavoj Z?iz?ek y la crítica del historicismo posmoderno.* Mexico D.F: Akal/Inter Pares.

Ceballos Ramírez, F.A. (2015) Fredy Chikangana: poesía, mito y negociación. MA thesis, Pontificia Universidad Javeriana, Bogota. Available at: http://repository.javeriana.edu.co/handle/10554/17052 (Accessed 8 January 2016).

Chikangana, F. (1995) Yo yanacona, palabra y memoria. BA thesis, Universidad Nacional de Colombia, Bogota.

Chikangana, F. (2010) *Samay pisccok pponccopi muschcoypa: Espíritu de pájaro en pozos de ensueño.* Bogota: Ministerio de Cultura. Available at: www.banrepcultural.org/sites/default/files/89046/07-Samay-Espiritu-de-pajaro.PDF (Accessed 15 September 2014).

Chikangana, F. (2016) Interview with Sandra Camelo. 29 March, Río Blanco. [Audio Recording].

Chomsky, N. (1995) Language and nature. *Mind* New Series 104 (413): 1–61.

Coulthard, G.S. (2014) *Red Skin, White Masks: Rejecting the Colonial Politics of Recognition.* Minneapolis: University of Minnesota Press.

De Sousa Santos, B. (2010) *Para descolonizar Occidente: más allá del pensamiento abismal.* Buenos Aires: Consejo Latinoamericano de Ciencias Sociales; Prometeo Libros.

Dussel, E. (1999) Más allá del eurocentrismo: el sistema-mundo y los límites de la modernidad. In S. Castro-Gómez, O. Guardiola-Rivera and C. Millán de Benavides (eds) *Pensar (en) los intersticios: teoría y práctica de la teoría poscolonial.* Bogota: Instituto Pensar; CEJA, pp. 147–161.

Dussel, E. (2000) Europa: modernidad y eurocentrismo. In E. Lander (2000) *La colonialidad del saber: eurocentrismo y ciencias sociales: perspectivas Latinoamericanas.* Buenos Aires: Consejo Latinoamericano de Ciencias Sociales, pp. 11–40.

Escobar, A. (1999) *El final del salvaje: Naturaleza, cultura y política en la antropología contemporánea.* Bogota: CEREC, Instituto Colombiano de Antropología.

Fabian, J. (1983) *Time and the Other: How Anthropology Makes its Object.* New York: Columbia University Press.

Gandhi, L. (2006) *Affective Communities: Anticolonial Thought and the Politics of Friendship.* Durham: Duke University Press.

González-Stephan, B. (1996) Economías fundacionales: diseño del cuerpo ciudadano. In B. González-Stephan (ed) *Cultura y tercer mundo.* Caracas: Nueva Sociedad, pp. 17–47.

Goody, J. (1987) *The Interface between the Written and the Oral.* Cambridge: Cambridge University Press.

Gros, C. (1991) *Colombia indígena: identidad cultural y cambios sociales.* Bogota: CEREC.

Gros, C. (2000) *Políticas de la etnicidad: identidad, estado y modernidad.* Bogota: Instituto Colombiano de Antropología e Historia.

Gros, C. (2002) Un ajuste con rostro indígena. In J.M. Blanquer and C. Gros (eds) *Las dos Colombias.* Bogota: Grupo Editorial Norma, pp. 323–358.

Grosfoguel, R. (2013) The structure of knowledge in Westernised universities: Epistemic racism/sexism and the four genocides/epistemicides of the long 16th century. *Human Architecture: Journal of the Sociology of Self-Knowledge* 11(1): 73–90.

Grosfoguel, R. (2014) La descolonización de la economía política y los estudios poscoloniales: transmodernidad, pensamiento descolonial y colonialidad global. In B. De Sousa Santos and P. Meneses (eds) *Epistemologías del Sur (perspectivas)*. Madrid: Akal, pp. 373–405.

Grosfoguel, R. (2016) "Decolonising the academy": Session 1. University of Edinburgh, Global Development Academy. Edinburgh, Scotland. 24 February 2016.

Guattari, F. (2000) *The Three Ecologies*. I. Pindar and P. Sutton (trans.). London: Athlone Press.

Gunew, S. (2004) *Haunted Nations: The Colonial Dimensions of Multiculturalism*. London: Routledge, pp. 33–50.

Gutmann, A. (ed) (1994) *Multiculturalism: Examining the Politics of Recognition*. Princeton: Princeton University Press.

Jamioy Juagibioy, H. (2010) *Bínÿbe oboyejuayëng: Danzantes del viento*. Bogota: Ministerio de Cultura. Available at: www.banrepcultural.org/sites/default/files/89044/06-Binybe-oboyejuayeng-Danzantes-del-viento-Hugo-Jamioy-Juagibioy.pdf (Accessed 15 September 2014).

KAS (2009) *Situación de los pueblos indígenas de Colombia. KAS PAPERS* 4. September. Konrad Adenauer Stifund, pp. 1–23. Available at: www.kas.de/wf/doc/23098-1442-4-30.pdf (Accessed 3 June 2014).

Kidd, S. (2000) Knowledge and the practice of love and hate among the Enxet of Paraguay. In J. Overing and A. Passes (eds) *The Anthropology of Love and Anger: The Aesthetics of Conviviality in Native Amazonia*. New York: Routledge, pp. 114–132.

Kusch, R. (2000) *Pozo de América*. I *Obras Completas*. Book. 4. Córdoba, Argentina: Editorial Fundación Ross, pp. 1–184.

Landaburu, J. (2009) La diversidad de lenguas nativas de Colombia y su situación actual. In Ministerio de Cultura (ed) *La fiesta de las lenguas nativas*. Bogota: Ministerio de Cultura, pp. 4–5.

Lemaitre Ripoll, J. (2009) *El derecho como conjuro: fetichismo legal, violencia y movimientos sociales*. Bogota: Siglo del Hombre Editores; Universidad de los Andes.

Maldonado-Torres, N. (2007) Coloniality of being: Contributions to the development of a concept. *Cultural Studies* 21 (2–3): 240–270.

Mignolo, W. (1992) On the colonisation of Amerindian languages and memories: Renaissance theories of writing and the discontinuity of the classical tradition. *Comparative Studies in Society and History* 34(2): 301–330.

Mignolo, W. (1994a) Signs and their transmission: The question of the books in the New World. In E. Boone and W. Mignolo (eds) *Writing without Words: Alternative Literacies in Mesoamerica and the Andes*. London: Duke University Press, pp. 220–270.

Mignolo, W. (1994b) Afterword: Writing and recorded knowledge in colonial and postcolonial situations. In E. Boone and W. Mignolo (eds) *Writing without Words: Alternative Literacies in Mesoamerica and the Andes*. London: Duke University Press, pp. 293–313.

Mignolo, W. (2000) *Local Histories/Global Designs: Coloniality, Subaltern Knowledges and Border Thinking*. Princeton: Princeton University Press.

Mignolo, W. (2005) Espacios geográficos y localizaciones epistemológicas: la *ratio* entre la localización geográfica y la subalternización de conocimientos. *GEOgraphia* 7(13): 7–28.

Minh-ha, T. (1989) *Woman, Native, Other*. Indianapolis: Indiana University Press.

Mmembe, A. (2001) *On the Postcolony*. Berkeley: University of California Press.

Mmembe, A. (2003) Necropolitics. *Public Culture* 1(15): 11–40.

Mora Púa, F.G. (2010) Mito y ética: una lectura del pensamiento mítico de los Uitoto y Muinane. *Franciscanum* 53(154): 115–149.

Mouffe, C. (2013) *Agonistics: Thinking the World Politically*. New York: Verso.

Nancy, J. (2000) *Being Singular Plural*. Stanford: Stanford University Press.

Ong, W. (2002) *Orality and Literacy: The Technologizing of the Word*. London: Routledge.

Overing, J. and Passes, A. (2000) Introduction: Conviviality and the opening up of Amazonian anthropology. In J. Overing and A. Passes (eds) *The Anthropology of Love and Anger: The Aesthetics of Conviviality in Native Amazonia*. London: Routledge, pp. 1–30.

Pineda Camacho, R. (1997) La Constitución de 1991 y la perspectiva del multiculturalismo en Colombia. *Alteridades* 7(14): 107–129.

Rama, A. (1998) *La ciudad letrada*. Montevideo: Arca.

Restrepo, E. (2004) Ethnisation of blackness in Colombia: Toward de-racializing theoretical and political imagination. *Cultural Studies* 18(5): 698–715.

Restrepo, E. (2007) El 'giro al multiculturalismo' desde un encuadre afro-indígena. *Journal of Latin American and Caribbean Anthropology* 12(2): 475–486.

Restrepo, E. (2014) Articulaciones coloniales, modernidades plurales: aportes al enfoque decolonial. In J. Gandarilla Salgado (ed) *América y el Caribe en el cruce de la modernidad y la colonialidad*. México: Universidad Nacional Autónoma de México, Centro de Investigaciones Interdisciplinarias en Ciencias y Humanidades, pp. 303–325.

Rocha Vivas, M. (2012) *Palabras mayores, palabras vivas: tradiciones mítico-literarias y escritores indígenas en Colombia*. Bogota: Taurus Pensamiento; Fundación Gilberto Alzate Avendaño.

Spivak, G.C. (1987) *In Other Worlds: Essays in Cultural Politics*. New York: Methuen.

Spivak, G.C. (1988) Can the subaltern speak? In C. Nelson and L. Grossberg (eds) *Marxism and the Interpretation of Culture*. London: Macmillan, pp. 271–313.

Truscott de Mejía, A. (2007) Realidades de la educación bilingüe en Colombia: implicaciones para el desarrollo del bilingüismo en América Latina. In C.L. Ordóñez (ed) *Segundo simposio internacional de bilingüismo y educación bilingüe en América Latina*. 5–7 October 2006. Bogota: Centro de Investigación y Formación en Educación, Universidad de los Andes, pp. 30–45.

Ulloa, A. (2007) La articulación de los pueblos indígenas en Colombia con los discursos ambientales, locales, nacionales y globales. In M. De la Cadena (ed) *Formaciones de indianidad: Articulaciones raciales, mestizaje y nación en América Latina*. Popayán: Envión, pp. 287–326.

Urbina Rangel, F. (2010) Los mitos. In F. Urbina Rangel (ed) *Las palabras del origen: Breve compendio de la mitología de los uitotos*. Bogota: Ministerio de Cultura, pp. 43–252.

Uzendoski, M. (2015) Los saberes ancestrales en la era del antropoceno: hacia una teoría de textualidad alternativa de los pueblos originiarios de la Amazonía. *Revista de Investigaciones Altoandinas* 17(1): 5–16.

Velasco Tumiña, E.A. (2015) Interview with Sandra Camelo. 11 December, Bogota. [Audio Recording].

Vivas Hurtado, S. (2011) Constitución y multiculturalismo. Jikanotikai. *Desde Abajo* 1(10) 21 June. Available at: www.desdeabajo.info/suplementos/item/17610-constituci%C3%B3n-y-multiculturalismo-jikanotikai.html (Accessed 25 July 2013).

Vivas Hurtado, S. (2013) Ñuera uaido: la palabra dulce o el arte verbal minika. *Devenires* 28: 89–120. Available at: http://filos.umich.mx/portal/wp-content/uploads/2014/06/89-120.pdf (Accessed 2 February 2014).

Woolard, K. (1998) Introduction: Language ideology as a field of inquiry. In B. Schieffelin, K. Woolard and P. Kroskrity (eds) *Language Ideologies: Practice and Theory*. Oxford: Oxford University Press, pp. 3–50.

Zibechi, R. (2006) La emancipación como producción de vínculos. In A.E. Ceceña (ed) *Los desafíos de las emancipaciones en un contexto militarizado*. Buenos Aires: CLACSO. Consejo Latinoamericano de Ciencias Sociales, pp. 123–149. Available at: http://bibliotecavirtual.clacso.org.ar/ar/libros/grupos/cece/Raul%20Zibechi.pdf (Accessed 20 May 2014).

Zimmermann, K. (1997) Introducción: apuntes para la historia de la lingüística de las lenguas amerindias. In K. Zimmermann (ed) *La descripción de las lenguas amerindias en la época colonial*. Madrid: Iberoamericana, pp. 9–20.

Zuleta Álvarez, E. (1997) *Lengua y cultura de Hispanoamérica en el pensamiento de Miguel Antonio Caro*. Bogota: Instituto Caro y Cuervo.

14

SURPASSING EPISTEMIC HIERARCHIES

A dialogue between expanded art practices and human scale development

Maricely Corzo Morales

Introduction

After completing undergraduate studies in fine arts and postgraduate studies in international cooperation, the question about the place of artistic processes in development interventions has continuously fed my creative projects and research. At that time I could barely find experiences of local development where culture and art had an important role to play, and I felt disappointed because it seemed that from my field of knowledge, art, I could not contribute much with respect to the development of a community, since nothing of this sort was part of the contents of the course. Since then I have studied how each of these fields of knowledge has been renewed and has expanded its frontiers and I have witnessed how the number of artistic interventions working with communities and in a more collaborative way has grown.

The university structure is very specialized and hierarchical, so that the arts are considered inferior to the social sciences and these are considered to be inferior to the physical sciences. These hierarchies have intensified in recent years, as the westernized university is tending to promote degrees that contribute to economic productivity and to the market in general. As a result, the arts and humanities find themselves in a more precarious position. This structure of epistemic hierarchies is a global one that has its origins in a Eurocentric university model. I draw on Ramón Grosfoguel's (2013) work to reveal how this model and its various historical configurations lead to a series of epistemicides. I also draw on research carried out in Colombia, to indicate how in this context the colonial heritage and the intervention of the State have put the study of the arts in an inferior position within the university structure. I then go on to discuss one of the efforts to overcome such epistemic hierarchies and to achieve changes in the structure of universities through the establishment of interdisciplinary dialogues. This chapter brings artistic practices into dialogue with development models as a decolonial move.

To introduce this particular interdisciplinary dialogue, I show how both fields of knowledge have experienced "expansion" processes during the second half of the 20th century, which generated common points of encounter. The question that I ask myself after observing the overflowing movements that have occurred in both fields, namely human development and artistic practices, is what kind of dialogue can result from their convergence? And, on the other hand, since artistic practices are directly linked to creativity, is it possible that artistic practices that deal with communal and collaborative aspects could be considered synergic satisfiers of fundamental human needs?

In response to these research questions, I argue that some conditions that shape artistic practices that work collaboratively and in community contexts give them great potential to be synergistic satisfiers with respect to the human need for creation. Finally, I present a proposal for a method of research technique that includes a visual work in its process, and which I have been using in several projects developed in the last few years.

At a recent conference in Bogotá, Boaventura de Sousa Santos (2017) warned that the increasing commodification of knowledge constitutes one of the principal threats to our societies in these times of post-democracy and post-truth, where we are witnessing the establishment of politically democratic, but socially fascist societies that run the risk of not being differentiated from a dictatorship. The seven threats he mentioned are: 1) the reconfiguration of the State (where sovereignty is increasingly conditioned by economic powers), 2) the emptying of democracy, 3) the destruction of nature, 4) the devaluation of labour, 5) the commodification of knowledge, 6) the recolonization and resexualization of difference 7) and the criminalization of protest. Despite long decades of changes and advances in research and experience in the field of human development models, States are under pressure from economic powers to put financial interests above human rights and sovereignty. Within this global context, westernized universities tend to cement hierarchical structures to get quality international certifications and become competitive and economically profitable.

Human development models have evolved, so it is very clear now that economic development is only one of many variables to take into account. Politics for environmental protection, gender equality, equal access to education and health services, and food sovereignty are crucial aspects that determine levels of development and are analysed every year by United Nations Development Programme (UNDP) publications. But the influence of transnational economic power means that most governments fail to act upon these recommendations and so tend to put capitalist interests ahead of those of their citizens.

While many universities continue to be public institutions, a growing number of them, especially in Latin America, are both private and for-profit. Rama (2012) indicates how the "for profit" dynamic is one of the new expressions of commodification and internationalization of higher education and in some countries it is becoming the fastest growing university sector.

Given this state of affairs, substantial efforts are needed so that knowledge production in universities becomes a real source of emancipation and critical thinking. In particular, various forms of engaged, participatory and community-based art practice can play a fundamental role in the promotion of epistemological diversity and the decolonial option. Art projects that achieve such goals include the process for the mural known as The Great Wall of Los Angeles between 1976 and 1984 coordinated by Judy Baca, a landmark pictorial representation of the history of ethnic peoples of California from prehistoric times to the 1950s; and the Cabanyal Portes Obertes project, a cultural project that claims the rehabilitation of the Cabanyal neighbourhood in Valencia (Spain) since 1998. Organized by the Platform Salvem El Cabanyal (Save El Cabanyal), this project uses the houses threatened by the projected prolongation of a great avenue as an artistic space to oppose the demolition of the neighbourhood. In the city of Bogotá, the group Mapa Teatro led an artistic interdisciplinary project around a degraded and conflict-ridden neighbourhood known as El Cartucho/Santa Inés, which was being dismantled as part of an ambitious urbanistic renovation plan for this part of the city. Between 2001 and 2005, among other artistic interventions, two plays –Prometeo I and Prometeo II – were performed with the participation of the inhabitants of El Cartucho, as well as an interactive exhibition and installation within an old house downtown, as a metaphor for the space of the disappeared neighbourhood, made from the vestiges of its demolition (Abderhalden, 2010).

Adolfo Albán's (2012) notion of the aesthetics of re-existence is useful to describe this kind of artistic intervention:

> The aesthetics of re-existence are those of the decentralization, those of vanishing points that allow us to visualize different, divergent, disruptive life scenarios against the narratives of cultural, symbolic, economic, socio-political homogenization.
>
> *(Albán, 2012: 292, my translation)*

In this sense, any attempt to de-hierarchize of the relations of power existing between different fields of knowledge constitutes an oppositional move that disrupts existing inequalities and provides opportunities for resistance. This is important given that the dominant analysis of social and historical realities has been based on the points of view of men from five "Global North" countries (Grosfoguel, 2013). As Grosfoguel (2013) writes, this knowledge is highly provincial and localized but thanks to the university system, it is posited as universal and forms the foundation of the social sciences and humanities in today's westernized universities where Eurocentric epistemologies are privileged. In order to explain the historical processes of the world that produced structures of knowledge based on epistemic racism and sexism, Grosfoguel (2013) analyses four key moments of what he calls the "long sixteenth century". 1) The Cartesian philosophy, 2) The conquest of Al-Andalus, 3) The conquest of the American continent by Europe and what it implied for the Muslims and Jews of sixteenth-century Spain and for the African population kidnapped on their land to be enslaved in America, 4) The epistemicide carried out against Indo-European women who were burned on bonfires, accused by the Inquisition of witchcraft. It is important to note that the acts of epistemicide committed against Indo-European women took place before the conquest of the American continent, and methods and arguments used to condemn "witches" were then transferred to the persecution of Indians in America (Federici, 2004). Nonetheless, these four genocides/epistemicides are interlinked, since all of them are "constitutive of Western men epistemic privilege" (Grosfoguel, 2013: 77). In this way, "Westernized universities internalized from its origin the racist-sexist epistemic structures created" (Grosfoguel, 2013: 87).

This analysis clarifies how epistemic hierarchies have historically been constructed and perpetuated and are still present in the production and legitimation of knowledge in universities around the world.[1]

However, this context of epistemicides presented by Grosfoguel constitutes a historical process which evidences how an important number of populations were decimated and dispossessed from their rights, and at the same time a lower human condition, or not even a human one, was attributed to them. In this way, these populations were relegated as subjects incapable of scientifically valid knowledge production. The emergence and development of the European university system were influenced by this historical context of epistemicides, so that the knowledge produced by Arabs, Jews, most European women and American Indians was not considered valid, and instead the theoretical production of white European men was considered not only valid but universal. This European university system thus formed an academic structure and a set of contents that were then exported to the rest of the world, that were also universalized. The fact that the model followed by university education institutions at the global level is a model based on epistemic racism is a central issue to understand the characteristics of current societies and their development models, bearing in mind that these institutions have been, and still are, responsible for establishing the requirements for access to university studies, in addition to training a sector of the population, granting it with greater social and cultural legitimacy within their community and, in addition, more possibilities to accede to instances of power within the society.

In the specific case of Colombia and the situation of art studies within the university model, the genealogical analysis of Santiago Castro-Gómez (2015) is a very illustrative example of how racism and colonial hierarchies took part in the shaping of the structure of universities. In his analysis of "the problematic of the arts in its relationship with science, the university and the Colombian State" (Castro-Gómez, 2015: 210), from the colonial times to the present, he shows how art studies were excluded from or included in the structure of university education, depending on the interests of the State and the perceived usefulness of such studies to their development policies. At the end of the 19th century, the arts related to handicrafts were finally banished from the university space and instead the Fine Arts (painting, sculpture, engraving, music) were accommodated. However, after the violent events of 1948[2] and the government crisis, the idea emerged that the barbarism of the population is a result of economic underdevelopment not of a lack of civilized habits and behaviours. So the arts had to struggle to maintain their legitimacy within university structures. Nowadays with the emergence of neoliberal government technologies and within a globalized economy, universities are considered a strategic place to promote the type of knowledge that the productive sector needs to boost economic activity. In this context the teaching of art within most Colombian universities is seen as an obstacle for technological development and innovation. Castro-Gómez concludes this genealogical analysis with a critical diagnosis summarized in three points.

First he emphasizes the persistence of colonial heritages today, which manifest themselves in the hierarchical distinction between the logos of science and the techné in arts, which accords an epistemic superiority to the scientific *logos* against what he calls the "ways of doing" in arts. This situation leads to the teaching of arts having to be part of the research system of the formal sciences. Castro-Gómez (2015: 210) affirms that the way in which the State values the knowledge produced by the arts perpetuates "a type of" "epistemic violence"' that humiliates and ignores the work of artists and art teachers in Colombia". Ivonne Pini (1991) notes how the idea that the arts and humanities are not necessary for society results in a lack of material resources. Moreover, the existing sources of economic support, namely Colciencias (Administrative Department of Science, Technology and Innovation of Colombia) and ICFES (Colombian Institute for the Promotion of Higher Education), are designed for knowledge areas which have nothing to do with arts. Consequently, social scientists and humanists find it difficult to meet the requirements for funding from Colciencias. In 2015, a significant group of the most important researchers of the humanities in the country decided not to participate in the call to audit and evaluate research groups in Colombia (Navarrete, 2015).

Second, he recognizes the difficulty of the situation, since the unilateral intervention of the State in universities has been constant, while the idea of "university autonomy" has remained exceptional. In this sense, Castro-Gómez suggests that universities have to preserve autonomy spaces to discuss and debate between different forms of knowledge production, and also around what science means. For Castro-Gómez (2015: 211), "The unification of all Colombian universities under a single 'system' can only generate impoverishment of democracy and depredation of epistemic multiplicity." Finally, he highlights the need for the arts to be decolonized and less influenced by government priorities. For this purpose, Castro-Gómez points out that the arts would have to stop appealing to their "radical otherness" (claiming to be the "other" of reason or the "other" of institutions so that they occupy a completely different and isolated sphere in society), in addition to abandoning the separation that puts fine arts above handicrafts, a distinction inherited from the colonial past. He asserts that the arts should take on two tasks: they must be willing to carry out a creative dialogue with other research forms, particularly those of the social sciences, and show the State that there are many ways to produce and transmit knowledge and that there are no criteria that establish the superiority of some types of knowledge over others.

In the same sense that Castro-Gómez talks about challenging the "radical other-ness" of the arts, Grant Kester (2015), in addressing the relationship between the theory and practice of "socially committed" arts, also urges us to recognize the ways that they co-exist with other types of cultural practices.

> In order to develop a more substantive theoretical analysis of socially engaged art, however, I do believe it's necessary to challenge the singular privilege we've been taught to assign to art and the personality of the artist, and to acknowledge that art exists along a continuum with a range of other cultural practices that hold the potential to produce disruptive or counter-normative insight.
>
> *(Kester, 2015)*

In order to make a decolonial intervention into these calls for greater creative dialogue with other research methods, I propose a dialogue between a concrete type of artistic practice – namely, the New Genre Public Art – and a specific theory of human development known as Human Scale Development.

The arts' expanded field and the utopian pulse

In a recent article, Javier Gil (2012) shows us the relevance of the roles of new artists with respect to what he calls "consolidation of another place and sense of the arts", when he states that:

> Such a situation appeals for the need of an extended conception beyond the objects to insert themselves in relationships, situations and contexts and therefore, demands comprehension schemes that surpass the mere artistic readings.
>
> *(Gil, 2012: 110)*

In the 1990s José Luis Brea (1996) published his theory of critical utopian pulse – an extension of Rosalind Krauss' (2002) theory of the expanded field. Krauss' text aimed to explain how new artistic productions such as land-art, or ephemeral interventions in public spaces, connect with the traditional notion of sculpture. This theory which explains the 60s and 70s art forms according to the modification of the sculptural forms is titled "Sculpture on the Expanded Field", since it theoretically places the new territories that the proposals of these decades occupied. Artistic productions such as land-art exceeded the classical parameters of what is considered a work of art – and the places where it is located – as artists started doing open-air performances as well as transforming country landscapes or urban buildings.

The fact that the art practice could be defined not only by the use of a given medium but by its relation with the occupied place opened the door to new possibilities of exploration and occupation of the emplacements by the artists.

Traditionally, the artists worked in their workshops and the produced objects were transferred to the place of exhibition. Nowadays, however, many artistic works are performed in situ – that is to say, that the artists directly intervene in the place where the work is located. That is why the word emplacement, used in geology, as the process of setting something in place or being set in place, was used to describe this kind of intervention. The relationship between art and place was transformed into a decisive aspect in the configuration of new ways of making and producing artworks. In the 1980s and 90s, art's action field was further enlarged through practices that included interactions with the body, appropriation of the media and the internet, as well as urban, activist and social transformation art. Brea's proposal gives place to a more complex scheme in which those new artistic practices could be included, including the public uses of artistic forms.

According to Brea's analysis, artistic practices have been advancing through a movement of utopian-critical expansion, which goes from what we traditionally understand as monument (a sculptural object placed in a public place, as an object of historical commemoration or in order to embellish the environment) to categories as recent as urban art or art linked to social practices. This study, based on art-extended territories, leads Brea to differentiate between ornamental-oriented art and the utopia-oriented art, between art which is concerned with aesthetics, and a visual impact, and art which is more concerned with the process and its social impact.

To Brea (1996), the advantages of the scheme that he proposes, besides the possibility of including many more practices than those allowed in Krauss' model, is that of assigning them a mutual relation, not only under the perspective of a formal analysis but "in relation to the public use of those forms, to their contextual relation with the social network, to their condition of communicative, significant practices, carried out by individuals of praxis and experience" (Brea, 1996: 36, my translation). As examples of the above I refer to two cases from Bolivia and Spain. Mujeres Creando (Women creating) is an anarchist feminist movement that was created in 1992, first with the name of Comunidad Creando (Community creating), in a suburb of La Paz, Bolivia. This group began with an anti-racist, feminist proposal that questioned the elite privileged women who separated the public from the private, and manual labour from intellectual work. They have used graffiti and actions with a strong presence in the streets, understanding creativity as a tool of social struggle; however, they do not consider themselves artists but street agitators (Álvarez, 2009). In Spain, the architect Santiago Cirugeda is known for his subversive architectural projects. Most of his works try to address the architectural demands of people who are denied political, legislative or real estate representation. Thus, the Spanish architect creates and installs extensions in public spaces, roofs or buildings, generating urban prostheses that, despite their "illegal" character, work and solve problems of housing and design of spaces. In 2003, after seven years of individual work, he created the architectural studio Recetas urbanas (Urban Recipes) and continues to carry out this kind of subversive project collectively.

New Genre Public Art

With the relevance acquired by the singularity of emplacements, the matter of context becomes crucial and many of the artistic propositions of the second half of the 20th century acknowledge the social, economic, cultural and religious reality of the community. I believe that the notion of *New Genre Public Art* (NGPA) synthesizes the distinct forms of hybridization that have arisen in recent decades, where the importance of the processes of production of an artistic project emerges. The term New Public Art refers to public art, often activist in nature, and created outside institutional structures in order to engage directly with an audience.

NGPA emerged as a result of a publication edited by the artist Suzanne Lacy in 1994, which is based on a public event realized in 1991 at the Museum of Art of San Francisco, with the name *Mapping the Terrain*. NGPA covers practices that range from cultural activism to communal and relational aesthetics; hybrid practices that serve as tools such as the photomontage and that are connected to different social demands such as the work of the AIDS activist artist collective from New York City Gran Fury (1988–1995) based on appropriation of commercial language for political ends, or the feminist activist artists group Guerrilla Girls, focused on fighting sexism and racism within the art world using culture jamming tactics.

This concept formulates a difference between all the spectrum of practices contained in public art and provides a more critical approach in relation to these kinds of practices. NGPA gathers various practices that have been developing throughout the 20th century, from contra-monuments to feminist and activist art practices that are committed to the revitalization and reactivation of the social forces that accompany such processes. In this manner, according to Lucy R. Lippard (2001: 68) "the redefinition of art and of the artist can help to purge a society that is alienated from its vital forces". Apart from being linked to sociocultural, relational and multicultural practices, among others, NGPA also has a highly transdisciplinary vocation, because when proposing a dialogue in specific contexts with attitudes and contents that are more committed and closer to reality, it needs to draw on perspectives from other knowledge fields.

Extensions in the field of development: Human Scale Development (H-SD)

If collaborative and socially engaged art practices seek a dialogue with other forms of research within the social sciences, one possible critical step to take is to connect these practices with a development model. Community-embedded artistic practices that seek to transform social reality can learn a lot from research practices in the social sciences and vice versa. And in order to have a better understanding of the role or impact of such practices in a given social and economic context, it is necessary to know the dynamics of that context and its relation to economic and social policies, since they set up the type of development model in which that community is enrolled.

Since the 1980s when the notion of development became linked to the question of human rights, disrupting the economic understandings of development that were dominant in the 1970s, development studies now works with much broader characterizations of what it is to enjoy wellbeing or to have an adequate standard of living. Accordingly, this process has resulted in the design of more diverse indicators to assess the level of development of a territory. I draw here on the proposal of Max-Neef *et al.* (1994) as an integral proposal and because it includes "aspects that have been too neglected in development practice: affection, leisure, creation" (Boni, 2005: 122) and that are relevant to the linkages that I present here.

The H-SD was first published in 1986 and proposed that human development should focus more on people and their basic needs. It basically stipulates that development is about people and not about objects, which means that transforming an object-person into a subject-person is a question of scale, as people cannot be active participators within gigantic systems and hierarchical organizations if decisions always flow from the top down to the bottom. Quality of life depends on people being able to adequately satisfy their fundamental human needs, so development processes should be geared towards those outcomes (Max-Neef, 1994).

To achieve the greatest improvement in people's quality of life H-SD indicates that, as a key principle, it is imperative to differentiate between basic human needs and satisfiers of those needs, since, against the general belief that human needs tend to be multiple and interminable, it argues that these needs can be counted and can be classified. As a result, there is a first group of needs: Being, Having, Doing and Interacting (Ser, tener, hacer, estar), and a second one that includes the need for: Subsistence, Protection, Affection, Understanding, Participation, Leisure, Creation, Identity and Freedom. H-SD thus affirms that everyone has the same fundamental human needs irrespective of culture or life context. What changes, however, is the way those needs are satisfied and the deployment of economic assets to do so.

There are at least five types of satisfiers. The first are violators or destroyers. These are applications intended to satisfy a need in a top-down manner. Not only do they fail to do so, but they often inhibit the adequate satisfaction of other needs due to their collateral effects. Exile is an example of a violator satisfier, since aiming to satisfy the need of protection it prevents the satisfaction of other needs such as subsistence, affection, participation and freedom. The second are pseudo-satisfiers. These are elements that stimulate a false sense of satisfaction of a given need. They are generally induced through propaganda, advertising or other means of persuasion. An example of this kind of satisfier is the overexploitation of natural resources, giving the sense of satisfying the human need of subsistence. The third are inhibiting satisfiers. These satisfy a particular need, often adequately, but seriously hinder the satisfaction of other needs. They often materialize in sets of rituals that emerge from ingrained habits. Commercial television can be considered among these kinds of satisfiers, since satisfying the need for leisure inhibits the satisfaction of understanding, creation or identity. The fourth are singular satisfiers. These aim to satisfy a single need and have no impact on the satisfaction of other needs, which is the case of curative medicine, which satisfies only the need for subsistence. They tend to be institutionalized and are common in development projects and programmes. The final are the synergic satisfiers. These satisfy a given need but they also stimulate and contribute to the simultaneous satisfaction of other needs. Among these kinds of satisfiers we find cultural television, which besides leisure satisfies the need for understanding; or preventive medicine, which allows for the satisfaction of understanding, participation and subsistence.

Max-Neef's definition of the synergy concept highlights the importance of these kinds of satisfiers for the transformation and overcoming of the hierarchies present in society, since they privilege collective and interdisciplinary processes over individual ones. For him: "synergy denotes a form of potentiality, a process in which the potential of the associated elements is bigger than the summed potential of each of those elements" (Max-Neef, 1994: 64).

Establishing a relationship between needs and satisfiers helps to develop processes that lead to a model of development that is genuinely humanistic. The cultural and political context must, however, be analysed, as it has the potential to repress, tolerate or stimulate the existing possibilities within a given environment. It is therefore necessary to differentiate between satisfiers and economic goods. Traditional political economy suggests a direct relationship between needs and economic goods, however, according to Max-Neef, including in this relationship an analysis of satisfiers allows for the addition of a subjective factor. The alienation of this subjective factor has contributed to the fact that our societies determine their means of production and consumption according to economic goods, through which "life, then, is placed at the service of artifacts, rather than artifacts at the service of life" (Max-Neef, 1994: 51).

Art practice as traditionally understood is a source of economic goods that exalts individual creativity. Based on the idea of an artist as a genius whose manual or intellectual skills outstrip those of common people, the value given to their artistic creativity is in most cases instrumentalized to inflate the possible value of the produced artworks. I would say then that art practice driven by galleries and art markets could be considered, at best, a singular satisfier of the fundamental need of creativity. Bearing in mind how the overvaluing of some particular art productions distorts the social imaginary about the role that artists have in their societies, I would also affirm that this instrumentalization and speculation made around art objects could even be considered a pseudo-satisfaction of the need for creativity. In these cases most art productions are the result of repeating a formula that is successful for sale rather than a result of experiment and research. Understanding the dialectic relationship between needs, satisfiers and economic goods poses a challenge to social planning and development policies, since it is about "conceiving forms of economic organisation in which goods empower satisfiers to meet fully and consistently fundamental human needs" (Max-Neef, 1994: 51). These movements of expansion, in the field of the artistic practices as well as in the models of human development, made it possible to find a contact zone between both fields, where synergic ways of satisfying the human need of creativity are achievable. This would not be possible in a context where it is considered that social transformations and local development are to be obtained only by economic policies, on the one hand, and on the other hand where the single function of artists is the production of ornamental objects suitable for gallery exhibitions and auctions.

Synergic satisfaction of needs from artistic practices

In his work on the redefinition of artistic practices, José Luis Brea (2003) emphasizes the growing responsibility that individuals who generate artistic practices have in relation to image and visual cultural power. According to him, the weight that current visual imagery places on the individual and social construction of people is decisive nowadays, and artistic practices have assumed that responsibility through three different routes. The artist as producer is a) a creator of narratives of mutual recognition; b) a generator of intensified situations of encounter and of socialization of experience; and c) a producer of mediators for their exchange in the public sphere (Brea, 2003: 164, my translation). Correspondingly, many of the practices belonging to New Genre Public Art could be considered a synergic satisfier of fundamental human needs as far as they tend to put the focus on processes which lead to a visual result or an artistic intervention rather than on the final result. As this kind of practice favours forms of collective creation and collaboration between artists and communities as well as between artists and professionals from other fields, it also facilitates new spaces for meeting and participation.

At the same time, the modus operandi of the aforementioned artistic practices is increasingly characterized by questioning its processes based on three key axes: the contextual, the relational and the representational. Besides approaching the concepts gathered from *Relational Aesthetic* (Bourriaud, 2004) and *A Contextual Art* (Ardenne, 2002), I am interested in observing the different ways of approaching those axes, and which of these ways would indicate a shift towards a more humane scale, within artistic practices.

By way of example I address the latest artistic intervention coordinated by the Colombian artist Doris Salcedo in Bogotá named Sumando Ausencias (Adding Absences), an enormous installation in a historical and strategic place in the heart of Bogotá, the Plaza de Bolívar. After the result of Colombia's referendum for peace, Salcedo produced and coordinated this project in a week to support the demands for a definitive peace agreement. The installation consisted of covering the whole area of the Plaza de Bolívar with a white blanket made of 2300 sheets, each one with the name of a conflict's victim written with ashes. Thousands of persons came voluntarily the day of the covering of the Plaza to help with sewing the 2300 sheets. The enormous installation required the closure of the square throughout the day and the help of the police to control the access, in addition to the temporary transfer of the Peace Camp that had been installed there for several days. This Peace Camp was moved to a side of the square where it did not visually interfere with the artistic installation, a decision made after the artist rejected the campers' demand to leave the camp in its place, because it affected the visual neatness that she always strives for in her projects.

This artistic project spoke to the importance of the political moment and attracted massive national and international media coverage. However, hundreds of people had to be hired by the artist for the process of writing the names with ashes, many of whom were art students, and some of them expressed their frustration, as they were motivated in participating and working with the artist, but the relation established was that of subordinates, given the speed and competence with which they were expected to work.

A woman passing through the square during the sewing process asked the artist Doris Salcedo how she could know if her son's name was there. The answer was that there was no way to know that because she had used a list of victim's names that was provided by the government. I heard this while I was there sewing two sheets and I questioned myself about the place for the victim's family members in this project and concluded that due to the urgency of the action, the artist needed only the victims' names and thousands of anonymous people collaborating with the installation. But reading the headlines that followed the artistic intervention, for the most part one could read the artist's name and that she had covered the square with thousands of sheets but only after reading the content was one aware that the action was made possible by the work of thousands of volunteers. Indeed, Colombia is a country full of victims' associations, of social movements linked to the search for disappeared family members, and most of them use the space of the Plaza de Bolívar to carry out artistic interventions to make their struggle and their demands for justice visible. The peace negotiations put them in a more privileged place, but there is no comparison between the media coverage and institutional and logistic support given to their interventions and that given to Doris Salcedo's projects.

Although this project addressed the Colombian political context, I tried to point out that human relations established between the artist and the collaborators were hierarchical, and hired people and volunteers were just following instructions, while the artist figure was presented in media and social networks as someone that facilitates the space for memory recognition and whose monumental art piece is of a participatory nature. From my point of view this kind of art project is ornamental-oriented and not utopian-oriented, since it takes the place of the victims to speak for them instead of giving voice to them. Volunteers, hired people, campers and victims' associations were passive participants, because the art process was not intended to connect or give continuity to their particular social processes, but to draw a huge and institutional visual gesture to honour the victims.

Visual translation of imaginaries as a research tool

From my experience having collaborated with different non-government organizations in Valencia between 2010 and 2013, I proposed some artistic processes linked to codevelopment projects. Assuming the role of an artist more in the line of the definition made by Brea, I carried out two projects with groups of women of different geographical origins: *Bellas Durmientes* (Sleeping Beauties 2010), *Soñadoras* (Dreamy Women 2011), and later the *Paisajes para la Paz* project (Landscapes for Peace 2013), which was carried out with a group of Colombian immigrants in the city of Valencia, Spain.[3] These experiences are gathered in my doctoral research presented at the beginning of 2016, where I described the methodological process conceived during the three projects (Corzo, 2016).

Inspired by the principles of Participatory Action Research (PAR), which considers that it is necessary to involve people in the process of researching their own reality in order to gain a clearer understanding of how they internalize and interpret natural and social phenomena, I developed a method I called Visual Translation of Imaginaries (VTI), put into practice for the first time during the Paisajes para la Paz project. It comprised three stages: 1) Production of narratives workshop, 2) Visual translation workshop, search/creation of images and 3) Visual creation workshop.

One of the key objectives of the Production of Narratives phase was to get closer to the participants' reality and context. The process consists of determining together with the participants a theme of common interest, establishing a series of questions from which each person expresses their point of view in writing, and then sharing those ideas with the group to finally group them by concepts. The Visual Translation Process starts with a process of a written description of the images through which everyone would represent each concept agreed in the first phase, followed by the collection or creation of images that corresponds to the written translation. This seeks to provide the concepts that result from the production of narratives with a visual body. In the final phase, that of Visual Creation Process, the visual composition is approached through the creation of photomontages put together from the images to which the different concepts, agreed upon in the narrative production phase, were translated. To illustrate these steps, the following figure shows the process of one of the participants of the *Paisajes para la Paz* project. First the concepts and the description of the elements imagined are given. Then the images chosen to represent those elements described and finally the same images are used to compose a collage.

A detailed monitoring of each stage in this process enables the elaboration and study of the three key axes mentioned above that enhance the synergy levels of the research-creation practices in the arts. On the one hand, I developed the process placing the element of desire as central, as a necessary aspect of activation for the definition of those personal and collective transformations that each one of us wants to see in our environment, seeking to strengthen the development of critical views and participatory action. On the other hand, there is a fundamental component in the construction of imaginaries of place, since these photomontages can be considered to be biographical and utopian landscapes. This is an exercise of thinking with images, however, besides the relevance of such an act, this methodology finds much of its meaning in constructing significant images that are opposed to the visuality that surrounds us in contemporary societies.

Finally, each of the workshops are sensemaking sites, since they are designed as meeting and exchange spaces for personal and collective reflection and work to

Example of the visual translation process
Paisajes para la Paz project

1. Coexistence	=	Street, neighbourhood
2. Peace	=	T-shirt with Colombian motto
3. Respect	=	Silence
4. Life	=	Jungle, seed
5. Development	=	Health card, doctor
6. Equality	=	The difference between equality and justice, with people standing on boxes
7. Environment	=	Clear river
8. Memory	=	Map of Colombia in a brain
9. Tranquility	=	A person reading while looking at a Colombian landscape
10. Human rights	=	The Constitution held in hands

FIGURE 14.1 Example of the visual translation process

promote the satisfaction of the fundamental human need of understanding. It is worth considering how these artistic practices that are linked to the New Genre Public Art and collaborative and community art could synergically satisfy the basic human needs of the beneficiaries of the process and as result complement and enrich the quality of the impact that these practices seek in society. It is then critical to redefine the role of artists from an individual one into a collective one that surpasses their mission of representing reality to become agents of critical reflection and social transformation.

All this facilitates interdisciplinary collaboration and therefore spaces of dialogue, as I have experienced within these projects. A group of Spanish and migrant women came together to talk about their dreams and describe the places where they would like to be in the future during the *Sleeping Beauties* project. The same issue was addressed between artist women and entrepreneur women for the *Dreamy Women* project. Finally, to carry out the *Landscapes for Peace* project and the process of Visual Translation I collaborated with the association of Colombian migrants to which I belonged and also with a group of Spanish associations committed to the human rights situation in Colombia. There were no artists in those associations but it was necessary for me to take part in the spaces of reflection they opened up regarding the political participation and contribution of Colombian migrants to the peace negotiations in La Habana to integrate those reflections to the Production of Narratives workshop.

The dialogue between artistic practices and social sciences is a step forward towards the utopian-critical movement described by José Luis Brea, and an advancement towards the expansion of possible realities, since it invites us to give more relevance and, therefore, visibility to the artistic process and not just to the impact of the outcome. The visibility of processes within artistic practice aims to prevent the invisibilization of the participants' diversity of experience, which is characteristic of individual and ornamental-oriented art practices.

The westernized university structure is a result of historical conformations that date back several centuries. The challenge of its transformation requires interdisciplinary dialogues to broaden the research perspectives of the different fields of knowledge. Regarding artistic practice, the redefinition of its ways of doing, assumed by the New Genre Public Art as well as by community and collaborative experiences, may have a synergistic potential, which is very difficult to evidence without extra-artistic criteria. It would be necessary, besides the design of indicators that can account for the degree to which different human needs are met during the processes, that the experiences and observations made up to the moment could be analysed by experts from other knowledge disciplines, not only by professionals in the field of art.

All projects mentioned were done as independent projects; the first two, with codevelopment funding, and the third was self-managed with the support of the Asociación Entreiguales Valencia. They were not part of an official university research programme or project, so knowledge creation at Universidad Politécnica de Valencia and Universidad de Valencia was not decisively impacted by these projects, although it did provide spaces where I could present or exhibit the results, including the exhibition room in the Social Sciences Faculty and the hall in the Psychology Faculty at the Universidad de Valencia and this material was also part of my doctoral thesis completed at Universidad Politécnica de Valencia. I believe that the westernized university must begin to embrace and support such collaborative artistic initiatives more fully as a part of decolonial efforts to dismantle epistemic hierarchies, provoke urgently needed shifts in knowledge production and contribute to transformative social change.

Notes

1 Within the structure of powers in the academy, there is a lower status for the arts and humanities than for the physical sciences, partly due to the progressive focus of universities on careers that meet market needs (Cordua, 2012). For example, in 2015 in Colombia, of the 189 doctoral programmes that competed to receive Colciencias (Administrative Department of Science, Technology and Innovation of Colombia) scholarships to finance their students, only 40 passed the preliminary evaluation. None of them were in the humanities and social sciences (Correa and Navarrete, 2015).
2 In 1948 the events known as *El Bogotazo* took place in the capital city of Colombia. *El Bogotazo* refers to the massive riots that followed the assassination in this city of the Liberal leader and presidential candidate Jorge Eliécer Gaitán on 9 April 1948. Much of downtown Bogotá was destroyed after the riots.
3 More information about these projects can be retrieved from: http://bellasdurmientesproject.blogspot.com.co/ and http://paisajesparalapaz.blogspot.com.co/.

References

Abderhalden, R. (2010) El artista como testigo: testimonio de un artista. *Utopías de la proximidad en el contexto de la globalización: La creación escénica en Iberoamérica*. Universidad de Castilla – La Mancha, pp. 295–302.

Albán, A. (2012) Estéticas de la re-existencia: ¿Lo político del arte? *Estéticas y Opción decolonial*. Bogotá: Editorial Universidad Distrital Francisco José de Caldas, pp. 281–295.

Álvarez, H. (2009) Mujeres creando: Feminismo de luchas concretas. Available at: www.mujerescreando.org/pag/articulos/2009/06-junio/mujerescreando.htm (Accessed 27 October 2017).

Ardenne, P. (2002) *Un arte contextual: Creación artística en medio urbano, en situación, de intervención, de participación*. Murcia: Cendeac.

Boni, A. (2005) La educación para el desarrollo en la enseñanza universitaria como una estrategia de la cooperación orientada al desarrollo humano. Unpublished doctoral thesis. Valencia: Universidad Politécnica de Valencia.

Bourriaud, N. (2004) *Esthétique relationnelle*. Les presses du reel.

Brea, J.L. (1996) Ornamento y utopia: Evoluciones de la escultura en los años 80 y 90. Available at: www.academia.edu/10070030/Marina_Pastor_-_Fia._Escultura_s.XX_-_2_Jose_Luis_Brea_-_Ornamento_y_Utopia.

Brea, J.L. (2003) *El tercer umbral: Estatuto de las prácticas artísticas en la era del capitalismo cultural*. Murcia: CENDEAC. Centro de Documentación y Estudios Avanzados de Arte Contemporáneo de la Región de Murcia.

Castro-Gómez, S. (2015) Descolonizar las artes: Una genealogía del modelo de la universidad-empresa en Colombia. *Intervenciones en estudios culturales* (2): 197–212.

Cordua, C. (2012) La crisis de las humanidades. *Revista de Filosofía* 68: 7–9.

Correa, P. and Navarrete, S. (2015) ¿El fin de las humanidades? Available at: www.elespectador.com/noticias/educacion/el-fin-de-humanidades-articulo-591959 (Accessed 14 October 2017).

Corzo, M. (2016) Paisajes y utopias: Traducción visual de imaginarios en contextos de migración e interculturalidad: Tres casos de estudio. Unpublished doctoral thesis. Valencia: Universidad Politécnica de Valencia. Available at: https://riunet.upv.es/handle/10251/61963 (Accessed 14 October 2017).

De Sousa Santos, B. (2017) *Democracia en tiempos inciertos.* Conference. 30th Bogotá Book Fair.

Federici, S. (2004) *Calibán y la bruja: mujeres, cuerpo y acumulación originaria.* Madrid: Traficantes de sueños.

Gil, J. (2012) De luces y sombras: a propósito de las estéticas comunitarias y colaborativas. *Errata # Revista de artes visuales* 7. Bogotá: Fundación Gilberto Alzate Avendaño e Instituto Distrital de las Artes, pp. 108–127. Available at: http://issuu.com/revistaerrata/docs/errata_7_creacion_colectiva_practic/7?e=0/8147026 (Accessed 14 October 2017).

Grosfoguel, R. (2013) The structure of knowledge in westernized universities: Epistemic racism/sexism and the four genocides/epistemicides of the long 16th century. *Human Architecture: Journal of the Sociology of Self-Knowledge* 11(1): 73–90.

Kester, G. (2015) On the relationship between theory and practice in socially engaged art. *Fertile Ground.* Available at: www.abladeofgrass.org/fertile-ground/on-the-relationship-between-theory-and-practice-in-socially-engaged-art/ (Accessed 27 October 2017).

Krauss, R. (2002) La escultura en el campo expandido. In H. Forster (ed) *La posmodernidad.* Barcelona: Ediciones Kairós, pp. 59–74.

Lippard, L.R. (2001) Mirando alrededor: Dónde estamos y dónde podríamos esta. *Modos de Hacer: arte crítico, esfera pública y acción directa.* Ediciones Universidad de Salamanca, pp. 51–71.

Max-Neef, M. (1994) *Desarrollo a escala humana: Conceptos aplicaciones y algunas reflexiones.* Barcelona: Icaria Editorial, S.A.

Navarrete, S. (2015) Colciencias y los humanistas en orillas diferentes. *El Espectador.* Available at: www.elespectador.com/noticias/nacional/colciencias-y-los-humanistas-orillas-diferentes-articulo-545541 (Accessed 15 October 2017).

Pini, I. (1991) La investigación en arte. In A. Gutiérrez and B. Carlos (eds) *La investigación en Colombia en las artes, las humanidades y las ciencias sociales.* Bogotá: Ediciones Uniandes, pp. 25–41.

Rama, C. (2012) El negocio universitario "for-profit" en América Latina. *Revista de la Educación Superior* 41(164). Available at: www.scielo.org.mx/scielo.php?script=sci_arttext&pid=S0185-27602012000400003.

Toro, J. (2016) Las dudas que levantó la obra de Doris Salcedo en la Plaza de Bolívar. *Vice magazine.* Available at: www.vice.com/es_co/article/nnp87g/doris-salcedo-plaza-bolivar-victimas-sumando-ausencias (Accessed 15 October 2017).

15

"LIBERTÉ, EGALITÉ, FRATERNITÉ"

Debunking the myth of egalitarianism in French education

Olivette Otele

Philosophy is part of the French education system. It starts in primary school and is further developed in secondary school. For decades, pupils were expected to be familiar with key texts that dealt with French history and philosophy. They learnt about Rousseau, Montaigne, Montesquieu, and others, and by the end of secondary school, specifically the year of the baccalaureate, philosophy was compulsory for all sections, even if students had chosen economics, biology and mathematics as their core subjects. In 2013, the curriculum was re-assessed in order to introduce pupils to other philosophical principles while adding elements of history, linguistics, psychology, law, sociology, and so on. A subject entitled Moral and Civic Education was created in 2013. It started in secondary schools and according to the Ministry of Education (MEN), "moral education is about linking the individual to the common, to re-articulate the moral with the civic, the person with the citizen, and to find adequate conditions to make a community" (MEN, 2013). Pupils before 2013 learnt about philosophy as a subject but also how to philosophize (Agostini, 2010). They are now taught about civic responsibility through an elaborate programme divided into four main parts (MEN, 2015). Part one, Sensitivity: Self and Others, looks at emotions and the language used to convey one's feelings. It also examines the question of respect and topics such as racism, anti-Semitism, homophobia, and sexism. Interestingly enough it does not look at Islamophobia. Part two, Laws and Rules: Principles to Live with Others, delves into behaviour in the classroom as well as one's relationship with the wider community. *Laïcité*, or the French interpretation of secularism is also looked at. Part three, Self-reflection and Thinking with Others, examines the question of choice and interprets the notion of *laïcité* as the "freedom to think and to believe or to not believe" (MEN, 2015). Part four covers the topic Individual and Collective Activism. This part is aimed at teaching pupils about solidarity and moral activism based on principles such as trust and faithfulness, while collective activism is based on the notion of fraternity. The principles of the French Republic, Liberty, Equality and Fraternity are interwoven into these four parts. Students are taught that these values form the founding principles of French and European citizenship and that they are at the basis of democracy.

The purpose of this chapter is to study the discrepancy between this "citizenship education" based for the most part on the philosophy of the French Enlightenment, which widely supported the colonization of the Americas on the one hand, and on the other, pervasive xenophobia, racism, and failed attempts to integrate minority ethnic groups in contemporary France (Agostini, 2010). French education is still based on colonial stances and the far-right ideology still influences French society and politics in spite of a thriving community of intellectuals from the African diaspora. This chapter will be articulated around three case studies: firstly former French president Nicolas Sarkozy's Dakar discourse of 2007, then the 2005 dispute about the "positive role of colonization" and finally the removal and quick reinstatement of the history of Africa in secondary schools in October 2015 paired with the question of French post-colonial citizens and immigration in the 2017 presidential elections.

The notions of citizenship, equality, and even the French form of secularism are based on the quest for freedom. Freedom of speech, for example, is one of the pillars of French education. It implies critical thinking and debating societal issues in the classroom. Later on, engaging with politics, history and literature in the public sphere is expected from young adults. This form of active citizenship has been inherited from French Enlightenment thinkers. In 1721, Charles de Montesquieu wrote *The Persian Letters*. Several themes, such as how to articulate freedom and reason with the question of happiness, were developed in this volume. In his seminal work *The Spirit of the Laws*, Book 12, chapters 12 and 13, he examined the question of freedom of expression and the ways in which criticism was presented. The philosopher noted that "Words do not constitute an overt act; they remain only in idea. When considered by themselves, they have generally no determinate signification; for this depends on the tone in which they are uttered" (Montesquieu, 1748). Montesquieu (1748) argued in favour of satirical writings: "They may amuse the general malevolence, please the malcontents, diminish the envy against public employments, give the people patience to suffer, and make them laugh at their sufferings". Philosopher Jean-Jacques Rousseau on the other hand challenged the idea of corruption in society. According to Rousseau (2009), human beings were in essence good but society corrupted them and that led to a state of inequality. In *The Social Contract*, first published in 1762, he defended the idea that individual freedom was crucial but it needed individuals to work towards the good of the community. It implied abandoning part of our liberty for the greater good. This freedom was to be transferred to legislators whose role was to protect society from individual corruption and personal gain. Each one of us, he contended, was invited to establish a "social contract". Fellow philosopher Voltaire criticized his contemporaries' issues and aspirations. Using satire in *Candide*, Voltaire noted that people needed to learn to question authorities and power, and to exercise caution towards all forms of religion. Voltaire's distaste for Islam was particularly salient in the play *Fanaticism, or Mahomet the Prophet*, published in 1736. He condemned what he saw as inconsistencies in the Qur'an. Yet twenty years later, in 1756, he published *Essai sur les Moeurs et l'Esprit des Nations* in which he argued for strong connections between Europe and the colonies in the East. As far as race and slavery were concerned, Voltaire believed in polygenism, which implied a hierarchy amongst human beings. Yet, in *Candide's* chapter about Surinam, Voltaire ridiculed slave-owners and attacked their treatment of slaves.

Montesquieu, Diderot and Voltaire's essays have contributed to the contemporary debates about France's involvement in overseas colonies. Voltaire's financial investments have been questioned over the last twenty years. It has been argued that Voltaire's anti-slavery position contrasted with his financial participation in overseas trade. Nelly Schmidt ironically noted that the philosopher's outrage at the way enslaved Africans were treated in *Candide* was at odds with his investments in Nantes' slave trading ventures (Schmidt, 2001). Opposing Schmidt's views, Jean Erhard (2008) argued in his volume that Voltaire, Montesquieu and Diderot had to be understood in the context of the 18th century economic, social changes. Voltaire, for instance, invested in trading ventures that brought him profits but did not necessarily seek to invest in the slave trade according to Erhard. These discussions bring to the surface the uneasy legacies of 18th-century French intellectuals. Beside financial gains, the question of prejudice and racism is another bone of contention related to colonial legacies. Francoise Vergès (2006) demonstrated that one could not understand the issues of race, racism, hierarchy and inequality in contemporary France, without re-evaluating key documents such as the Declaration of Human Rights and the Black Code or the bill about the abolition of the slave trade and the decree turning Algeria into a French colony. Vergès' point is that the laws regulated the movement of indigenous populations and enslaved Africans whilst legally denying them the means to combat subjugation. These laws legitimized inequality and promoted racial hierarchy in the colonies. Verges therefore contended that contemporary racism and Islamophobia were a legacy of 18th-century laws and 19th- century measures against indigenous populations in Algeria.

Enlightenment philosophers' writings all highlighted the importance of the freedom of individuals. Freedom of expression is again inscribed into the 1789 Declaration of the Rights of Men and of the Citizen (Conseil Constitutionnel, 1789). Articles 10 and 11 specify that freedom of expression is guaranteed by the law but should not jeopardize public order. The constitution made it clear that the country had a political project based on equality between citizens. Equality was also to take precedent over individual freedom, as previously seen in Rousseau's idea about the broader community's good over individual gains. Those rights were refined over the next centuries. In 1881, for example, a bill was ratified which stated that the law guaranteed the freedom of the press but the press could not "insult, vilify or falsely accuse someone" (Legifrance, 2017). The nation's founding principles were also based on the notion of individual rights and the idea of various communities' identities coming together. Common identity was supported by common values. School was there to transmit these values. Member of Parliament Jules Ferry's school reforms of 1880–81 were to promote moral values rather than religious ones (Mougniotte, 2006). The relationship between the values defended by society and state schools' programmes was close in 1881. It has been argued that one of the roles of the school was to "reinforce social homogeneity" (Mougniotte, 2006: 176). This could be achieved by firstly teaching cultural values through literary and artistic masterpieces; secondly by transmitting social values represented by Republican institutions seen as symbols of freedom; and thirdly by supporting moral values considered to be universal values. Problems arose when there was a clear discrepancy between what these values encompassed and people's lived experiences. In the 19th century these values were still accepted by the vast majority of the French population. Centuries of colonialism, decolonization, World Wars and post-colonial immigration with its cohort of racism, discrimination and oppression widened the gap between these values taught in school and the experiences of post-colonial children of the Republic.

State school has had a political function since Jules Ferry's reforms. It had been contended that schools were to promote France as a nation. It meant relegating individual social practices and an individual sense of belonging, including ethnicity, to the private sphere (Xypas, 2006: 224–225). The aim was to create "a common historical memory" and "fundamental values" that would use the past to build the present and shape the future (Xypas, 2006: 224). Constantin Xypas (2006: 225) noted that because school was "a place in the material and metaphorical sense that stood against the inequalities of social lives and that resisted social civil movement" [...] "the combined action of the school on the past, present and futures of younger generations, makes its main contribution to the consolidation of the citizen and of the nation". The limit to these considerations appears when schools reproduce social inequality and when that unifying environment is perceived as being hostile. Laurent Mucchielli (2016) has demonstrated that delinquency of post-colonial young men, for example, was the result of a number of key factors such as fragile family structures, society's failings, and identity crisis. Exclusion from school and rapid isolation at a young age, Mucchielli (2016: 236–238) argued, participated in the "identity construction of young delinquents". These points regarding the close link between the French curriculum and the political, social and socializing role of the state school help us understand why certain topics such as history and history programmes have become tools used by politicians to convey specific messages to the population.

The debate in France echoes similar discussions in European countries that had overseas colonies. In the Netherlands, the debate about Black Pete has produced a vast number of scholarly volumes. The Netherlands' involvement in the slave trade has been widely documented (Nimako and Willemsen, 2011) and started in the 15th century. When the character created by Jan Schenkman appeared in a children's book in 1850, the Netherlands had not yet abolished slavery in their colonies. That took place thirteen years later. Supposedly a Moor from Spain, Black Pete, was allegedly an amiable character. In 2016, a study based on a sample of 201 pupils aged 5–7 showed that Black Pete's skin colour and his lower status in society were well known to children. Their perception of Black Pete did not imply that all Black people were or should be considered in a similar way. In fact Black Pete was associated with a clown loved by most children. The research argued that adults in the Netherlands have shared similar perceptions since the book was published. As the study demonstrated, that was the reason why it was difficult for most Dutch people to acknowledge that racial stereotypes could be attached to the character of Black Pete. Every year Black faces parades still take place during the children's holidays at the Sinterklaas Festival. The links between Black Pete, slavery, racial hierarchy born during colonial times, received ideas about Black people and education in the Netherlands are still subject to controversy to this day. Scholars such as Goddeeris (2015) and others have shown that similar debates about colonial legacies, education and racism are also taking place in Belgium. In France, the story took on a new turn when Nicolas Sarkozy decided to use these issues in the political arena.

In 2006, the then Home Secretary Nicolas Sarkozy pushed forward a bill that was aimed at supporting what is known as "chosen immigration" or white colour immigration (Ministère de l'Intérieur, 2006). Before an assembly of senators, Sarkozy presented his objectives, which were to further reduce immigration and thus foster, according to him, a stronger social cohesion. He explained that he had met with several African heads of state and civil servants to work together on measures to reduce migration to France. Sarkozy was eager to explain that he was simply answering to public expectations and was determined to save the "sinking ship" – in other words, a France burdened by a high number of immigrants. He quoted a poll done by a right-wing newspaper, sympathetic to his party, *Le Figaro* as a reliable source for public opinion. The spectre of immigration largely used by the far-right party Front National since 1972 became one of the most debated topics in the following presidential campaign. Sarkozy decided to run for the presidency and used his work and what he saw as "successful immigration measures" as Home Secretary to present a more restrictive immigration programme.

As far as links with Africa and Africans were concerned, he claimed that it was time to end the "Francafrique", an informal network that comprised French and African heads of states and business partners. The network's aims had always been to protect France's geopolitical interests overtly and covertly (Thomas, 2012). Galvanized by opinion polls about his popularity, Sarkozy decided to deliver a speech at Université Cheikh Anta Diop in 2007. African intellectuals would later debunk the speech in 2008 (Gassama *et al.*, 2008 and Chrétien, 2008). The audience was made up of students, academics, and politicians. He talked about common ties between France and Africa but used backhanded compliments. He acknowledged that it was wrong of Europeans to destroy African languages and customs during colonization. However, the most problematic part of the speech was when he talked about the African man, who had not really entered History (*L'Obs*, 2012). He was not part of History with a big H Sarkozy claimed. The African peasant, continued the politician, "was following the rhythm of seasons", which meant that he was unaware of what was happening around him (*L'Obs*, 2012). "The African man was still living in harmony with nature" and was still longing for a "lost paradise" (*L'Obs*, 2012). The speech was racist in essence and presented the well-used trope of the African as the noble savage or as a child. Dominique Wolton (2008: 667) noted that when people travelled they were more likely to revise their views about the local population and their position about colonial stereotypes. Yet, Sarkozy, a citizen of the world himself (his father was a Hungarian aristocrat and his mother was of Greek Jewish descent), did not take advantage of his trip to Africa to challenge those colonial views. Instead, he used it as a platform to promote ideas that perpetuated negative stereotypes about Africans. Sarkozy's advisor, politician Henri Guaino, wrote the speech. Highly criticized, Guaino refused to apologize. Instead, he argued that some colonists built bridges, roads, hospitals, schools, and so on (Guaino, 2008). Sarkozy's speech and Guaino's subsequent media appearances provoked strong reactions. An ideological battle was taking place about the ways in which the history of colonial encounters was to be taught if Sarkozy was to win the election.

The question about what colonization brought to the colonized had always been in the background of education in France. The charity Survie asked teacher Raphael Granvaud and other educators to look at the way colonialism was presented, by analysing the twelve most used history textbooks in secondary schools in 2006 (Cahiers d'Histoire, 2006). Sarkozy, the media and other politicians never expanded on the context in which infrastructures such as bridges, schools and hospitals that supposedly brought civilization were set up. Nothing was said about the fact that bridges and roads were planned by colonial administrators but built by the local population as Granvaud's study demonstrated. The study also found that these infrastructures were not built for the benefit of the said population but for the exploitation of natural resources (Cahiers d'Histoire, 2008). Regarding schools, only a small minority of the local population was allowed to attend these places prior to the twentieth century. As for hospitals, they were few and "medical campaigns were limited and often oppressive towards women" (Cahiers d'Histoire, 2008). Yet, none of these textbooks mentioned these facts. They presented French intervention as a benevolent act. These textbooks had shaped the views of school children for decades.

Sarkozy's ludicrous statement about the African man not being part of history draws attention to the question of the distribution of knowledge. In order to promote education, the World Bank issued an open call to publishers able to provide books in French-speaking countries in Africa. French publisher Hachette was successful in the bid. Hachette ended up holding the lion's share (85 per cent) of the market of textbooks in francophone Africa. Except for Cameroon and the Ivory Coast, most countries did not have the funds to produce and then widely distribute textbooks. The World Bank and the African Development Bank bought the textbooks for these countries from Hachette (Perucca, 2010). The publisher's position has been challenged in the last few years. As the CEO of Hachette International lamented, the World Bank increasingly preferred to pay local publishers rather than turn to France, thus disregarding "quality and content" according to Hachette (Perucca, 2010). In relation to textbooks transmitting knowledge, Sophie

Lewandowski analysed twenty-four textbooks used by French students preparing for the baccalaureate between 1945 and 1998 (Lewandowski, 2001). She noted that the titles and contents reflected political and maybe societal changes. More specifically they reflected the ways in which France viewed and wanted to teach the history of Africa. She contended that there were three distinct periods that could be summarized as follows: "colonised Africa", "Africa and civilisations", and "Africa: an economic disaster". Each period saw subtle changes that had a more and more pronounced Franco-centric stance. However, one period (1945 to 1957), according to Lewandowski, was characterized by an opening towards diversity. The books attempted to show that beside Western civilization, other cultures were worth studying. However, within fifty years France had moved towards a single educational model. In the meantime, the number of young people in the age group taking the baccalaureate had moved from 5 per cent to 50 per cent between 1945 and 1998. In other words, an increasing number of young people had been exposed to these textbooks. Quoting historian Alain Choppin, Lewandowski (2001) added,

> The school textbook is the repository of knowledge and techniques, the acquisition of which is deemed necessary by the society. It is, in that respect, an incomplete or staggered reflection, but always revealing in its schematization of the state of knowledge of an era and the main aspects and stereotypes of a society.

Regarding the Dakar speech, Sarkozy presented the so-called low level of education of the African man as a choice and a way of life. Nonetheless, Sarkozy became president of the Republic. His views greatly influenced the way Africa was to be presented and taught in the school curriculum. In fact, the history of France became a fiercer locus for an ideological battle under his presidency. France was to be the centre point from which knowledge about the world was to be accessed. French textbooks did not focus on Africa but rather on the links between France and Africa. African history was simply not deemed as relevant a subject for pupils as the history of France. This case showed how difficult any decolonization of the mind would be, especially when articulated around Frenchness, legacies of the past, and immigration. Sarkozy's appearance in Senegal had been not simply about reconnecting with geopolitical partners. He had been campaigning for the general elections and the speech was aimed at long-term Conservative party supporters. Using history for political aims was not, as Giovanni Levi (2001) noted, a new phenomenon. Historians have always been aware that their work could serve the political sphere. What has changed over the last ten centuries according to Levi (2001: 37) is that "History is manipulated and used while the voices of historians have become veiled and far off."

On 23 February 2005, a law was promulgated and it related to French citizens' contribution overseas during the colonial period. It specifically stated, "The nation expresses its gratitude towards women and men who helped France in former colonies of Algeria, Morocco, Tunisia and Indochina" (Legifrance, 2005). Point Two of Article Four was particularly problematic, as it included changes in the curriculum to "acknowledge the positive role of French presence overseas" (Legifrance, 2005). Historians, lawyers and novelists petitioned to have the law removed. Several articles were published in national newspapers (*Le Monde*, 2005). In politics, several ministers, including the high-profile Member of Parliament Christiane Taubira, demanded that the bill be withdrawn. Taubira had been instrumental in presenting and convincing Parliament to ratify a bill recognizing slavery as a crime against humanity in 2001. During the controversy, many historians also demanded the suppression of what they called "memorial laws". The Taubira Law of 2001 was a memorial law. A complaint was made in June 2005 when historian Olivier Pétré-Grenouilleau said that slavery had indeed been a crime against humanity but could not be considered a genocide. He was accused of revisionism. The debate was about historians' interpretation of the past. Most French historians involved in the debate argued that they were beyond dogmas and taboos. Having these laws was infringing on their intellectual freedom. The petitioners argued that it was not the government's place to do the work of historians. The debate in 2005 was based on two major laws that led to discussions about redefining the way colonial history should be written. On the one hand, the Taubira Law of 2001 acknowledges the past but does not hold the colonizers responsible for that past. The government's bill of February 2005, on the other hand, intended to have a very specific reading (positive) of France's colonial past. On the 15th of February 2006, France's High Council repealed the law of February 2005 on the basis that it imposed a detrimental reading of history. The debate surrounding France's colonial history continued. It further merged with the question of national identity, Frenchness and immigration in France.

Academics and teachers in France fought hard to have the history of Africa from the 8th to the 16th century become part of the national curriculum for Year 8 classes. They succeeded in 2010 but also provoked the ire of French nationalists, including known historians, novelists, and various intellectuals. A campaign of denigration of the new curriculum rested on two main arguments: first, it was disgraceful to remove key French figures such as Napoleon, Clovis and Louis XIV from the curriculum. Second, it was important for young minds to know "their own history first" (Estelle Pech, 2010). In other words at such a young age it was argued that learning about the history of France would help create a sense of belonging and national pride. Salient to these views was the question of the origins of the young learners, some of whom were indeed of African descent. The programme was also aimed at providing them with the opportunity to understand the world. The history of China and India was equally important in the new curriculum. The debate continued but the programme remained unchanged until 2015 despite vocal opposition from right-wing politicians. In 2015, under the left-wing socialist government, academics and teachers discovered after the summer holidays that the history of Africa had been removed from the curriculum. The Ministry of Education had asked a consulting body of educators and senators to reform history programmes. There was no national consultation. The consulting body (Conseil supérieur des programmes) is a group of experts appointed by the Ministry of Education. The body does not always have the support of teachers' unions. In this case, the experts on the history of Africa who had fought for that history to be included in the curriculum in 2010 were not consulted. Teachers only found out in September, at the beginning of the academic year, that changes had been made to the history curriculum. A new controversy was ignited and it threatened to create chaos in the academic world with strikes already being planned by politicians, teachers, educators, and students. Open letters were written and petitions started (Chrétien, 2015). The reason behind the changes, according to the education minister's advisors, was that students were already looking at the history of the transatlantic slavery and the history of colonialism, so they did not need to be burdened with the history of Africa. Historians objected that presenting only these two histories, slavery and colonialism, showed the colonized as victims and with no past prior to these periods. French historians Jean-Pierre Chrétien and Pierre Boilley specializing in the history of Africa noted that to understand the transatlantic slavery one needed to know the pre-colonial history of Africa. One also needed to understand that Africa was made of small tribes as well as powerful kingdoms that fought against each other, just like Europe did during the same periods. They explained that to understand the disruption created by colonialism it was crucial to learn about certain facts, such as the clearly structured courts of African kingdoms and their wealth. It was also noted that the teaching was not beneficial only for children of African origins. The majority group also benefitted from this input. Finally, a wider perspective of the history of people of African descent that examines the "longue durée" as well as various kingdoms and countries could help fight received ideas about uncivilized Africans living in an ahistorical continent. The debate served as a reminder that the economic domination of Europe over the last few centuries should not mean intellectual superiority. Under the weight of these arguments, the Ministry of Education had to cancel the new changes.

The 2016–17 academic year proved to be taxing for history teachers in favour of a global and inclusive history. The part dedicated to the history of Africa was completely removed this time, leaving only the history of the transatlantic slavery. Mobilization amongst left-wing politicians to keep the history of Africa was less virulent. The Socialist Minister of Education, Najat Vallaud-Belkacem, supported the new curriculum. In fact, the 2014 cabinet of president Francois Hollande led by Prime Minister Manuel Valls had increasingly adopted measures that reflected a shift towards centre-right ideas. History teacher Jean-Riad Kechaou, paraphrasing Nicolas Sarkozy's statement "The African man has not entered History", added "the African man has left history classes" (Kechaou, 2016). This was a sign of France's stance on the colonial past. The 2017 presidential campaign was dominated by the question of national identity and the role of French history in bringing people together. Each presidential candidate tried to outbid the other by using and quoting key French figures who supposedly epitomize Frenchness and French values. Sarkozy, who had lost the presidential elections against socialist François Hollande in 2012, came back in 2016 to try and win the Conservative primaries. During his campaign he was relentless in using the term "national narrative". For Sarkozy, textbooks had to be revised to inscribe a coherent national narrative based on men, women, places and monuments that made France. He added "whatever your parents' nationality, young French people, when you become French, your ancestors, are now the Ghouls and Vercingetorix" (Durand, 2016). It was noted that these views were based on those of French positivist historian Ernest Lavisse, who argued that the history of France should be centred on conquest and big names such as Vercingetorix, Joan of Arc, Napoleon, and others. Lavisse's approach had shaped French textbooks from 1884 to 1950 (ibid., 2016).

Fifteen days before the first round of the elections, Conservative candidate François Fillon compared himself to Vercingetorix, a Celtic king who rose against Julius Caesar's army. Concluding his speech in Auvergne in central France, he talked about the role of the state school that was aimed at "teaching how to be French and proud to be French". He added it was "up to foreigners to take the necessary steps to integrate into France and not France's role to bow down before customs that were not conforming to the Republican pact" (Auffray, 2017). This was a clear reference to a theme that has been dear to most of the main political parties: immigration bringing people whose customs are deemed too different to be accommodated. It was a euphemism to talk about the Muslim community in France. It was also a discourse that was born from far-right ideology. This was exemplified by the frenzy caused by the Burkini and the reaction of political parties. The dispute had ended when the Constitutional Council stated that it was illegal to prevent Muslim women from wearing the Burkini. The debate about so called key historical figures continued. Front National candidate Marine Le Pen kept Joan of Arc as the symbol of the party but went further regarding the question of national identity. She stated that there would be no dual citizenship for those born from non-European parents. According to Le Pen, those willing to be French had to renounce any other citizenship and by extension any other culture that shaped their views (Chazan, 2017). After pressure from politicians at home and abroad, Marine Le Pen gave up the idea a couple of weeks later. The measure seemed to be aimed at Africans but ended up also targeting the Jewish community. These questions became the focus of Israeli newspapers (Bar, 2017). The debate surrounding dual citizenship had already provoked anger in France when president François Hollande decided to promote a bill designed to strip French citizenship from people convicted for acts of terror. He was then accused of using the Front National's ideas and ideology and his position led to the resignation of his justice minister Christiane Taubira, author of the previously mentioned Taubira Law. The bill was dropped by Hollande but the debate about national identity and re-centring the "national narrative" around French-born figures continued, even though it meant excluding populations' histories who shaped the French Empire and who had contributed to the nation's wealth.

France's education has, since its inception, inscribed critical thinking and free speech as the bases to shape pupils' knowledge acquisition. Young minds were to follow in the footsteps of Enlightenment thinkers by promoting individual freedom and France's values. Jules Ferry's school became, in 1881–84, a vehicle for values that were dear to the nation. The "grandeur" of France also rested on its colonial hegemony and on its so-called civilizing mission. It is only in the second half of the 20th century that a re-writing of textbooks was deemed necessary. The population of metropolitan France was changing. The post-war and the decolonization periods were characterized by a rise in the volume of populations from former colonies. France's demographic and ethnic landscape had changed but that was not reflected in the education system because school was, from the end of the 19th century, supposed to be a safe place that relegated specificities such as religion, ethnicity and social practices to the private sphere. School, however, also proved to be a terrain that sometimes reflected and even reproduced these inequalities. The debate about textbooks and the national curriculum merged with the political agenda of various parties. The history of France had to focus on French citizens and not on the colonial past. Faithful to Jules Ferry's project about an education system that did not reflect diversity but rather presented a uniformed and carefully constructed idea of the nation, textbooks in France have shunned debates about legacies of the imperial past. It has been argued that reflecting on the competition between the "victims of the past" meant that the "traditional part of the history of France taught up to that point" would also have to be reduced. That was deemed an arbitrary decision (Falaize and Lantheaume, 2010).

On the one hand, various parties were in favour of following a Lavissien way of studying history and that implied focussing on key names, monuments and places specific to mainland France. On the other hand, it was stated that minority ethnic groups, French or foreign nationals had to accept this history as theirs. The salient question of multiple identities was relegated to the notion of "communitarism". Educators and politicians stated that communitarism characterized by the exclusion of the majority populations' values was a threat to social cohesion. In fact, communitarism, it was argued, led to isolation for these populations who were then at risk of falling victim to dangerous ideologies. Communitarism could lead, both conservative and liberal politicians and intellectuals predicted, to acts of terror. It therefore threatened the Republic. The 2017 presidential campaign brought a new element to the debate. Newcomers from Africa as well as Muslim communities in France were viewed with suspicion. France was "under attack" declared then prime minister, socialist Manuel Valls, from populations within and outside. Immigration had to be restricted and communitarism had to be dismantled. One of the tools used was to re-write a national narrative that was to celebrate a unique and unifying vision of the history of France. The narrative relegated part of the French population to the periphery of education. Sarkozy's Dakar speech, the debate about the positive role of colonization, the campaigns to introduce and remove the history of Africa in the curriculum, the links with immigration and the rise of far-right ideology all appeared to be attempts to re-colonize minds, but closer analysis has revealed that a decolonization of minds had never taken place in the French curriculum.

French Enlightenment thinkers continue to be used as models in primary and secondary schools. Contemporary universities do not advocate their removal from French textbooks because they are supposed to constitute pupils' basic knowledge.

Yet, universities have tended to add post-colonial scholarly work produced by people of African descent rather than removing colonial references. History students in higher education have benefitted from a variety of schools of thought, such as Ernest Renan's views on so-called racial hierarchy, March Bloch's dynamic approach to deconstructing French history in the Annales School, Pierre Nora's work on memory and history of France or Elika Mbokolo's seminal publications on African history, etc. The only historical debates that unsettle Eurocentrism in higher education are linked on the one hand to the parts of France's colonial history that should be taught in primary and secondary schools, and on the other hand to the legacies and the political uses of that past.

References

Agostini, M. (2010) L'Apprentissage du Philosopher à l'École Primaire: Analyse d'une Expérience d'un Atelier de CM2 sous l'Éclairage de la Pensée de Montaigne. PhD Thesis, Université d'Aix-Marseille I. Available at: https://halshs.archives-ouvertes.fr/tel-00561515/file/Volume_I.pdf.

Auffray, A. (2017) Fillon, sur les Pas de Vercingétorix et de Sarkozy. *Libération* 8 April. Available at: www.liberation.fr/france/2017/04/08/fillon-sur-les-pas-de-vercingetorix-et-de-sarkozy_1561411 (Accessed 8 April 2017).

Bar, R. (2017) Le Pen: French Jews will have to give up Israeli citizenship. *Haaretz* 10 February 2017. Available at: www.haaretz.com/world-news/europe/1.770915 (Accessed 10 February 2017).

Cahiers d'Histoire (2006) Colonisation et décolonisation dans les manuels scolaires de Collège en France. *Relecture d'Histoire Coloniale* 99. Available at: https://chrhc.revues.org/792?lang=en (Accessed 29 March 2016).

Chazan, D. (2017) French Jews 'will have to give up dual Israeli citizenship' if Marine Le Pen wins presidential election. *The Telegraph* 10 February. Available at: www.telegraph.co.uk/news/2017/02/10/french-jews-will-have-give-dual-israeli-citizenship-marine-le/ (Accessed 10 February 2017).

Chrétien, J.P. (ed) (2008) *L'Afrique de Sarkozy: Un Déni d'Histoire*. Paris: Karthala.

Chrétien, J.P. (2015) Enseignement de l'Afrique à l'École: Un Oubli Coupable. *Libération* 9 October 2015. Available at: www.liberation.fr/debats/2015/10/09/enseignement-de-l-afrique-a-l-ecole-un-ououpable_1400534 (Accessed 29 March 2016).

Conseil Constitutionnel (1789) Déclaration des Droits de l'Homme et du Citoyen, 1789. Available at: www.conseil-constitutionnel.fr/conseil-constitutionnel/francais/la-constitution/la-constitution-du-4-octobre-1958/declaration-des-droits-de-l-homme-et-du-citoyen-de-1789.5076.html (Accessed 14 April 2016).

de Montesquieu, C. (1748) *The Spirit of the Laws*. Available at: http://press-pubs.uchicago.edu/founders/documents/amendI_speechs3.html (Accessed 29 March 2016).

Durand, A.-A. (2016) 'Roman National', 'Récit National': De quoi parle-t-on? *Le Monde* 28 September. Available at: www.lemonde.fr/les-decodeurs/article/2016/09/28/roman-national-recit-national-de-quoi-parle-ton_5004994_4355770.html (Accessed 28 September 2016).

Erhard, J. (2008) *Lumières et Esclavage: L'Esclavage Colonial et l'Opinion Publique en France au XVIIIe Siècle*. Brussels: André Versailles Editeur.

Falaize, B. and Lantheaume, F. (2010) Entre pacification et reconnaissance: les manuels scolaires et la concurrence des mémoires. In P. Blanchard and I. Veyrat-Masson (eds) *Les Guerres de Mémoires*. Paris: La Découverte, pp. 177–186.

Gassama, M., Diagne, M. Diop, D. and Lamko, K. (2008) *L'Afrique Répond à Sarkozy: Contre le Discours de Dakar.* Paris: Philippe Rey.

Goddeeris, I. (2015) Colonial streets and statues: Postcolonial Belgium in the public space. *Postcolonial Studies* 18(4): 397–409.

Guaino, H. (2008) L'homme africain et l'histoire par Henri Guaino. *Le Monde* 26 July. Available at: www.lemonde.fr/idees/article/2008/07/26/henri-guaino-toute-l-afrique-n-a-pas-rejete-le-discours-de-dakar_1077506_3232.html (Accessed 21 March 2016).

Kechaou J.-R. (2016) L'Homme Africain est Sorti des Cours d'Histoire. *Politis* 29 June. Available at: www.politis.fr/blogs/2016/06/lhomme-africain-est-sorti-des-cours-dhistoire-34073/ (Accessed 29 June 2016).

Legifrance (2005) Loi n° 2005–158 du 23 Février 2005 Portant Reconnaissance de la Nation et Contribution Nationale en Faveur des Français Rapatriés. 24 February. Available at: www.legifrance.gouv.fr/affichTexte.do?cidTexte=JORFTEXT000000444898&categorieLien=id (Accessed 29 March 2016).

Legifrance (2017) Loi du 29 Juillet 1881 sur la Liberté de la Presse. Available at: www.legifrance.gouv.fr/affichTexte.do;jsessionid=73C2E9838829A2FE2B356D4E1E926C17.tpdila13v_3?cidTexte=JORFTEXT000000877119&dateTexte=20170428 (Accessed 12 December 2017).

Le Monde (2005) Colonisation: Non à l'enseignement d'une histoire officielle. 24 March. Available at: www.lemonde.fr/societe/article/2005/03/24/colonisation-non-a-l-enseignement-d-une-histoire-officielle_630960_3224.html (Accessed 29 March 2016).

Levi, G. (2001) Le passe lointain: Sur l'usage politique de l'histoire. In F. Haryog and J. Revel (eds) *Les Usages Politiques du Passé.* Paris: Éditions de L'École des Hautes Etudes en Sciences Sociales, pp. 25–37.

Lewandowski, S. (2001) L'Afrique dans les manuels scolaires français (1945–1998): Du colonialisme à l'économisme in *Cahiers de la Recherche sur l'Éducation et les Savoirs* 11. Available at: https://cres.revues.org/308 (Accessed 29 March 2016).

L'Obs (2012) L'intégralité du discours de Dakar prononcé par Nicolas Sarkozy. Available at: http://tempsreel.nouvelobs.com/politique/20121122.OBS0195/l-integralite-du-discours-de-dakar-prononce-par-nicolas-sarkozy.html (Accessed 29 March 2016).

Ministère de l'Éducation Nationale (2013) Morale Laïque, pour un Enseignement Laïque de la Moral. Available at: http://geographie-histoire.info/130422-rapport-enseignement-morale-laique.pdf (Accessed 29 March 2016).

Ministère de l'Éducation Nationale et de l'Enseignement Superieur (2015) École Élémentaire et Collège, Programme d'Enseignement Moral et Civique. Available at: www.education.gouv.fr/pid25535/bulletin_officiel.html?cid_bo=90158 (Accessed 29 March 2016).

Ministère de l'Intérieur (2006) Projet de Loi Immigration et Intégration. 6 June. Available at: www.interieur.gouv.fr/Archives/Archives-ministre-de-l-interieur/Archives-de-Nicolas-Sarkozy-2005-2007/Interventions/06.06.2006-Projet-de-loi-immigration-et-integration (Accessed 29 March 2016).

Mougniotte, A. (2006) Rôle et portée de l'école dans l'apprentissage de la citoyenneté. In Y. Lenoir, C. Xypas and C. Jamet (eds) *École et Citoyenneté: Un Défi Multiculturel.* Paris: Arman Colin, pp. 171–184.

Mucchielli, L. (2016) Immigration, délinquance et terrorisme: Erreurs et dangers d'une assignation identitaire persistante. In P. Blanchard, N. Bancel and D. Thomas (eds) *Vers la Guerre des Identités?* Paris: Éditions La Découverte, pp. 231–238.

Nimako, K. and Willemsen, G. (2011) *The Dutch Atlantic: Slavery, Abolition and Emancipation.* London: Pluto Press.

Pech, M-E. (2010) Polémique sur les Programmes d'Histoire au Collège. *Le Figaro* 27 August. Available at: www.lefigaro.fr/actualite-france/2010/08/27/01016-20100827ARTFIG

00586-polemique-sur-les-programmes-d-histoire-au-college.php (Accessed 29 March 2016).

Perucca, B. (2010) La France règne en maître sur le marché des manuels scolaires en Afrique francophone. *Le Monde* 10 June. Available at: www.lemonde.fr/planete/article/2010/06/10/la-france-regne-en-maitre-sur-le-marche-des-manuels-scolaires-en-afrique-francophone_1370530_3244.html (Accessed 29 March 2016).

Rousseau, J.J (2009) *The Discourse on the Origin of Inequality*. Oxford: Oxford University Press.

Schmidt, N. (2001) *Abolitionnistes de l'Esclavages et Réformateurs des Colonies 1820–1851*. Paris: Karthala.

Thomas, D. (2012) The adventures of Sarkozy in EuroAfrica. *Contemporary French and Francophone Studies* 16(3): 393–404.

Vergès, F. (2006) *La Mémoire Enchaînée: Question sur l'Esclavage*. Paris: Albin Michel.

Voltaire (1759) *Candide*. Available at: www.gutenberg.org/files/19942/19942-h/19942-h.htm (Accessed 15 April 2017).

Wolton, D. (2008) Des stéréotypes coloniaux aux regards post-coloniaux: L'indispensable évolution des imaginaires. In P. Blanchard, S. Lemaire and N. Bancel (eds) *Culture Coloniale en France: De la Révolution Français à Nos Jours*. Paris: CNRS Éditions, pp. 655–667.

Xypas, C. (2006) La fonction politique de l'école au défi des jeunes issus de l'immigration. In Y. Lenoir, C. Xypas and C. Jamet (eds) *École et Citoyenneté: Un Défi Multiculturel*. Paris: Arman Colin, pp. 223–235.

16

DISMANTLING EUROCENTRISM IN THE FRENCH HISTORY OF CHATTEL SLAVERY AND RACISM

Christelle Gomis

Eurocentrism and its French discontents

Eurocentrism serves as an interpretative framework to analyse any phenomenon through the lens of Western superiority (Araujo and Maeso, 2015). Immediately available, it shapes the most basic assumptions into affirming the supremacy of the West over the rest of the world. Eurocentrism acts like a filter that constantly privileges Western achievements (Mazrui, 2009). As sociologist Gurminder Bhambra (2007: 5) explains, Eurocentrism can be defined as: "the belief, implicit or otherwise, in the world historical significance of events believed to have developed endogenously within the cultural-geographical sphere of Europe". Bhambra contests the idea that Europe is special, that the West is exceptional, and that both entities are amenable to analysis on their own terms. She questions the notion that Europe has always been an identifiable, well-defined space. But Eurocentrism obfuscates this reality.

To follow the basic tenets of Eurocentrism, scholars have charted and documented an epistemological progression that starts in Ancient Greece moving to Rome to feudal Christian Europe and to capitalist Europe (Dussel, 1993). Present everywhere, always available, from school books to video games, it contaminates popular opinion across the world. The West is and has always been superior in every way. The West leads the good and only way towards the best of worlds for the good of human kind. Europe is repeatedly posited as an ideal model for other societies to follow in the arts, philosophy, science, technology, and governance, if they can. The particularity of European evolution is the best "expression of a general law "that will be inevitably reproduced elsewhere, even if delayed" (Amin, 2009: 256).

Accordingly, this Eurocentric teleology makes one consider other societies through the paradigm of European evolution. For instance, French President Emmanuel Macron embodied Eurocentrism during a press conference at the G20 summit in Hamburg in 2017 (Dearden, 2017). A journalist from Ivory Coast asked President Macron why there was no Marshall Plan for Africa. He then answered:

> The challenge of Africa, it is totally different, it is much deeper, it is civilizational, today. […] Development, security, and then there's a shared responsibility, the Marshall plan you want will also be supported by African governments and regional organizations. It is through a more rigorous governance, through fight against corruption, through struggle for good governance, a successful demographic transition. When countries still have seven to eight children per woman — you can decide to spend billions of euros, you will not stabilize anything.[1]

President Macron identifies each problem that plagues Africa against the backdrop of Europe's particular evolution, as if it was a natural law and as if racism and colonialism did not lie at the core of the historical processes affecting the entire world. In the wake of his discourse, the Economic Community of West African States hastily announced their aim to cut back the birth rate to three children per woman (Penney, 2017).

Within Eurocentric narratives, colonialism is routinely constructed as being outside the realm of French history, rarely related to the fundamental changes that affected France – a fortiori Europe – such as the French Revolution. Colonial histories allegedly do not bear any consequences for the present situations of Europe and Africa (Trouillot, 1995; Bancel, Blanchard and Thomas, 2017). In its attempt to repeatedly reproduce the idea of the superiority of the West, Eurocentrism warps visions of world societies and histories. The narratives are often coherent, but not always. Depending on the ideological needs of the moment, the bias leads to either emphasizing or eclipsing phenomena such as colonialism. Either way, the fundamental role played by colonialism in the making of Europe is underplayed.

The French have (re)assured themselves about their moral high ground because their imperial rule was civilizing "backward" people (Conklin, 2003). Modern civilization has justified the violence that characterized the second wave of colonization and, in retrospect, chattel slavery. Just as the colonized were tainted with "guilt" because they dared challenge the civilizing mission (Dussel, 1993) and thus cannot claim to be "victims", their descendants must proclaim today how grateful they are for the colonization of their ancestors and not look critically at the past actions of France in its "former" colonies.

This framework only considers the benefits of colonization or how much it has cost France to "invest" in the colonial pursuit. The violence of colonialism is never permanent or long-lasting. It is only perceived through specific moments, such as conquest and independence wars.[2] Usually put forth as symbols of benevolence, routes, hospitals and railroads were built by colonized people for the sole benefit of settlers. When in February 2017 presidential candidate Macron toured Algeria to hone his international profile and ensure continuing "cooperation" between France and Algeria, he declared that colonialism was a crime against humanity.

While Algerians reacted with satisfaction, Macron's statement provoked much out-cry in France, where a particular version of Eurocentrism euphemistically remembers colonialism and, especially, chattel slavery.[3]

Along with the Portuguese and the British, the French were the third largest chattel slave traders. Thousands of French ships transported captive Africans to Guadeloupe, Martinique and Saint-Domingue, the most lucrative colony of France's first empire. Despite the increase of contemporary scholarship on the historiography of Haiti and its past as France's most profitable colony – Haiti or Saint-Domingue can hardly find a place in the Eurocentric national narratives of France (Geggus, 2002; Fischer, 2005; Dubois and Garrigus, 2017). There is no French word to designate chattel slavery as a specific phenomenon. The widely used French term "*esclavage*" conflates every form of human bondage across the ages. Other terms like "transatlantic (slave) trade" or "triangular trade" make the enslaved persons disappear. This lack of conceptualization leads to a tendency to confuse chattel slavery with any form of exploitation or infringement of personal freedom and integrity.

The often-used phrase "*traite(s) négrière(s)*"[4] generalizes every instance of enslavement of Black people. Chattel slavery is equated with the intra-African slave trade, the Arab Muslim slave trade, and the Indian Ocean slave trade (Many Chroniques, 2015). All these phenomena are disconnected from their own contexts to be conflated within a Eurocentric timeline. In these narratives of chattel slavery, everybody is to blame, but especially Africans and Arabs. To conceal the specialness of chattel slavery, Eurocentric common accounts of chattel slavery insist that it is Africans' responsibility: they sold their own flesh and blood and they did not resist. Chattel slavery and its role in enriching the entire French nation, a fortiori Europe, disappears.

When chattel slavery does enter the public sphere, it is through a celebratory framework. The phenomenon sits so uneasily with French Eurocentric narratives that France manages to commemorate slavery and its abolition at the same time.[5] The French Republic remains unscathed because it abolished slavery twice. Indeed, France is the only nation that had to abolish chattel slavery[6] twice, while most French people still do not know how long it lasted. Under the pressure of descendants of the enslaved people in the West Indies, chattel slavery found its way to national recognition.

However, in the Eurocentric framework, chattel slavery is made foreign to Western modernity. After the mobilization of the French Antillean population in 2001, the first Taubira Law was adopted, but only after French parliamentarians emptied substantial amendments out of the bill (Khiari, 2009; Fleming, 2017). The law does recognize slavery as a crime against humanity but does so equally for every phenomenon of enslavement throughout the ages. But there is no talk of French or European responsibility. State politics of commemoration neutralized the political demands of descendants of the enslaved people.

The "abolitionist mythmaking" (Larcher, 2014; Schmidt, 2016) obscures the achievements of other societies. Enslaved people's persistent resistance against chattel slavery is routinely diminished or dismissed. While celebrating freedom born at home, France chastises other countries for tyranny and oppression and emphasizes its own role in bringing it to an end. Most French people do not know that the first abolition of French chattel slavery was due to the only known successful slave revolt, the Haitian revolution.

Consequently, racism as a long-held colonial practice is explained away from France. The notions of subjugation and domination are left out of Eurocentric definitions of racism. France was not really involved because there were supposedly no slaves on metropolitan soil, whereas chattel slavery was constitutive of countries like the United States, France's favourite scapegoat (Fleming, 2017). While racism and chattel slavery made Europe rich and predominant, the dominance of Eurocentric thinking means that Europeans have never been held accountable in any international court of justice. Eurocentric narratives preclude any possibility of mentioning and analysing the social structures based on race and created to make colonialism palatable since the beginning of the French colonial empire in 1534 (Quijano, 2007). It is one of the reasons why racism is currently considered as an exceptional phenomenon, usually the appanage of the United States or at "best" far-right sympathizers. Racism is located only in intersubjective relationships, never in structures or institutions, let alone the French state. It is not as widespread as it is in the United States because France is the country that invented universalism (Schor, 2006).

French universalism means the refusal of any particularism, be it sexual, racial, religious or ethnic.[7] In its effects, this French discourse of race-blindness is very similar to the US version, but the two ideologies have different genealogies (Bleich, 2004; Keaton, 2010). "Homeland of the human rights", the French Nation is irrevocably linked to universalism (race-blindness) thanks to *The Declaration of the Rights of Man and the Citizen* while other nations are undermined by their particularisms. Enlightenment philosophers and more radically the French Revolution transformed Catholic universalism into a more civic religion for the new Republic. The universal creature is a Republican citizen, supposedly devoid of all particularities: unsexed, ungendered, unraced, unclassed. The subject is emptied of all spatial or temporal determinations. It becomes ahistorical and apolitical. It is only as a supposedly abstract individual that the French citizen is entitled to all the rights the French Republic offers.

Eurocentric narratives are mobilized by the French state to defend the idea that race and racism cannot exist within republican structures and institutions. They present serious obstacles to demands for dignity, equality, and fairness from people of colour. French universalism (race-blindness), notably through the centrality of the French Revolution, contributes to uphold claims of French uprightness. But how could this discourse prosper while the institution of chattel slavery was still in place?

Constitutionalizing chattel slavery

Although its current crisis forces many nationalists to look for earlier origins of France even before the mythical Gaul to uphold the myth of a nation that preexists its constitution into a state, the French version of Eurocentrism still rests upon the Revolution of 1789. It remains the matricial event that created the Nation. The French Revolution went on to symbolize the universal revolution *par excellence* and precipitated the identification of France with universalism. It is still taught as such in French curricula (Vovelle, 2007).

In France, each pupil must learn that the French Revolution was, is and always will be the pinnacle of liberty and equality. Each pupil must learn that it is one of the most important events in the history of humanity. Everything leads to it; everything stems from it. The traditional periodization of French history suggests that the Revolution marks the beginning of the modern era (Le Goff, 2014). Also, its advent allegedly signalled the birth of the modern French civilization, superior to the rest of the world in every way and thus poised to offer it to the "backward and immature" people who need it.

However, the very French revolutionaries who came up with the generous motto Liberty! Equality! Fraternity! constitutionalized chattel slavery on the 13th May 1791 – that is, during the very first stages of the revolution, well before counter-revolutionaries won and managed to impose their narratives (James, 1938; Césaire, 1962). It revealed the true nature of the motto as Liberty! Equality! Property! How come French pupils rarely learn about this fateful decree? More importantly, how can we explain the fact that French revolutionaries declared human rights for "all" then and decided to constitutionalize slavery in the very same breath?

> Colonial authority was constructed on two powerful but false premises. The first was the notion that Europeans in the colonies made up an easily identifiable and discrete biological and social entity … The second was the related notion that the boundaries separating colonizer from colonized were thus self-evident and easily drawn.
>
> *(Stoler, 2002: 42)*

On the eve of the French Revolution, phenotypical characteristics had become of the utmost importance both in the colonies and the metropole as tensions flared in the French colonies, especially in Saint-Domingue, later (re)known as Haiti. These race struggles turned into one of the most important issues facing the Constituents during the very first stages of the Revolution.

The crown jewel of the French Empire, Saint-Domingue, was characterized by growing tensions within the free population. Wealthy planters allied themselves with arriving land-hungry European settlers against free people with darker skin. The latter were progressively designated as "coloured" people (*gens de couleur*) while the former became white (respectively *grands blancs* and *petits blancs*). Creating discriminatory laws, they redrew racial lines to separate people of apparent European parentage from those with both apparent European and African ancestry within the class of planters. *Grands* and *petits blancs* primarily targeted wealthy slave owners with darker skins. Thus, they could dispossess and redistribute land and enslaved people between themselves (Garrigus, 1996).

From the 1770s white and light-skinned slave owners were increasingly defined through their privileges to become known as white people, while freedom was reattached more firmly to whiteness, thus restricting the meaning of freedom for all Blacks, free and enslaved. For example, free Black people could no longer use European names. Instead, they were forced to adopt African names. They now had to prove their freedom with legal documents, which now stipulated that they belonged to the *gens de couleur*. Any union between white and black people was forbidden. Social behaviours were progressively controlled. The *gens de couleur* could not sit with white people in public places. Free Blacks could not dress or wear their hair in ways that would outrival Whites. They were progressively banned from exercising certain professions, carrying weapons and serving in the colonial militias. Incidentally, these militias exerted terror to enforce these newly crafted legal definitions of race. This racial obsession also existed in the metropole. The French government created the *Police des Noirs* to stave off the presence of free and enslaved Blacks (Peabody, 2002).

The publication of the Declaration of the Rights of Man and Citizen, right after the Bastille was stormed, gave the opportunity to both the enslaved and free Blacks to challenge this hierarchy. Conversely, sentences such as "Men are born and remain free and equal in rights" *terrorized*[8] white slave owners, who tightened their rule in Saint-Domingue. They quashed revolts of both enslaved and free Blacks, while a delegation hurried to the National Constituent Assembly in Paris to fight any threat to their authority. Free *gens de couleur* quickly followed and were the first to test white revolutionaries' determination to realize their own principles. They demand voting rights for free *gens de couleur*, regardless of their wealth. They also insisted that abolishing discriminatory laws would strengthen their resolve to uphold the slave system.

But white slave owners and their supporters attached blackness to enslavement and whiteness to liberty, arguing that the separation needed to be clear between who could be enslaved and who could be free. Otherwise, enslaved people would never "understand" their fate and continue to revolt, especially because the ratio was of ten enslaved people for one free person. Antoine Barnave, one of the most influential revolutionary leaders at that time, explicitly separated again the humanity into two "races", the free Whites and the enslaved Blacks (Gauthier, 2007).

Given the opportunity to apply the Declaration of the Rights of Man and Citizen to all slave owners, revolutionaries sided with white people only and tied universalism to whiteness. On 13th May 1791, they gave full control over colonial matters, including voting rights, to white slave owners and made slavery an integral part of the French Constitution. They denied the rights of Black people. They chose to betray their principles and instead maintained slavery. White revolutionaries deliberately and explicitly created a legal order based on anti-blackness that would pretend to be universal. Learning about the constitutionalizing of chattel slavery during the French Revolution in schools and universities is urgently needed to dispel the Eurocentric dogmas that plague the struggles of people of colour against racism.

What has to change

Saint-Domingue's history dispels a first myth: the primacy of economic class. It is rather a Eurocentric construct, insulated from the colonial histories that show how race and class have been deeply intertwined for a longer time than the post-World War II era. Despite the claims of free Blacks, despite their own principles, white revolutionaries remained committed to the institution of slavery to keep extracting resources from their colonies through continued dehumanization and dispossession. Race has been "the modality in which class is lived" for much longer than any French person can currently realize (Hall *et al.*, 1978).

Although intersectional theory has made a noteworthy foray into French social sciences, the embeddedness of race and class is still analysed in a static manner, as if the history of French/European class relations remained unaffected by racial considerations. The narrative of the primacy economic class is the following in France: The white proletariat created by the industrial revolution was lured into racism after World War II because of massive immigration. Following, it would suffice to show white and non-white workers that they have a common interest in overthrowing capitalism. While this might be a valid perspective, because of Eurocentrism French workers cannot possibly understand their possessive investment in whiteness or the substantial social benefits they have reaped from living in Europe. A critique of capitalism without a critique of white supremacy – and vice versa – is incomplete and ahistorical.

Second, the careful consideration of colonialism and chattel slavery makes racism an unavoidable feature of French history. Giving a longer genealogy of race-making is crucial in France where the nineteenth century is always favoured because it reinforces certain Eurocentric beliefs. Chattel slavery's role in accumulating riches that financed the industrial revolution is erased. France started to colonize because it was already powerful, thus legitimizing the civilizing mission. The histories of French slave societies – especially on the eve of Revolutions – show how racism can be understood as a social construct, a social relationship and a mechanism of structural domination.

Racism transforms differences into the basis of domination because of the relationship between modernity and coloniality (Quijano, 2007). The popular belief in France is that preexisting differences inevitably produce racism. Thus, resistance is futile or changing hearts and minds is just a lengthy but inevitable process. But differences are constructed and have been legitimized through biology over time. The case of colonial Saint-Domingue reveals how these differences were ascribed to divide the class of plantation owners. Resistance of both enslaved and free Blacks render these differences very unstable. These contestations are erased and thus contribute to the essentializing of the slave as a Black person and the slave owner as white.

Biological definitions of race supposedly disappeared from France after World War II. The scientific invalidity of biological races is the cornerstone of any French discourse or policy, but in many cases it is the only one. The priority is not to fight racism but to fight racial categories. The structural dimension of this systemic domination is completely absent.

Race, defined as a social construct,[9] has to be evoked with much caution in France. In order to keep French colour-blindness intact, one will twist language not to talk about race. For instance, a Black child born and raised in France is – theoretically – French before anything else, albeit conditionally. Invited to talk about genital mutilations in France and in Africa, a surgeon struggled to describe one of her Black patients. After some hesitation, she evoked "a little French girl with Malian genetic origins". Even French social scientists contort themselves not to use race as an analytical lens. They designate people of colour as first, second, third generation, even when they are born in France. They are at a loss as how to account for the situations of Caribbean-born citizens.

As Tricia Keaton (2010: 108) argued, "'Frenchness' and 'blackness' in the discourse of 'race-blindness', reinforced by the Republic's myths, become an impossible equation because 'blackness' is demographically[10] rendered non-existent". Therefore, it is impossible to be a black French person, a fortiori a Black European. Because of Eurocentrism, French or European necessarily means white and superior. Incidentally, the English adjective "Black" instead of the French adjective "noir" is routinely used to designate people of African descent living in France, suggesting an inherent extraneity and strangeness to the Black presence in France (Ahouansou, 2013; Gay, 2015).

Anti-racist activists who define racism as a system of domination and race as a social construct – especially when they are on the receiving end of racism – are deemed suspicious or guilty of importing the worst of US sociology, as if it was a disease (Fassin, 2010). The simple uttering of the definition of race as a social construct can effectively end any discussion, because it equals conjuring racism (Keaton, 2010; Fleming, 2017). As a result, the French Minister of Education can threaten to sue a teacher's union for using phrases such as "state racism", "whiteness" because these notions supposedly produce racism (Silga, 2017; Durupt, 2017).

Third, the constitutionalizing of chattel slavery by the French Revolution denaturalizes the Eurocentric separation between colonies and metropole. It exposes it as an artificial demarcation meant to obscure how racial considerations took place during the crucial event (Boulbina, 2008). In French historiography, colonies are generally construed as distinct peripheral entities that affected France's internal developments less than England or Germany. But colonies and metropole are irrevocably intertwined and have evolved in a "shared but differentiated space of empire" (Cooper and Stoler, 1997: 3). In this colonial context, universal claims came with the price of particularism. To uphold them, French politicians consciously distinguished "between what was happening on 'their' soil and belonged to the history of rights and freedom, and what was happening in the colonies where the exception could be explained by the notion of race and a hierarchy of civilizations" (Vergès, 2010: 104). Human rights were explicitly conceived for Whites only. Eurocentric narratives exclude colonies from the realm of European history and, therefore, myriad of histories get lost while remaining constitutive of developments in Europe like an invisible chain of events. The constitutionalizing of chattel slavery is one of the most understudied moments where the colour line is explicitly put forward, well before the nineteenth century.

Today, the separation metropole/colonies lives on in the expression "Hexagon" or even metropole, used to designate the French territory located in the Eurasian peninsula. The term "outre-mer" is employed for overseas territories and was coined in the 1930s to replace the word "colonial". This Manichaean dichotomy continues to be preserved because it effectively conceals how Eurocentric grammars of difference have been contested. Carefully honed by white revolutionaries, the separation metropole/colonies allowed for the obfuscation of white supremacy in France.

Finally, the French ideal of universalism is to be understood as historically mediated by the context of colonial insurrections in the Caribbean. When white revolutionaries constitutionalized chattel slavery and claimed universality for themselves only, they created a racialized citizen characterized by its proximity to whiteness and masculinity instead of an abstract creature. The lighter the skin, the more human, the more universal. Conversely, a darker skin inevitably meant enslavement and the exclusion from the human realm. The French state hates particularities, but the universalism it built and has revered for such a long time is particular. It is white and masculine by default because Eurocentrism makes the abstract white because the rest of the world is actively excluded. French universalism is historically built to be discriminatory. As Maldonado-Torres (2007: 244) aptly argued, "Modernity, usually considered to be a product of the European Renaissance, or the European Enlightenment, has a darker side, which is constitutive of it". For France, it means that exclusionary practices were assimilated into the French Republican fabric and then invisibilized by the separation metropole/colonies.

This critique of French universalism is crucial to unsettle Eurocentrism. Today, it is invoked to reinforce the narratives of French people as part of an imagined community of whiteness. France is imagined as phenotypically white because Eurocentrism exists to safeguard whiteness as Eurocentric narratives erase the presence of people of colour again and again. Therefore, they are constructed as perpetually arriving. Eurocentrism renders the citizenship rights of French people of colour precarious. They cannot become full members of the French nation because Eurocentric narratives legitimized the imagined whiteness of the French community. Consequently, French people of colour are always constructed as undeserving of the welfare state, as tearing apart the French nation when they assert their demands for equality. For instance, "failures" to achieve economic success in the overseas territories is solely their fault. It is difficult for them to confront past and present structural barriers and develop informed critiques of racial injustices. As soon as French citizens of colour develop analyses based on racism, they are perceived as threatening, unpatriotic and disloyal. French universalism must be understood against the backdrop of the "basic tensions of empire": white revolutionaries' unwillingness to deal with colonialism as well as the resistance of enslaved and free people of colour.

So French anti-racist paradigms must change. With the exclusion of the long-lasting history of racism, the French ideal of universalism promoted by the state cannot be a generous intention. Currently, common narratives consider the French Revolution and its aspirations as being incomplete. Thus, the struggle against racism is simply about enforcing the law, changing the minds and hearts of ignorant people, and building harmonious race relations in keeping with the supposedly universal principles of 1789. But rarely do good-intentioned policymakers, activists or ordinary citizens consider that these principles are an apparatus devised to be compatible with racism, sexism and classism. For the moment, French conceptions of citizenship are steeped in whiteness and used to maintain it. Within a Eurocentric framework, French universalism is built to make people of colour disappear.

Conclusion

Thanks to Eurocentrism, France can refashion its own identity as a democratic, anti-racist world saviour *ad infinitum*. This epistemology affirms "the universal validity of the European trajectory and the ultimate necessity for others to imitate the same experience" to achieve progress (Tansel, 2014). It conceals people and territories that have been dominated, exploited to enrich Europe, and ultimately leads to the whitewashing of colonialism and chattel slavery.

Discovering the constitutionalizing of chattel slavery by the French Revolution was a way of "excavating the ruins of the marginalized, suppressed or silenced traditions upon which Eurocentric modernity built its own supremacy" (De Sousa Santos, 2001: 193). Looking for what has not been permitted to exist in Eurocentric curricula, it is possible to understand that conceptions of progress and human rights were designed to be the universal privilege of only a few white men located in the French Hexagon.

Although the decree of 13th May 1791 was repealed two days later, the idea of separating humanity did not disappear and has remained at the heart of the French Republic, as chattel slavery was reestablished by Napoleon and colonialism reprised its course during the subsequent political regimes. The constitutionalizing of chattel slavery was voted on again in September 1791, but by then, enslaved people had revolted in Saint-Domingue and would eventually abolish it. The disappearance of such stories helps to uphold claims of European victory over racism. But racism in France is alive and well. It is grounded in a relationship to land and geography marked by coloniality. Its inhabitants of colour cannot fully belong to the nation because it is understood as eternally white.

In the late 18th century, the enslaved people of Saint Domingue would lead the only successful slave insurrection that very few French people know as the Haitian Revolution, among the ones that had the greatest universal potential to this day (Ehrmann, 2015). Teaching and researching Haiti's history is, then, a crucial starting point to challenge the epistemic violence carried by Westernized universities. Incorporating the Haitian Revolution into mainstream narratives is not the territory but it is certainly a part of the map. Telling these stories is a means to humanize people of colour and to break away from Eurocentrism. It constitutes a first step towards (re)constructing pluriversal imaginaries.

Notes

1 See Dearden (2017) for the original quote. The English translation is mine.
2 The Algerian war of independence is the best known, while others – such as the Indochina or the Cameroonian wars – are disconnected from French history. Some colonial massacres (Sétif and Guelma, for instance) are now better known thanks to mobilizations of victims' descendants.
3 Four days later, Macron apologized for his statement to the Pieds-Noirs. The community lobbied for the 2005 law that required the teaching of "positive" aspects of colonialism.
4 Literally trade of Black people.
5 On the 10th May every year in mainland France. Some activists celebrate it on May 23. Overseas territories celebrate the abolitions on different dates according to their respective histories.
6 In 1794 the Convention declared the universal emancipation of slaves, but it did not actually outlaw the slave trade. Napoleon reestablished chattel slavery in 1802. It was abolished again in 1848 but Africans were transported to the French West Indies well into the 1870s.
7 In the sense of regional differences. See Bretons, Corsicans, Catalans and Basques and their respective separatist movements.
8 They used this exact term in their correspondence. See Gauthier (2007).
9 Biological racism remains politically correct. For example, mixed marriage is still hailed as a measure of good integration, of successful French multiculturalism.
10 The adverb refers to the official rejection of the ethno-racial classifications because of French universalism (race-blindness).

References

Ahouansou, K.M. (2013) De l'anthropologie des Français noirs aux politiques quotidiennes: Transversalité, genre, race. Unpublished paper presented to the TransOceanik international seminar, Florianópolis, 27–29 May.

Amin, S. (2009) *L'eurocentrisme: Critique d'une idéologie*. 2nd edn. Paris: Anthropos.

Araujo, M. and S. Maeso (2015) (eds) *Eurocentrism, Racism and Knowledge: Debates on History and Power in Europe and the Americas*. Basingstoke: Palgrave Macmillan.

Bancel, N., P. Blanchard and D. Thomas (2017) *The Colonial Legacy in France – Fracture, Rupture, and Apartheid*. Bloomington: Indiana University Press.

Bhambra, G. (2007) *Rethinking Modernity: Postcolonialism and the Sociological Imagination*. Basingstoke: Palgrave Macmillan.

Bleich, E. (2004) Anti-racism without Races: Politics and policy in a 'color-blind' state. In H. Chapman and L. Frader (eds) *Race in France: Interdisciplinary Perspectives on the Politics of Difference*. New York: Berghahn, pp. 162–188.

Boulbina, S.L. (2008) *Le singe de Kafka et autres propos sur la colonie*. Lyon: Parangon.

Césaire, A. (1962) *Toussaint Louverture: la Révolution française et le problème colonial*. Paris: Présence africaine.

Conklin, A. (2003) *A Mission to Civilize: The Republican Idea of Empire in France and West Africa, 1895–1930*. Stanford: Stanford University Press.

Cooper, F. and A.L. Stoler (1997) (eds) *Tensions of Empire: Colonial Cultures in a Bourgeois World*. Berkeley: University of California Press.

Dearden, L. (2017) Emmanuel Macron claims Africa held back by 'civilisational' problems and women having 'seven or eight children'. *The Independent*, 11 July. Available at: www.independent.co.uk/news/world/europe/emmanuel-macron-africa-development-civilisation-problems-women-seven-eight-children-colonialism-a7835586.html (Accessed 20 November 2017).

De Sousa Santos, B. (2001) Nuestra América: Reinventing a subaltern paradigm of recognition and redistribution. *Theory, Culture and Society* 18 (2–3): 185–217.

Dubois, L. and J. Garrigus (2017) *Slave Revolution in the Caribbean, 1789–1804: A Brief History with Documents*. Boston: Macmillan.

Durupt, F. (2017) Blanquer porte plainte contre un syndicat qui a utilisé l'expression 'racisme d'Etat'. *Libération* 21 November. Available at: www.liberation.fr/france/2017/11/21/blanquer-porte-plainte-contre-un-syndicat-qui-a-utilise-l-expression-racisme-d-etat_1611537 (Accessed 21 November, 2017).

Dussel, E. (1993) Eurocentrism and Modernity (Introduction to the Frankfurt Lectures). *Boundary* 2 20(3): 65–76.

Ehrmann, J. (2015) Konstitution der Rassismuskritik: Haiti und die Revolution der Menschenrechte. *Zeitschrift für Menschenrechte* 9(1): 26–40.

Fassin, D. (2010) Ni race, ni racisme: Ce que racialiser veut dire. In D. Fassin (ed) *Les Nouvelles Frontières de la Société Française*. Paris: La Découverte, pp. 147–172.

Fischer, S. (2005) *Modernity Disavowed: Haiti and the Cultures of Slavery in the Age of Revolution*. Durham: Duke University Press.

Fleming, C.M. (2017) *Resurrecting Slavery: Racial Legacies and White Supremacy in France*. Philadelphia: Temple University Press.

Garrigus, J. (1996) Colour, class and identity on the eve of the Haitian revolution: Saint-Domingue's free coloured elite as colons américains. *Slavery and Abolition* 17(1): 20–43.

Gauthier, F. (2007) *L'Aristocratie de l'Epiderme: Le combat de la Société des Citoyens de Couleur 1789–1791*. Paris: CNRS Editions.

Gay, A. (2015) Deny and punish: A French history of concealed violence. *Occasion* 9: 1–11.

Geggus, D. (2002) *Haitian Revolutionary Studies*. Bloomington: Indiana University Press.

Hall S., C. Critcher, T. Jefferson, J.N. Clarke and B. Roberts (1978) (eds) *Policing the Crisis: Mugging, the State, and Law and Order*. London: Macmillan.

James, C.L.R. (1938) *The Black Jacobins: Toussaint L'Ouverture and the San Domingo Revolution*. London: Secker & Warburg.

Keaton, T.D. (2010) The politics of race-blindness. *Du Bois Review* 7(1): 103–131.

Khiari, S. (2009) *La Contre-Révolution Coloniale, de De Gaulle à Sarkozy*. Paris: La Fabrique.

Larcher, S. (2014) *L'Autre Citoyen: L'Idéal Républicain et les Antilles après l'Esclavage*. Paris: Armand Colin.

Le Goff, J. (2014) *Faut-il Vraiment Découper l'Histoire en Tranches?* Paris: Seuil.

Maldonado-Torres, N. (2007) On the coloniality of being: Contributions to the development of a concept. *Cultural Studies* 21(2–3): 240–270.

Many Chroniques (2015) Transmettre la mémoire de l'esclavage: un défi politique primordial – Partie 1, *Many Chroniques* [blog]. Available at: https://manychroniques.wordpress.com/2015/05/24/transmettre-la-memoire-et-lhistoire-de-lesclavage-un-defi-politique-primordial-partie-1/ (Accessed 20 November 2017).

Mazrui, A. (2009) Foreword – The seven biases of Eurocentrism: A diagnostic introduction. In R. Kanth (ed) *The Challenge of Eurocentrism: Global Perspectives, Policy, and Prospects*. New York: Palgrave Macmillan, pp. xi–xix.

Peabody, S. (2002) *"There are no Slaves in France": The Political Culture of Race and Slavery in the Ancien Régime*. Oxford: Oxford University Press.

Penney, J. (2017) West African governments want to cut population growth in half, but for whose benefit? *Quartz* 1 August. Available at: https://qz.com/1042602/west-african-governments-want-to-cut-population-growth-in-half-but-for-whose-benefit/ (Accessed 20 November 2017).

Quijano, A. (2007) Coloniality and Modernity/Rationality. *Cultural Studies* 21(2): 168–178.

Schmidt, N. (2016) Teaching and commemorating slavery and abolition: From organized forgetfulness to historical debates. In A. Araujo (ed) *Politics of Memory: Making Slavery Visible in the Public Space*. New York; London: Routledge, pp. 106–123.

Schor, N. (2006) Universalism. In L. Kritzman (ed) *The Columbia History of Twentieth-Century French Thought*. New York: Columbia University Press, pp. 344–348.

Silga, J. (2017) La protection juridique contre la haine raciale: Le point de vue de la victime. In H. Bentouhami and M. Möschel (eds) *Critical Race Theory: Une Introduction aux Grands Textes Fondateurs*. Paris: Dalloz.

Stoler, A.L. (2002) *Carnal Knowledge and Imperial Power: Race and the Intimate in Colonial Rule*. Berkeley: University of California Press.

Tansel, C.B. (2014) For a Marxist critique of Eurocentrism, or refusing to throw the baby out with the bath water. *Progress in Political Economy* [blog]. Available at: http://ppesydney. net/for-a-marxist-critique-of-eurocentrism-or-refusing-to-throw-the-baby-out-with-the-bath-water/ (Accessed 20 November 2017).

Trouillot, M.-R. (1995) *Silencing the Past: Power and the Production of History*. Boston: Beacon Press.

Vergès, F. (2010) "There are no Blacks in France": Fanonian discourse, "the dark night of slavery and the French civilizing mission reconsidered". *Theory, Culture and Society* 27 (7–8): 91–111.

Vovelle, M. (2007) *1789. L'Héritage et la Mémoire*. Toulouse: Privat.

17

BEYOND THE WESTERNIZED UNIVERSITY

Eurocentrism and international high school curricula

Marcin B. Stanek

Colonialism, Eurocentrism and schooling

In this chapter I focus on the epistemic politics of an international high school curriculum. I do so because I believe that an effective decolonization of westernized universities and, indeed, the wider world, is not possible without a thorough understanding of how Eurocentrism and other epistemic injustices permeate high school curricula (see also Otele, this volume). There is a well-established consensus that schooling has often been used by colonial powers and postcolonial elites to discipline and control colonized populations (Apple and Buras, 2006; Rizvi, Lingard and Lavia, 2006; Larsen, 2011; Ticona Alejo, 2011). There are countless examples from all colonized nations, which shed light on the brutal epistemic violence implicit in compulsory education. One of them can be found by looking at reforms implemented in early twentieth century Bolivia (Larsen, 2011). Designed by the country's establishment in order to ensure indigenous peoples' obedience in the light of increasing rural-urban migration, the first Bolivian national education system was based on segregation between rural and urban schools. Indigenous people in rural schools were indoctrinated in nationalist values and taught specialized labour skills, while children of the urban elite were prepared to become future lawyers, politicians, journalists and professors (ibid.).

Another example, which demonstrates that epistemic injustice is deeply rooted in the politics of schooling, can be found by looking at school curricula in my own country, Poland. Beginning in 1772, the country was gradually partitioned by Russia, Prussia and Austria to eventually disappear from the map of Europe for 123 years. At that time teaching programmes were designed to ensure a smooth incorporation of Poland into the political and cultural systems of the occupying countries. The Polish language was banned from use and education focused on simple maths, German or Russian languages and faith-related subjects (Lukowski,

1999). Some of the documents from that time are quite explicit in stating that education was not meant to make Poles brighter and open opportunities for them to enter university-level education, but to make sure that they become efficient and obedient servants to the new foreign authorities (Meissner and Potoczny, 2013). While Poland was not colonized per se, and indeed at different times had its own aspirations to become a colonial power (Arpad Kowalski, 2006), the political aim behind schooling in times of partitions shows clear parallels with that in the colonies. Accounts describing the brutality with which Russian and German languages, histories and values were forced on students (Zeromski, 2017 [1897]) inspire me to speak out in solidarity against any act of using school curricula to colonize, universalize and impose ways of thinking of those economically and politically stronger on others. Up until the present day, discourses that posit the exclusion and inferiority of Eastern Europe with respect to the Western part of the continent are frequent in Polish media and popular debates. Whether we, as Poles, are as modern, advanced, smart, efficient or innovative as Western Europeans is constantly subject to a heated discussion, both in the sphere of politics and everyday life. Consequently, to me as a Pole, Eurocentrism has a very complex geography. It is as much about a set of parochial ideas from a few European metropolises being falsely introduced as universal to people on other continents as it is about these ideas being imposed on people in other parts of Europe.

The examples from Bolivia and Poland show that schooling played a crucial role in the maintenance of the epistemic and political hegemony of dominant groups. Both demonstrate that Eurocentric high school curricula actively reproduced the privilege of the powerful. They did so by introducing subordinate groups of people to selected elements of hegemonic epistemology, so that they could be governed effectively by the elites. They also ensured that the group of people who had a thorough understanding of Eurocentric epistemology and exercised power over the rest of the population was composed exclusively of children of already privileged, white, European elites. Moreover, the cases of Bolivian and Polish education reforms show how particular high school curricula were designed to ensure that universities remain epistemically, socially and culturally exclusive enclaves. It is also important to keep in mind that the number of people who attended universities was significantly smaller than those who finished their education at primary or secondary level. As such, not only did high school curricula instruct future university students on what is valuable knowledge, but they also played an important role in the reproduction of Eurocentric epistemology as the dominant, institutionalized way of knowing within wider society.

If one considers such historical connections between schooling and maintenance of hegemonic, colonial power through a decolonial lens, a series of important questions emerges. Inspired by Grosfoguel (2007: 220), who argues that despite the official end of colonial rule, modernity is underpinned by coloniality, or "colonial situations", defined as "cultural, political, sexual, spiritual, epistemic and economic oppression/exploitation of subordinate racialized/ethnic groups by dominant racialized/ethnic groups", I ask whether contemporary schooling still plays a key role in the reproduction of Eurocentric privilege. It is in the context of this question that I set out to analyse the geopolitics of knowledge in what is considered to be one of the most innovative and progressive secondary school curricula in the world, the International Baccalaureate (hereafter the IB). Through this discussion, I check for evidence of epistemic violence in the IB curriculum and discuss possible connections between this secondary education programme and Eurocentrism which persists at westernized universities. I do so by analysing interviews with fifteen Eastern European IB teachers. Due to ethical commitments to my research partners, I cannot provide further details about their geographical location. However, based on the interview material it is clear that the vast majority of the interviewees do not feel in a position of epistemic privilege or power. Quite the contrary, most of the teachers, all of whom are fully qualified from Eastern European universities, feel that their pedagogic skills and knowledges are inferior to those of Western European and US American teachers teaching the same curriculum. Through an analysis of our conversations, I sketch a rather dark picture of the epistemic politics of the IB. Eastern European teachers who work with the curriculum clearly underline that the content of the programme, the values, teaching methods and evaluation techniques it promotes all represent a Western European approach to schooling. Ironically, this context-specific pedagogy is adopted as universal by an education system which claims to run truly international "world schools", in which cultural diversity and intercultural understanding are at the heart of the curriculum. Not only do I argue that contemporary secondary international education is Eurocentric, but that it also plays a key role in cementing Eurocentrism at westernized universities. Most importantly, I underline that without tackling epistemic injustices present in secondary education, epistemic violence will continue to underpin wider structures of our society. In order to do so, I divide this chapter into three parts. First, I outline a theoretical framework for understanding the geopolitics of knowledge based on the work of scholars working within the Modernity/Coloniality/Decoloniality research programme (hereafter MCD). Second, I present evidence which demonstrates that the IB curriculum is indeed Western European in its content. Third, I discuss some of the consequences that Eurocentric curricula at high school level might have for the wider epistemic landscape of the contemporary world. However, before I begin I introduce the IB education system, its significance and main principles.

The International Baccalaureate

The IB is an international non-profit organization with headquarters in Geneva, Switzerland and major offices in The Hague, Washington D.C, Singapore and Cardiff (IBO, 2017). The organization designs and administers four different types of school curricula to be used in primary, middle, vocational and high schools. The four programmes are currently taught in 4,655 schools in 109 countries around the world, with the number of schools offering the IB increasing by almost 40% in the course of the last five years (ibid.). The beginnings of the IB can be traced back to the International School of Geneva in the late 1960s. The school was attended by children of employees of the numerous international organizations based in the city. Most of the parents wanted their children to attend universities in various English-speaking countries, which meant that apart from the Swiss national curriculum, students often had to study to pass British or American high school exams. As such, the IB was born out of the need for a universal preparatory school curriculum, which would be recognized by a wide variety of universities in the Western world (Tarc, 2009).

The founders adopted what they perceived to be a progressive educational model, as they wanted to make strategic use of the fact that unlike state-funded education systems, the new curriculum would not be constrained by any kind of national politics. They designed a child-centred education based on constructivist pedagogy and intercultural understanding (IBO, 2017). However, IB's understanding of the latter is not as rich and critical as the work on interculturality by decolonial scholars (Walsh, 2009; Aman, 2015); for the IB intercultural understanding is simply about "recognizing and reflecting on one's own perspective, as well as the perspectives of others" (IBO, 2015: 6). The curriculum is based on principles such as critical, independent thinking, reflection, international-mindedness and the relative autonomy of the teacher (Halic *et al.*, 2015). Its structure and content demonstrates its innovative nature quite effectively. Apart from successfully passing exams in each of the six subject groups (mother tongue literature course, foreign language, humanities and social sciences, experimental sciences, mathematics and arts), the high school programme graduates have to pass a course in the theory of knowledge. The latter includes a discussion of different kinds of knowledges, including indigenous epistemologies as well as the ethics of knowing. Moreover, students have to submit an essay which comprises an individual, original piece of research undertaken as part of the programme. Each of the IB students has to critically reflect on their involvement in after-school activities related to creativity, sports and service to the wider society. In terms of humanities and social sciences, schools can choose to offer subjects as diverse as social and cultural anthropology, philosophy, global politics or world religions. All of the IB assessments, apart from experimental sciences and mathematics, are based on essays and academic writing. The IB curricula are revised every five to ten years; they are always consulted with IB teachers and academics from chosen Western universities. Apart from the mother tongue literature course, all the subjects are currently offered in English, Spanish and French with German and Chinese versions being in a pilot stage (ibid.). When compared to most state-funded education systems worldwide, the IB stands out as a system which, at least in theory, gives its students tools necessary to become independent, critical and self-aware members of the society. Without doubt, it is widely considered to offer high school education of the highest quality.

For a long period, IB education was not only associated with quality, but also with financial elites, but over time the IB programmes became more accessible to the wider population. In 2015 approximately 50% of schools teaching the curriculum were state-funded, including 250 "Title 1" schools in the United States (Hill,

2012). The latter are state schools in which at least 40% of children are from low income families. The IB has also recently increased its cooperation with national governments, and plans to make their programmes available in the majority of state schools in countries such as Canada, Ecuador and Malaysia are well under way (IBO, 2017). There is no doubt that many Western national curricula and qualifications re-structure their programmes in order to remain competitive with IB education. For example, recent changes to A levels in England resemble some of the key IB features, such as compulsory, assessed fieldwork components or new assessment techniques which require students to reflect on their learning process. As such, the IB is setting trends that are followed by other curricula and it is quite likely that the programme will become even more widely available worldwide. It is also possible that the IB will replace a number of state-funded education systems in the near future.

Overall, the fact that the IB is a good indicator of the general direction in which most Western education systems develop makes the programme appropriate for the kind of analysis that is the aim of this chapter. The IB as an education system projects itself as a symbol of modern progressive pedagogy and as such is well-suited to show possible connections between modernity and coloniality in the field of secondary education. Moreover, the programme is taught in a wide variety of epistemic, cultural, social and political settings. It forms the secondary curriculum for a rapidly increasing group of students from the non-Western world, with the programmes being now available in the vast majority of South American, African and Asian countries. This allows for reflection on the extent to which European origins of the programme allow for the presence of non-Western epistemologies within the system. Furthermore, as a preparatory school for university and a curriculum frequently consulted with academics, the programme has clear links with westernized universities. The latter are the main reference point for a critique of knowledge production processes offered by scholars working within the MCD, whose work is adapted as a theoretical framework for my analysis.

Geopolitics of knowledge in secondary education

My understanding of geopolitics of knowledge as an analytical category is underpinned by the work of scholars who collaborate together within the MCD. The basic premise of the research programme is that "modernity and coloniality are two sides of the same coin" (Grosfoguel, 2007: 218). This has far reaching consequences for the nature of contemporary knowledge production processes. The ability to produce meaningful knowledge in modern times is reserved for Western metropolitan subjects. Dussel's (1994) work is particularly effective at explaining that the Cartesian thinking subject (ego cogito), which is symbolic of modern ways of knowing, is preceded 150 years by "I conquer, therefore I am" (ego conquiro) and conditioned upon a colonial "I exterminate, therefore I am" (ego extermino). Therefore, contemporary knowledge production processes are rooted in the brutal rejection of non-European knowledges as invalid, esoteric or irrelevant. According to MCD scholars, the dominance of European knowledges persists despite the official end of colonial rule. Grosfoguel (2013: 76) highlights that contemporary epistemic injustice is based on the fact that "any knowledge that [is] opposed to the myth of the unsituated knowledge of the Cartesian ego-politics of knowledge is discarded as biased, invalid, irrelevant, unserious and inferior". Elsewhere, he further explains that the modern world is built on a premise that "universal reason and truth can only emerge through a white-European-masculine-heterosexual subject" (Grosfoguel, 2012: 94). Not only is contemporary knowledge based on the work of white European philosophers seen as the only valid way of thinking, but also as the way of knowing which can be applied to everything everywhere. Walsh (2007: 225) argues that "the persistence of Western hegemony [...] positions Eurocentric thought as 'universal', while localizing other forms of thought as at best folkloric". While De Sousa Santos (2006) points to scientific knowledge as a Western hegemonic way of thinking about the world, which claims universality and superiority over other ways of thinking, Walsh (2007) suggests that the epistemic violence caused by European claims to universality extends beyond scientific knowledge to Western critical thought. She insists that an analysis of contemporary geopolitics of knowledge focuses on questions such as "who produces critical knowledge? With what recognition? Whose critical knowledge? For whom? Why and for what uses?" (ibid.: 227). Inspired by such an understanding of the contemporary geopolitics of knowledge production processes, in this article I position particular knowledges which circulate through the IB system against the colonial matrix of power. Based on interviews with IB teachers, I reflect on what is considered valid knowledge within this secondary education system. Who has the power to outline curricula, who decides on the mark schemes used to assess students' essays and what does the International Baccalaureate understand as international education are some of the key questions which guide this analysis. Given the theoretical context outlined here, I pay particular attention to possible parallels between colonial and contemporary power relations which epistemically underpin secondary education curricula.

In order to reflect on the extent to which high school curricula are epistemically rooted in coloniality, I structure my analysis around De Sousa Santos' (2006) notion of the ecology of knowledges and his larger project that he refers to as "the sociology of absences". De Sousa Santos (ibid.: xix) argues that "there is no global social justice without global cognitive justice". He insists that despite "a growing recognition of the cultural diversity of the world", there is no "recognition of the epistemological diversity" of our planet (ibid.). He suggests that capitalism is based on a "monoculture of scientific knowledge" enunciated from the Western world (De Sousa Santos, 2006). De Sousa Santos insists that one of the necessary conditions for a socially just world is a recognition and appreciation of an "ecology of knowledges" which implies a "promotion of non-relativistic dialogues among knowledges [and] granting equality of opportunities to the different kinds of knowledge". The "sociology of absences" project "consists of an enquiry that aims to explain that what does not exist is in fact actively produced as non-existent, that is − as a non-credible alternative to what exists" (ibid). De Sousa Santos (2006) insists that a sociology of absences should focus on five logics which constantly reproduce the monoculture of scientific knowledge. The five logics include the monocultures of: rigour of knowledge; linear time; the naturalization of difference; the universal and the global; and the criteria of capitalist productivity and efficiency. The first and the fourth logics are particularly important to this study. The monoculture of rigour of knowledge is understood by De Sousa Santos (ibid.) as "the most powerful mode of production of non-existence [which] turns modern science and high culture into the sole criteria of truth and aesthetic quality". The monoculture of the universal and the global implies that "the scales adopted [by Western modernity] as primordial determinant of the irrelevance of all other possible scales [are] the universal and the global" (ibid.: 17). Elsewhere, De Sousa Santos (2001) distinguishes between "knowledge-as-emancipation and knowledge-as-regulation"; he suggests that in order to stop the unconscious reproduction of the latter we must replace the five logics of monoculture with five ecologies: "the ecology of knowledge, the ecology of temporalities, the ecology of recognitions, the ecology of trans-scales and the ecology of productivities" (De Sousa Santos, 2006). In this manner the world of monoculture governed by the "epistemology of blindness" might move closer towards a utopian project of a world governed by "epistemology of seeing" (De Sousa Santos, 2001). Consequently, the analysis of interviews with Eastern European IB teachers that follows is structured according to what my interviewees saw as epistemic presences and absences within the IB programme. This format highlights how the IB actively produces epistemic absences.

The IB and epistemic presences

The interviewed IB teachers asserted that the content of subject-specific curricula within the IB system is defined by English and US American teachers and academics. One example is given by the history teacher who suggests that:

> The history curriculum is [...] it does not include French, Belgian, Polish or German historiographies. These are somewhat present while discussing grand topics such as the First or Second World Wars, where the work of German historians, for example, is taken into account, as they have a lot to say about these topics, but in general four fifths of the historiographic sources on which the curriculum is based are American and British.
>
> *(Interview 4, 7/06/2017)*

It is rarely the case that the work of scholars and teachers from outside of the English-speaking world is entirely absent. However, their presence in relation to US American and British knowledges is significantly limited. It is also interesting to note that the teacher points to a certain Anglo/American-centrism rather than to Eurocentrism. He actually considers the absence of other European knowledges as a certain injustice. Rather than an epistemic injustice, this relates to a concerning lack of geographical diversity amongst the authors of the programme. However, it is most important to underline the unquestioned lack of presence of non-European epistemologies in an international curriculum. The mindset here is so Eurocentric that the debate is centred on questions of which of the European historiographies should be treated as universal knowledge, rather than whether non-hegemonic ways of thinking about history should be included. The teacher goes on further to explain the dominance of Anglo-Saxon knowledges within the IB by the fact that the textbooks endorsed by the organization are written by teachers and academics from the English-speaking world:

> It is overwhelmingly Anglo-Saxons who write the IB History textbooks, which explains the prevalence of British and American historiographies; they [the authors of textbooks] source their knowledge from the books they find in libraries and bookshops in their countries.
>
> *(Interview 4, 7/06/2016)*

Similarly, a geography teacher underlines that:

> Despite the fact that we are meant to help students understand the world in its entirety, to support them in becoming global citizens, as it is written in all the IB policies, both the textbook and the exam questions narrow the world to the Western European perspective. [...] Practically speaking, the textbook which provides suggestions for specific case studies presents a Western outlook. [...] What's more, the exam questions afterwards are also most often based on content covered in the two most popular textbooks.
>
> *(Interview 2, 6/06/2016)*

What is at stake here goes beyond the economic power of British and US American publishers who play a dominant role in the IB textbook market. Many of the interviewed teachers repeatedly underline that it is only educators and academics in some Western European countries, Britain and the United States, who are considered to have enough expertise to decide on the curriculum or write textbooks. Only those positioned at the centres of Western metropolises, based at the most highly ranked and commercialized universities, are understood as thinking subjects capable of defining the curriculum. A literature teacher self-evaluates the level of her expertise as follows:

> I am fully dependent on the materials and training provided by teachers and IB specialists from the Netherlands, Switzerland, the UK and the US. I mean of course, I am familiar with all the documents, curricula and marking criteria, but they can really be interpreted in many different ways. In here [referring to the country] we do not have enough expertise, so we are dependent on the interpretations of the specialists from the West. It is them who are most knowledgeable; they are the only ones who can clarify what they mean. After all, the IB is a programme from the West and teaching it is

about understanding a particular vision of education from there. What I have learnt at the university in here [referring to the Eastern European country] does not really apply.

(Interview 13, 9/06/2016)

The number of interviewees underlining that pedagogical and subject-specific content of the programme comes only from the Western world was so large that I have noted the following in my field diary:

I have met so many excellent teachers in here, but almost all of them lack self-confidence in their knowledge and skills. Every day I keep hearing them refer to "their [Western specialists'] requirements", "their methods", to not knowing "what they want from us". The programme, in the way it is organised and in terms of policies and curricula makes teachers feel like instruction followers instead of fully knowledgeable subjects. They also keep pointing out to evaluation, [especially] the [negative] comments they get from examiners on their students' work, which exacerbates the feeling of insecurity about one's own expertise and knowledge.

(field notes 8/06/2016)

The feeling of uncertainty about the validity of Eastern European expertise and pedagogical knowledge is omnipresent at the school. Western European and US American textbook authors, examiners, moderators and curriculum designers are constantly referred to by the teachers as "IB experts". Given that the design of curriculum content, teaching methods, marking criteria, moderation of students' work and teacher training are all done by Western European and US American specialists, the IB actively reproduces the epistemic and political hegemony of the Western world. The programme reinforces the unequal geography of expertise, which apparently is only to be found in spaces that continuously act as centres of economic, political and epistemic power since the conquest of the Americas. Most "IB experts" are Western, metropolitan, white, heterosexual subjects.

The geography teacher also suggests that the subject syllabus prioritizes the so-called "peripheral" parts of the world which are of strategic importance to the Western society:

The way the curriculum is designed is good in a sense that it does not let students from the peripheries think that their countries are economically and politically disabled; that their only fate is to be taken over by the more powerful. The way the case studies are designed is to make students think that there is another way, a third way which leads to a more peaceful and sustainable world. However, the case studies focus entirely on peripheries which are of strategic importance to the Western world; for example, they challenge the assumptions that Africa or South East Asia are countries where nothing ever happens. It is a shame that this part of the world, Central and Eastern

Europe, is completely omitted, as if this part of the world did actually not exist [...] the so called alternative approaches to, say, development geography or globalization are present, but, again, this is all content that is coming from the West.

(Interview 2, 6/06/2016)

It is worth to note at this point that the Western-centric nature of the syllabus is not only about the presence of case studies from locations of particular importance to the authors of the programme, but also about the lack of diversity of critical perspectives included. A large number of the interviewed teachers assert that when critical perspectives are present in the curriculum they are almost exclusively those articulated in the Western world. There is no doubt that the presence of non-hegemonic knowledges – for example, those including critical approaches to development, elements of gender studies or queer theory – are of huge benefit to the programme. However, we must not forget that these represent only a small part of a large ecology of critical, non-hegemonic epistemologies articulated in a wide array of geographical locations. As such, geographical diversity of knowledges present in the programme is as important as the epistemic variety of the curriculum. The geography teacher recognizes the presence of some non-Eurocentric perspectives articulated from within Western Europe and sees the lack of critical perspectives from Eastern Europe as an important exclusion. This is consistent with Walsh's (2007) argument that the epistemic violence caused by European claims to universality extends beyond scientific knowledge to Western critical thought.

In this section I have focused on epistemic presences within the IB curriculum. I highlighted the unequal geography of expertise, which is based on the hyper-presence of Western European and US American ways of thinking and teaching. I have presented evidence to suggest that the international high school curriculum is dominated by pedagogical and subject-specific knowledges produced in Western Europe and the United States. While most of these knowledges are formulated from within the Eurocentric, hegemonic episteme, there are also some critical, non-hegemonic knowledges present. However, the content related to non-hegemonic ways of thinking is articulated from a limited number of geographical locations, as only critical perspectives articulated in the Western world find their way onto the syllabus. This highlights the importance of both epistemic and geographical diversity of knowledges for efforts to decolonize education. In the next section I highlight what Eastern European teachers see as absences within the curriculum.

The IB and epistemic absences

Most of the absences talked about by Eastern European IB teachers relate to important linguistic differences between Eastern and Western Europe as well as to difficulties in access to Western European books. There is no doubt that this is important, as it points to the already mentioned Western-centric nature of the programme.

However, it also highlights a deeper issue of a lack of concern for broader epistemic diversity within the curriculum.

In terms of linguistic differences between Eastern and Western European languages, one of the examples, which shows that the former are not given equal opportunities within the IB system, is given by a mathematics teacher:

> There are many differences between our [referring to the Eastern European country] and English Maths. The biggest difference is that they use the symbol "." in the place of our "," [The British 1,000,000.55 would be written as 1000000,55]. That is one thing. Another difference is in differential calculus. We teach that there are four trigonometric functions and in the English system, which makes more sense actually, they teach that there are three main trigonometric functions and that these have three inverse functions. There are also differences in the naming of large numbers, our billion is one thousand times more than an English billion. Yet another difference is in writing. For example the number 1 in the English system is written as the letter "I" in our system and the number 7 is not hyphened in England [Number 1 here is written as 1 and is distinguished from number 7 by a hyphen half way through the vertical line 7]. So sometimes our 1 can be interpreted by English examiners as a 7. That creates a bit of a problem.
>
> *(Interview 5, 7/06/2016)*

Similarly, a literature teacher comments that:

> in one of the recent exams the examiners have translated the word "novel" as "a short story". In our literary theory, the words "novel" and "a short story" define two entirely distinct genres, which have entirely different features. Some students who answered that question defined a specific novel as a short story, which given the distinctive features of our literature is a serious error. Similarly, there is a lot of confusion when students are asked to write an essay. The IB has incorrectly translated the word "essay" as "an opinion piece". In our country a literal translation of the word "essay" indicates the most flexible and creative form of writing which does not suggest a certain logical structure of the argument; its main characteristics are that it doesn't have any formal or structural requirements. In the English-speaking culture an essay is simply the equivalent of what we call "a short dissertation".
>
> *(Interview 13, 9/06/2016)*

Both the mathematics and the literature teachers conceptualize the absence of Eastern European terminology in exam questions and marking schemes as a technical problem that is specific to teaching the IB outside of the English-speaking world. Importantly, the absence of non-Western nomenclature is seen by teachers as something to be expected. The fact that one of the conditions for students to acquire an international high school diploma is that they need to show that they are capable of using a set of local, specifically English, terminologies as universal ones, is, according to teachers, a frustration but not a surprise. Moreover, the teachers are aware that they need to pay particular attention to the differences between the two ways of describing the world and make students extremely cautious about using terminology in their exam answers; however, this is seen as a technical issue rather than any kind of injustice. This shows how deeply ingrained Eurocentrism is in secondary international education. However, it also suggests that within the IB system students from the non-Western world have to develop an awareness and critical understanding of the differences between Western and non-Western ways of describing the world as well as the politics which underpin their uses. To an extent, this is a possible opening for a future decolonization of this education system. Apart from the prevalence of US American and British perspectives in exam questions, the teachers also point out to the lack of materials about Eastern European Geography or History in English. Moreover, the teachers express concern about the inequality in access to sources of knowledge between local students and those in Western European countries:

> My students have problems accessing many of the historical sources or books which are readily available in American or British bookshops and libraries. Here we are limited to the books which are available in our country. It would be unjust of me to force my students to purchase expensive books from abroad.
>
> *(Interview 4, 7/06/2016)*

The teacher highlights the inaccessibility of certain books which are available in the Western world. While the IB syllabi theoretically allow teachers to choose specific case studies and materials and adjust the curriculum to the local needs, a closer analysis reveals that such adjustment is much cheaper and quicker in the English-speaking world, where textbooks with context-specific case studies and additional teaching materials are readily available. From an economic and logistical perspective, the infrastructure of the IB system does not grant "equality of opportunities to the different kinds of knowledge" (De Sousa Santos, 2006: xix). The way in which curricula, textbooks, teaching materials and exam questions are prepared clearly favours content relevant to the Western world, actively producing pedagogical approaches and subject-specific materials relevant to Eastern European and other non-Western parts of the world as non-existent.

While the teachers point to an important lack in what is meant to be a truly international education system, there is also another absence present in this context; I call it a deep absence. What I am referring to is an absence of epistemic diversity which goes beyond a purely geographical variety of locations from which knowledges present in the system are articulated. It is an absence of an ecology of epistemes, which should be present given the international and apparently intercultural focus of the system. I call it a deep absence because it is absent from both the Eurocentric curriculum and the Eastern European teachers' vision for a more diverse programme. While the interviewed teachers do see the lack of Eastern European perspectives as an important fault of the programme, their understanding of a varied curriculum does not reach beyond diversity in a purely geographical sense. As such, the absence of desire for epistemic plurality is deeply ingrained in the monoculture of hegemonic epistemology. As the interviewed teachers did not mention other epistemologies when asked for absences, I asked directly about other ways of thinking. I referred to an ethnic group of people living in the country and known for their non-Cartesian understanding of the world. The teacher gave me a surprised look and said:

> No, no [...] no. I really do not think they need to know this kind of stuff to become global citizens. I mean, maybe in a form of a legend or to highlight the historical diversity, for students in primary school, but I am not convinced if this kind of content would be useful for them in this context. No, I don't think so.
>
> *(Interview 13, 9/06/2016)*

The absence of a desire for a curriculum based on an ecology of knowledges which would represent the epistemic, rather than the purely geographical diversity of the world is apparent here. Particular non-hegemonic ways of thinking are deemed unimportant to an international curriculum. Based on the absences identified earlier as important, the teachers show a desire for specific Eastern European ways of interpreting the world to be included in the Eurocentric curriculum, so that these could become an official part of the curriculum which would continue to represent a slightly modified, but still hegemonic way of thinking. However, they do not see it as necessary to include other non-hegemonic epistemologies, such as those excluded not only in the Western, but also in the Eastern European contexts in the programme. It is also interesting to recall that many of the teachers argued that some critical, non-hegemonic perspectives articulated from within the Western world are indeed present in the curriculum. The majority of the teachers showed a desire for other critical approaches to be included; however, it seems that they have quite a specific idea of which of the critical, non-hegemonic knowledges are worth including and which are not. As such, the interviewed teachers' vision of a better international education system is based on a slightly broader understanding of the monoculture of knowledge, but does not include a transformation of the very structures of the system to an epistemic ecology.

Both the epistemic absences and presences identified by Eastern European IB teachers clearly point to a Eurocentric nature of this seemingly international high school curriculum. Western, or more specifically British and US American knowledges are actively reproduced through textbooks, curricula, teaching materials and exam questions. Given the presence of some critical thought articulated from within the Western world in the programme, IB experiences of Western students and teachers are likely to suggest that the programme is progressive. However, an analysis of experiences of Eastern European IB teachers suggests that the programme is deeply Eurocentric, as it is constructed based on the expertise of specialists in Western European and US American schools and universities. The concluding section will outline some of the consequences of this for the wider epistemic landscape of the contemporary world.

Concluding remarks

The evidence presented in this chapter suggests that the international secondary education curriculum is indeed Eurocentric. It is designed and administered by academics and teachers from Western Europe and the United States, uses terminologies and ways of thinking typical of the hegemonic epistemology and makes most of the Eastern European teachers interviewed think that their pedagogical expertise is irrelevant. There are clear parallels between the Bolivian and Polish education reforms introduced at the beginning of this chapter and contemporary, progressive education. Just as with the historical reforms in Bolivia and Poland, contents of curricula, textbooks, exam questions, mark schemes and teaching guides in the IB are controlled by specialists from the Western world. The IB introduces students and teachers from the non-Western world to Eurocentric knowledges – including some critical perspectives developed therein – but does not necessarily give them the tools to feel the ownership of that knowledge. Acquiring elements of the Eurocentric curriculum is a basic requirement to gain a high school diploma recognized as international. As such, Eurocentric knowledges are treated as universal and the absence of non-Eurocentric epistemologies is seen as normal, as something to be expected. Ruptures resulting from this absence are seen as technical issues to be overcome by teachers and students, rather than signs of an injustice. While all this paints a rather dark picture of the geopolitics of knowledge within the IB curriculum, one might ask what does it mean for unsettling Eurocentrism at the westernized university? One of the key consequences is that an increasing number of high school students enter prestigious universities with a pre-conceived idea that Eurocentric knowledge is the valid, universal way of thinking, which defines quality in education and beyond. We must also remember that the majority of those students are very likely to enter departments in which critical theories are at best absent, if not ridiculed. When one takes the fact that the IB is often seen by parents, students and teachers as the entry ticket to the world's best universities, one can clearly see that a Eurocentric style of learning and teaching is at the heart of their pre-conceived idea about what high quality university education is about. While there are many differences between the Bolivian and the Polish education reforms mentioned in the introduction and the IB system, from a perspective of geopolitics of knowledge, all three programmes are part of a machine that uses education to divide populations and knowledges into those who are critical, independent and powerful knowledge producers and those who are merely taught to be obedient listeners and knowledge consumers.

While I have no doubt that some of the decolonial movements will have a trickle-down effect and will slowly be incorporated into some of the high school curricula (for example, the revised IB Geography curriculum to be taught from 2017/2018 onwards focuses on geographical relations between power and places), this is most likely going to be seen by Eastern European IB teachers as another imposition of Western critical thought on their classrooms. As critical scholars, who sometimes have an influence on the shape of high school curricula, we should always remember that unsettling Eurocentrism and decolonizing monocultural ways of thinking always need to be context specific. For if we do not wish to be remembered as the contemporaries of those who draw on schooling's colonial heritage and continue to design education systems which divide, exclude and maintain epistemic privilege, we must ensure that our decolonial efforts are open to an ecology of multiple, radical and continuous practices that are attentive to epistemic injustices and alternative epistemic systems.

References

Aman, R. (2015) The double bind of interculturality, and the implications for education. *Journal of Intercultural Studies* 36(2): 149–165.

Apple, M. and K. Buras (eds) (2006) *The Subaltern Speak Curriculum, Power, and Educational Struggles*. New York: Routledge.

Arpad Kowalski, M. (2006) *Kolonie Rzeczypospolitej*. Warszawa: Bellona.

De Sousa Santos, B. (2001) Toward an epistemology of blindness: Why the new forms of 'ceremonial adequacy' neither regulate nor emancipate. *European Journal of Social Theory* 4(3): 251–279.

De Sousa Santos, B. (2006) *The Rise of the Global Left: The World Social Forum and Beyond*. Zed Books: London.

Dussel, E. (1994) *1492: El encubrimiento del Otro: Hacia el origen del "mito de la modernidad"*. La Paz: Plural Editores.

Grosfoguel, R. (2007) The epistemic decolonial turn: Beyond political-economy paradigms. *Cultural Studies* 21(2–3): 211–223.

Grosfoguel, R. (2012) Decolonizing Western uni-versalisms: Decolonial pluri-versalism from Aimé Césaire to the zapatistas. *Transmodernity* 1(3): 88–104.

Grosfoguel, R. (2013) The structure of knowledge in westernized universities: Epistemic racism/sexism and the four genocides/epistemicides of the long 16th century. *Human Architecture* 11(1): 73–90.

Halic, O., Bergeron, L., Kuvaeva, A. and Smith, A. (2015) The International Baccalaureate's bilingual diploma: Global trends, pathways, and predictors of attainment. *International Journal of Educational Research* 69: 59–70.

Hill, I. (2012) An international model of world-class education: The International Baccalaureate. *Prospects* 42(3): 341–359.

IBO (2015) *What is an IB education?* Geneva: International Baccalaureate Organization.

IBO (2017) About the IB. Available at: www.ibo.org/about-the-ib/ (Accessed 28 July 2017).

Larsen, B. (2011) Forging the unlettered Indian: The pedagogy of race in the Bolivian Andes. In L. Gotkowitz (ed) *Histories of Race and Racism: The Andes and Mesoamerica from Colonial Times to the Present*. Durham: Duke University Press, pp. 134–156.

Lukowski, J. (1999) *The Partitions of Poland: 1772, 1793, 1795*. London: Longman.

Meissner, A. and Potoczny, J. (eds) (2013) *Oświata na ziemiach polskich pod zaborami wobec wyzwá cywilizacyjnych*. Lublin: WSEI.

Rizvi, F., Lingard, B. and Lavia, J. (2006) Postcolonialism and education: Negotiating a contested terrain. *Pedagogy, Culture and Society* 14(3): 249–262.

Tarc, P. (2009) What is the 'International' in the International Baccalaureate? Three structuring tensions of the early years (1962–1973). *Journal of Research in International Education* 8(3): 235–261.

Ticona Alejo, E. (2011) Education and decolonization in the work of the Aymara activist Eduardo Leandro Nina Qhispi. In L. Gotkowitz (ed) *Histories of Race and Racism: The Andes and Mesoamerica from Colonial Times to the Present.* Durham: Duke University Press, pp. 240–253.

Walsh, C. (2007) Shifting the geopolitics of critical knowledge: Decolonial thought and cultural studies 'others' in the Andes. *Cultural Studies* 21(2–3): 224–239.

Walsh, C. (2009) The plurinational and intercultural State: Decolonization and state refounding in Ecuador. *Kult* 6: 65–84.

Zeromski, S. (2017 [1897]) *Syzyfowe Prace.* Warszawa: Litres.

18

WHAT IS RACISM?

Zone of being and zone of non-being
in the work of Frantz Fanon and
Boaventura de Sousa Santos

Ramón Grosfoguel
Translation by Jordan Rodriguez

Introduction

Since Jean Paul Sartre there has not been a European thinker with the political and social commitment to the Global South like Boaventura de Sousa Santos. However De Sousa Santos surpasses Sartre to the extent that the latter was never influenced by, or took seriously, the epistemologies of the South. Sartre was a lifelong existentialist without rethinking his Eurocentric biases or without being influenced by the thought from the South. In contrast, Boaventura de Sousa Santos has posed as a priority in the production of knowledge in the social sciences thinking together and with the Global South. De Sousa Santos (2014: 19) shares the principle that "the understanding of the world is much broader than the Western understanding of the world". His sociology is a rupture with the Eurocentric universalism, calling for the production of an epistemology of the South through an ecology of knowledges that includes everything from social scientific knowledge to Other epistemologies and knowledges produced from the South. The ecology of knowledges is a fundamental epistemic principle in the work of De Sousa Santos that represents the dialogic point of departure enabling an escape from the Eurocentric monoculturalist monologue. In this intervention I wish to highlight the significance and fruitful encounter between Sousa Santos's and Fanon's work to understand the complexities involved in the concept of racism.

Fanonian conception of racism

For Fanon, racism is a global power hierarchy of superiority and inferiority along the line of the human that has been politically produced and reproduced for centuries by the "modern/colonial, capitalist/patriarchal imperialist/Western-centric" world system (Grosfoguel, 2011). People that are above the line of the human are socially recognized in their humanity as being human with rights and access to subjectivity and human/citizen/civil/labour rights. People below the line of the human are considered sub-human or non-human – that is to say, their humanity is in question and thus negated (Fanon, 2010). There are many important points to highlight with this definition.

First, the Fanonian definition of racism allows us to conceive of diverse forms of racism, avoiding the reductionisms of many definitions. Depending on the different colonial histories in various regions of the world, the power hierarchy of superiority/inferiority along the line of the human can be constructed with various racial markers. Racism can be marked through colour, ethnic, linguistic, cultural, or religious identity. While colour racism has been predominant in many parts of the world, it is not the only and exclusive form of racism. Many times we mistake the particular form of marking racism in one region of the world as the exclusive, universal form of marking racism everywhere. This has created a huge amount of theoretical and conceptual problems. If we collapse the particular form of racism that one region or country of the world adopts as if it were the universal definition of racism, we lose sight of the diversity of racisms that are not necessarily marked in the same way in other regions. Thus we adopt the false conclusion that in other parts of the world racism does not exist if the form of marking racism in one particular region or country does not coincide with the form of marking it in another region or country. Racism is a hierarchy of superiority/inferiority along the line of the human. This hierarchy can be constructed/marked with various markers. Westernized elites of the third world (Africans, Asians, or Latin Americans) reproduce racist practices against socially "inferiorized" ethnic/racial groups, where depending on the local/colonial history inferiorization can be defined or marked across religious, ethnic, cultural, or colour lines.

In Irish colonial history, the British constructed their racial superiority over the Irish not through markers of skin colour but through religious identity. What seemed in appearance like a religious conflict between Protestants and Catholics was in fact a racial/colonial domination. The same can be said of Islamophobia in Europe and the United States today. The Muslim religious identity constitutes today one of the most prominent markers of inferiority below the line of the human. I say "one of the markers", because in these two regions of the world colour racism remains very important and entangled in complex ways with religious racism. Nevertheless, while in many regions of the world the ethno-racial hierarchy of superiority/inferiority is marked by skin colour, in other regions it is constructed by ethnic and linguistic practices or religious and cultural identity. Racialization occurs through the marking of bodies with identities that are considered symbols of superiority or inferiority. Some bodies are racialized as superior and other bodies are racialized as inferior. The important point for Fanon is that those subjects located in the superior side of the line of the human live in what he calls the "zone of being", while those subjects that live in the inferior side of this line live in the "zone of non-being" (Fanon, 2010). This is not a geographical concept but a position in racial/ethnic hierarchies.

Differentiated intersectionalities: Zone of being and zone of non-being

In a colonial/capitalist/imperial world, race constitutes the transversal dividing line that crosses through relations of class, national, sexual, and gender oppressions at a global scale. This is what has become known as the "coloniality" or race as infrastructure (Fanon, 2010). The "intersectionality" or "interlocking" of power relations of race, class, sexuality, and gender, a concept developed by Black feminists (Davis, 1983; Crenshaw, 1991), occurs in both zones of the world that Fanon describes. However, the lived experience of the diverse oppressions and the particular way the intersectionality occurs is different in the zone of being in comparison to the zone of non-being. This is related to differences in the *materiality of domination* used by the system. In the zone of being, subjects, for reasons of being racialized as superior, do not live racial oppression but racial privilege. As it will be discussed later, this has fundamental implications in how they live class, sexual, and gender oppressions. In the zone of non-being, due to the subjects being racialized as inferior, they live racial oppression in place of racial privilege. Therefore, the oppression of class, sexuality, and gender that exist in the zone of non-being is qualitatively different and more devastating from those oppressions that exist in the zone of being. The point that should be emphasized is that there is a qualitative difference between the intersectional/interlocking of oppressions that exist in the zone of being and the zone of non-being in the "modern/colonial Christiancentric/Westerncentric patriarchal/capitalist" world system (Grosfoguel, 2011). But while in the zone of being oppressions of class, gender, and sexuality are *mitigated* due to racial privilege, in the zone of non-being these oppressions are *aggravated* due to racial oppression.

For Fanon, neither of these zones is homogenous. Both zones are heterogeneous spaces. Within the zone of being, there exists continuous conflict between what the Hegelian philosophy characterizes as the dialectics of recognition between "I" and the "Other". In the dialectic of the "I" and the "Other" within the zone of being there are conflicts but they are not racial conflicts, because the humanity of the oppressed other is recognized by the oppressor "I". All of the subjects living above the line of the human, even those oppressed, are *hyper-humanized* as superior beings. The "I" in a patriarchal/capitalist/imperial system are the Western, heterosexual, masculine, metropolitan, capitalist elites and the Westernized heterosexual masculine elites in the periphery of the world economy. The "Other" are Western populations of the metropolitan centres or the Westernized people within the peripheries whose humanity is recognized but at the same time live non-racial oppressions of class, sexuality, or gender at the hands of the imperial "I" in their respective regions and countries. The zone of being and zone of non-being is not a specific geographical location but a positionality in racial power relations that occurs on a global scale between centres and peripheries, but that also occurs at a local and national scale against diverse racially inferiorized groups everywhere in the world. Zones of being and non-being exist on a global scale between Westernized centres and non-Western peripheries (global coloniality) but also zones of being and non-being exist within the peripheries (internal colonialism). Internal colonialism exists in both the centre as in the periphery. The zone of non-being within a (core or peripheral) country would be the zone of internal colonialism. In the zone of non-being, all of the subjects are *dehumanized* as inferior beings. However, it is here where the decolonial critical theory of Boaventura de Sousa Santos (2010) helps to clarify the difference in the materiality of domination between the zone of being and the zone of non-being.

Fanonian zones and De Sousa Santos's abyssal line

For De Sousa Santos (2010), in modernity an abyssal line exists between the people above this line and the people below it. If we translate this line as the Fanonian line of the human and assign the zone of being to those who reside above the abyssal line and the zone of non-being to those who reside below this line, we can enrich our understanding of modernity and its colonial/racial/patriarchal/imperial/capitalist world system of oppressions that we inhabit. For De Sousa Santos, the way conflicts are managed by the system in the zone of being (above the abyssal line) is through what he calls mechanisms of regulation and emancipation. There exist codes of labour/civil/women/human rights, relations of civility, spaces of negotiation, and political actions that are recognized and extended to the oppressed "Other" in its conflict with the oppressor "I" within the zone of being. Emancipation refers to concepts of liberty, autonomy, and equality that form part of the discursive, institutional, and legal ends of the management of conflicts in the zone of being. As a trend, conflicts in the zone of being are materially managed through non-violent methods. Violence is always an exception and used in exceptional moments. The latter does not deny that violent moments exist in the zone of being. However, they exist more as an exception than as a rule.

On the contrary, as De Sousa Santos states, in the zone of non-being of the abyssal line, where the populations are dehumanized in the sense of being considered below the line of the human, the methods used by the heterosexual/masculine/capitalist/imperial "I" and its institutional system to manage and administer conflicts is by means of violence and open and blatant appropriation/dispossession. As a trend, conflicts in the zone of non-being are managed by perpetual violence and only in exceptional moments are peaceful methods of regulation and emancipation used. Given that the humanity of the people classified in the zone of non-being is not recognized, given that they are treated as non-humans or sub-humans, which is to say, without norms of rights and civility, then acts of violence, violations, and appropriation/dispossession are allowed that in the zone of being would be unacceptable. For De Sousa Santos, both zones are part of the project of colonial modernity. On the other hand, for Fanon, the dialectic of mutual recognition of the "I" and the "Other" that exist in the zone of being collapses, it falls apart in the zone of non-being by failing to recognize the humanity of the other. In short, the materiality of domination is different in the zone of being as opposed to the zone of non-being. In the zone of being we have forms of managing conflicts of perpetual peace with exceptional moments of war, while in the zone of non-being we have perpetual war with exceptional moments of peace.

Intersectionality and stratification in the zones marked by the abyssal line

The oppression of class, gender, and sexuality lived within the zone of being and within the zone of non-being are not the same. Since the conflicts with dominant classes and dominant elites in the zone of being are by nature lived with racial privilege, in the class, gender, and sexual conflicts the oppressed "Other Being" shares the privileges of imperial rights, Enlightenment emancipatory discourses, and processes of negotiation or conflict resolution. Thus, in the zone of being, the multiple oppressions are lived mitigated by racial privilege. On the contrary, since in the zone of non-being, class, gender, and sexual conflicts are aggravated by racial oppression, the conflicts are managed and administered with violent methods and perpetual appropriation/dispossession. The oppression of class, sexuality, and gender lived by the "Other Non-Being" is aggravated due to the articulation of these oppressions with racial oppression. For example, while workers in the zone of non-being risk their lives when they try to organize a union, earning one or two dollars a day while working 10 or 14 hours a day, workers in the zone of being enjoy labour rights, higher wages per hour, and better working conditions. Although a worker in a maquiladora in Ciudad Juarez who earns two dollars a day is formally a waged worker, her lived experience has nothing to do with that of a waged manual worker in Boeing Company in Seattle who earns 100 dollars an hour. The same principle applies to the oppression of gender and sexuality. Western women and Western gays/lesbians enjoy access to resources, wealth, rights and power, disproportionately greater than oppressed non-Western women or gays/lesbians in the zone of non-being. As a matter of fact, despite the gender oppression in the zone of being, Western women, while being a demographic minority in the world, have more power, resources, and wealth than the majority of men in the world that are of non-Western origins that live in the zone of non-being of the present system. In the order of Western-centric imperial things, it is not the same to be an "Other human" in the zone of being as it is to be an "Other non-human" in the zone of non-being. For Fanon and De Sousa Santos the zone of being is the imperial world that includes not only the imperial elites but also its oppressed Western subjects or Westernized subjects (elites in the Third World), while the zone of non-being is the colonial/neocolonial world with its oppressed non-Western subjects.

But the zone of non-being is also heterogeneous and stratified. What this means is that in the zone of non-being, besides the oppression that subjects live at the hands of the privileged subjects in the zone of being, there are in addition oppressions exercised within the zone of non-being between subjects belonging to this same zone that are also stratified. A non-Western, heterosexual man of the zone of non-being has privilege over non-Western heterosexual women and/or gays/lesbians within the zone of non-being. Despite the fact that the non-Western heterosexual man is an oppressed subject in the zone of non-being in relation to the zone of being, they oppress woman and/or a gay/lesbian in the zone of non-being. The problem is that the non-Western woman and non-Western gays/ lesbians in the zone of non-being are oppressed by not only Western people who inhabit the zone of being, but also by other subjects belonging to the zone of non-being. This is related to the multiple oppressions Black feminists always emphasize. This implies a double, triple, or quadruple oppression for oppressed non-Western subjects within the zone of non-being that has no comparison to those oppressed Western subjects within the zone of being with the access to labour/civil/human rights, norms of civility, and emancipatory discourses recognized and lived by those oppressed Western subjects within the zone of being.

This conceptualization is crucial to understand urban zones today. Cities are one of the most important spaces where racism and its materiality of domination of violence and dispossession are manifested. Although the zone of being and zone of non-being is not a geography but a position in global power racial/ethnic hierarchies, it has spatial manifestations. There are urban spaces that belong to the zones of non-being where the materiality of domination passes through violence and dispossession. In these spaces, the institutional practices are racist because they are violent and conducive towards the dispossession and premature death of the population. The question of premature death could happen by sudden and unexpected death because of being killed by police brutality or a sickness provoked by being exposed to the garbage of the city due to unhealthy or poisoned food/water/ buildings/communities by contamination, fast food, high levels of radiation, and so on. They are dispossessed from the services of the city in areas such as education, health, and transportation. By contrast, there are spaces that belong to the zone of being and that are ecological "safe heavens" because the materiality of domination is through methods of regulation and emancipation. Communities in the zone of being experience police protection instead of police brutality, insulation from the garbage and poisonous material of the city, and access to all of the city services, including high quality education, transportation, and health care.

In sum, the spaces in the city that have "symbolic capital" (Bourdieu, 1977) and spaces with "symbolic discredit" overlap with the racist power relations of zones of being and zones of non-being. Symbolic capital – that is, the capital of prestige and honour is possessed by inhabitants of urban spaces positioned as zones of being and dispossessed by the inhabitants of urban spaces located in the zone of non-being.

Colonial epistemology, decolonization of knowledge, and the critique of radical anti-essentialism

What is the relevance of the zone of being and the zone of non-being for the discussion about epistemic decolonization in combating Eurocentrism? Epistemic decolonization implies a delinking from Eurocentrism. However, the question is: delinking from what? What we know today as critical theory or critical thought is social theory produced from the social and historical experience of the "Other" oppressed within the zone of being. Marxism, critical theory, post-structuralism, psychoanalysis, and feminism are modalities of critical thought produced from the epistemic position of the "Other" within the zone of being. The racial inferiority of the zone of non-being occurs not only in relation to processes of domination and exploitation in economic, political, and cultural power relations, but also in epistemological processes. Epistemic racism/sexism refers to a hierarchy of colonial domination where knowledges produced by Western masculine subjects (both imperial and oppressed) within the zone of being are considered *a priori* as superior to knowledges produced by non-Western colonial subjects in the zone of non-being. The claim is that knowledge produced by subjects belonging to the zone of being from the rightist point of view of the imperial "I" or from the leftist point of view of the Westernized oppressed "Other" within the zone of being is automatically considered universally valid for all contexts and situations in the world. This leads to an imperial/colonial epistemology both of the right and of the left inside the zone of being that does not take seriously the theories produced from the zone of non-being and imposes its theoretical schemas conceived from realities very distinct from the situations of the oppressed others in the zone of non-being. Critical theory which is produced from the social conflicts that the oppressed "Other" lives within the zone of being with access to processes of regulation and emancipation, where racial domination is experienced as privilege and not as oppression, is taken as the conceptual criteria for understanding the social-historical experiences of those subjects that experience perpetual violence and appropriation/dispossession produced by racial conflicts in the zone of non-being.

The problem is that critical theory produced from the zone of being does not consider the social conflicts and colonial particularities of the oppressions lived in the zone of non-being. And when it does consider them, it is done from the perspective of the social-historical experience of the oppressed in the zone of being. The imposition of this critical theory from the zone of being onto the zone of non-being constitutes a coloniality of knowledge from the left. Critical theory of the left produced from the geopolitics of knowledge and the body-politics of knowledge of the "Other" within the zone of being is not sufficient to understand the lived problems nor the way in which to articulate the processes of violence and appropriation/dispossession of domination and exploitation in the zone of non-being. When the colonial subjects that reside in the zone of non-being adopt uncritically and exclusively social theory produced from the experience of the "Other" oppressed in the zone of being without taking seriously the critical theory produced from the social/historical experience of the "Other non-being" in the zone of non-being, they undergo a mental colonization subordinate to the Westernized left. Critical theories of the Westernized left in the zone of being, with very few exceptions, are blind towards the lived problems in the zone of non-being and to the qualitative difference between lived oppression in the zone of being in contrast with the zone of non-being. Epistemic racism/sexism in this critical theory is such that the claim is that the theory produced from the Global North should apply equally in a universal way to the Global South. But the theories produced by the "Others" oppressed in the zone of being tend to be blind towards the social/historical experience of those of the Global South, who live within the zone of non-being. This blindness leads to the invisibility of the experience of domination and exploitation lived in the zone of non-being, such as perpetual violence that is ignored or sub-theorized by the critical theory produced from the zone of being. Therefore, an important consequence derived from this discussion is that the project of epistemic decolonization implies a delinking from both the theory of the right and the theory of the left, which is produced from the social experience of those of the zone of being, who are blind towards the social experience of the zone of non-being. But decolonization has to be produced from a decolonial critical theory that makes visible the wasted and invisibilized experiences by the Northern-centric critical theories produced from the zone of being.

At the political level, what happens is that critical theories produced from the geopolitics of knowledge and body-politics of knowledge of those living the materiality of domination characterized by methods of violence and appropriation/dispossession in the zone of non-being are not as well known and are considered inferior to the critical theories produced by the Westernized left in the zone of being. The problem is not only one of epistemic colonization but also of political misunderstanding. The political challenge is how to construct coalitions and political alliances between oppressed subjects as "Others" in the zone of being and oppressed subjects in the zone of non-being against the Western/capitalist/masculine/heterosexual/military "I". The lack of communication and misunderstanding of the oppressed in the zone of being about the situation lived in the zone of non-being leads to an impasse to achieving political alliances. How can a politics of solidarity develop a way that can go in both directions and not entail a unilateral, paternalist, colonial, racist act of the Westernized left towards the people racially oppressed in the zone of non-being? If the oppressed people in the zone of being produce critical theory that they consider to be the only one valid and exclusive to understand, critique, and transform the world, making invisible and inferior other ways of critical theorization that are produced from the experience of the zone of non-being, then there are no conditions for the possibility of a political alliance on equal terms. The future rests on building political projects that are epistemically pluriversal and not universal, where there is room for critical epistemic diversity. For this, the oppressed in the zone of being has to take seriously the critical theories and knowledges produced from the zone of non-being and, therefore, be capable of building political alliances against the imperial "I" in the zone of being. This implies a decolonization of the subjectivity of the "Other" in the zone of being. However, decolonization in the zone of being is not equivalent to decolonization in the zone of non-being. It is here where Boaventura de Sousa Santos proposes another key concept: translation (De Sousa Santos, 2010). Translation is fundamental to building bridges between diverse social movements. Without translation it is not possible to understand or respect differences. It is not that there is absolute commensurability in translation processes. There exist many untranslatable things that are incommensurable. However, this does not discredit spaces of translation, negotiation, and respect that enable acting politically together despite the difference of situations and political projects. But translation is not only to be used in our political work but also in our intellectual work. As De Sousa Santos (2010) says, translation complements the sociology of absences and the sociology of emergences to enrich the number and multiplicity of available experiences. The latter is fundamental to have methods of intelligibility, coherence, and articulation. Translation represents in De Sousa Santos's work a fundamental methodological mechanism to account for intelligibility and coherence facing the increase of possible and visible experiences that the sociology of absences and the sociology of emergences produce.

However, the same decolonial methods cannot be applied in the same way in the zone of being and the zone of non-being if we want to avoid falling into another form of colonialism from the left. In the zone of non-being, inflating and building strong identities and epistemologies with strong metanarratives is necessary in the process of reconstruction and decolonization. Rebuilding strong identities and epistemologies is a requisite for rebuilding in the zone of non-being what coloniality has destroyed and reduced to inferiority through centuries of European colonial expansion. Many postmodernists, poststructuralists, and Marxists, apply in a reactionary way the method of radical anti-essentialism against indigenous and aboriginal people, aboriginal, Afrodescendants, immigrants of the South, non-Western citizens, and other colonial subjects that produce decolonial metanarratives from the zone of non-being. What the Westernized left do with its radical anti-essentialism – in place of translating the proposals, visions, and conceptions of colonial subjects – is discredit them. The anti-essentialist methods of the Westernized left end up being accomplices in the historical colonial racism of inferiorizing the knowledge and epistemologies produced by colonial subjects. After centuries of destroyed epistemologies, knowledges, and identities, decolonization in the zone of non-being is undergoing a necessary process of reconstructing its own thinking and identities. The Westernized left has difficulties understanding these processes. The radical anti-essentialism of the Westernized left has now become an instrument of colonial silencing, epistemological inferiority, and political underestimation of the critical voices producing critical thought from the zone of non-being. This is due to the fact that in the zone of being the opposite has happened – that is, epistemologies and identities have been inflated as superior. Therefore, in the zone of being, radical anti-essentialism is crucial to decolonize inflated egos, identities, and epistemologies conceived as racially superior. But when applied to the zone of non-being, where identities and epistemologies have been considered racially inferior for centuries, radical anti-essentialism becomes a colonial tool to keep silencing their voices and impeding their identity and epistemic decolonial reconstructions.

The work of Boaventura de Sousa Santos is an important antidote to this radical anti-essentialism. His anti-essentialism is accompanied by a great respect for and caution against the cultures and epistemologies of the South. This is implied in his calling for the ecology of knowledges, recognition of multiple temporalities, trans-scales, and productivities. For De Sousa Santos, epistemic diversity involves a dialogue of knowledges that allows the incorporation of thoughts and experiences of non-Western cultures as a point of departure to recover the wasted experiences discarded by Western reason. Hence he defends a moderate anti-essentialism.

Radical deconstruction, de-essentialization, and de-totalization are fundamental methods of decolonization within the zone of being, which always, and when not extrapolated to discredit the critical thought of the colonized subjects, represent an important step for imperial subjects beginning a decolonial process. Since historically the identities and epistemologies within the zone of being have been over-inflated as superior by the West, radical anti-essentialism becomes an important decolonial strategy for the oppressed in the zone of being. But this is only a first step in the process of decolonization of the zone of being. Decolonizing the West and the privileges of "whiteness" is something that entails many other things. For example, to decolonize the privileged positionality of a Western "white" subject further entails, among other things, fighting for the following demands:

1. Loss of privilege (social, political, economic, epistemological, etc.) facing the colonial subjects searching for egalitarian/equal relations.
2. Transferring material and symbolic resources from the zone of being to the zone of non-being.
3. Radically oppose the imperial/military aggressions as well as police violence in the zone of non-being.
4. Radical anti-racism.
5. Take serious the critical thought produced by and from the Global South.

There are many other demands that can be taken today in a decolonizing direction for Western/"white" subjects. However, to overcome this, we need to create a world with a new structure of power where the West stops the domination or exploitation of the rest of the world. But in reality the question "what does the decolonization of Westernized subjects signify?" is just beginning to find answers. It constitutes a challenge and still there is no clear idea about it. The decolonial sociology of Boaventura de Sousa Santos contributes a fundamental step in this direction.

References

Bourdieu, P. (1977) *Outline of a Theory of Practice*. Cambridge: Cambridge University Press.

Crenshaw, K. (1991) Mapping the margins: Intersectionality, identity politics, and violence against women of color. *Stanford Law Review* 43: 1241–1279.

Davis, A. (1983) *Women, Class and Race*. New York: Vintage Books.

De Sousa Santos, B. (2010) *Epistemologías del Sur*. Mexico: Siglo XXI.

De Sousa Santos, B. (2014) *Epistemologies of the South: Justice Against Epistemicide*. Boulder: Paradigm Publishers.

Fanon, F. (2010) *Piel Negra, Máscaras Blancas*. Madrid: Akal.

Grosfoguel, R. (2011) Decolonizing post-colonial studies and paradigms of political-economy: Transmodernity, decolonial thinking and global coloniality. *Transmodernity: Journal of Peripheral Cultural Production of the Luso-Hispanic World* 1(1): 1–38.

INDEX